Symmetry in Particle Physics

Symmetry in Particle Physics

Editors

Michal Hnatič
Jaroslav Antoš
Juha Honkonen

MDPI • Basel • Beijing • Wuhan • Barcelona • Belgrade • Manchester • Tokyo • Cluj • Tianjin

Editors
Michal Hnatič
Pavol Jozef Šafárik University
Institute of Experimental
Physics SAS
Slovakia
Joint Institute for
Nuclear Research
Russia

Jaroslav Antoš
Institute of Experimental Physics
of Slovak Academy of Sciences
Slovakia

Juha Honkonen
National Defence University
University of Helsinki
Finland

Editorial Office
MDPI
St. Alban-Anlage 66
4052 Basel, Switzerland

This is a reprint of articles from the Special Issue published online in the open access journal *Symmetry* (ISSN 2073-8994) (available at: https://www.mdpi.com/journal/symmetry/special_issues/Symmetry_Particle_Physics).

For citation purposes, cite each article independently as indicated on the article page online and as indicated below:

LastName, A.A.; LastName, B.B.; LastName, C.C. Article Title. *Journal Name* **Year**, *Volume Number*, Page Range.

ISBN 978-3-03943-801-3 (Hbk)
ISBN 978-3-03943-802-0 (PDF)

© 2020 by the authors. Articles in this book are Open Access and distributed under the Creative Commons Attribution (CC BY) license, which allows users to download, copy and build upon published articles, as long as the author and publisher are properly credited, which ensures maximum dissemination and a wider impact of our publications.

The book as a whole is distributed by MDPI under the terms and conditions of the Creative Commons license CC BY-NC-ND.

Contents

About the Editors vii

Preface to "Symmetry in Particle Physics" ix

Stanislav Dubnička, Anna Zuzana Dubničková, Mikhail A. Ivanov and Andrej Liptaj
Dynamical Approach to Decays of XYZ States
Reprinted from: *Symmetry* **2020**, *12*, 884, doi:10.3390/sym12060884 1

Roman Lysák
Charge Asymmetry in Top Quark Pair Production
Reprinted from: *Symmetry* **2020**, *12*, 1278, doi:10.3390/sym12081278 45

Andrej Arbuzov, Serge Bondarenko and Lidia Kalinovskaya
Asymmetries in Processes of Electron–Positron Annihilation
Reprinted from: *Symmetry* **2020**, *12*, 1132, doi:10.3390/sym12071132 93

Svetlana Belokurova and Vladimir Vechernin
Long-Range Correlations between Observables in a Model with Translational Invariance in Rapidity
Reprinted from: *Symmetry* **2020**, *12*, 1107, doi:10.3390/sym12071107 109

Amina Khatun, Adam Smetana and Fedor Šimkovic
Three Flavor Quasi-Dirac Neutrino Mixing, Oscillations and Neutrinoless Double Beta Decay
Reprinted from: *Symmetry* **2020**, *12*, 1310, doi:10.3390/sym12081310 121

Dmitry V. Naumov, Vadim A. Naumov, Dmitry S. Shkirmanov
Rephasing Invariant for Three-Neutrino Oscillations Governed by a Non-Hermitian Hamiltonian
Reprinted from: *Symmetry* **2020**, *12*, 1285, doi:10.3390/sym12081285 133

Oleg Teryaev
Energy-Momentum Relocalization, Surface Terms, and Massless Poles in Axial Current Matrix Elements
Reprinted from: *Symmetry* **2020**, *12*, 1409, doi:10.3390/sym12091409 149

About the Editors

Michal Hnatič (Prof., Dr.Sc.) is a Full Professor of Physics in the Faculty of Science of P.J. Šafarik University. He received his Master's and Ph.D. degrees in physics at the Leningrad State University in Leningrad (1983 and 1987, respectively, supervisor prof. A.N. Vasil'ev) and the highest scientific degree, Doctor of Science (Dr.Sc.), at the Comenius University in Bratislava (2007). His scientific interests are in quantum field theory methods, including their application in complex systems of classical physics (critical and stochastic dynamics, percolation, developed turbulence). He supervised 12 successfully defended Ph.D. students. He is a co-author of more than 150 scientific publications, including more than 80 in WOS or SCOPUS indexed international journals. He is participating in long-term successful international collaboration with colleagues from Helsinki University, Finland; Sankt-Petersburg State University, Russia; Joint Institute for Nuclear Research Dubna, Russia; and Uzhhorod National University, Ukraine.

Jaroslav Antoš (Dr.) graduated in 1974 in physics (Safarik University, Košice, ČSFR). From 1976 until now he has worked at the Institute of Experimental Physics (IEP), SAS, in Košice (Slovakia). In 1976, he was delegated as a visiting scientist to the Joint Institute for Nuclear Research in Dubna, USSR. He did work there until 1981 and was involved in an experiment called HYPERON. In 1986, he went to DESY for a short visit. This short visit helped to establish collaboration between Czecho-Slovak institutes and H1 experiment at DESY. This was the first official collaboration of HEP Czech and Slovak groups in former Czecho-Slovakia on an experiment at DESY! In 1989, he took a position for one year as a scientific associate at CERN. He did work in the group of Professor Willis on the analysis of data from the HELIOS/1 experiment. Established contacts later helped him to start the official collaboration of the group in IEP with the HELIOS/3 experiment. This was the first official collaboration of the HEP group in former Czecho-Slovakia on an experiment at CERN! The aim of the HELIOS/3 experiment was to study dimuon production at 200 GeA in pW and SW reactions. His analysis proved the existence of a nontrivial effect in SW which cannot be explained by simple extrapolation from pW. This effect was quoted in support of the existence of quark–gluon plasma announced at CERN in the year 2000. In 1992, he was invited by Academia Sinica (Taiwan) to help to organize an experimental physics group at the Institute of Physics, Academia Sinica, Taipei. The group was quickly established and recognized. In January 1993, this group was accepted to the CDF collaboration at FNAL, USA. The Taiwan group was active in the search for the top quark in dilepton decay mode; after the discovery of the top quark, the main focus was on the determination of top quark mass in this channel. In mid-1995, Dr. Antoš returned to his home institute (IEP). Even from Slovakia, he continued to collaborate with colleagues at CDF in the framework of the Taiwan group. He developed several original methods for top quark mass determination in the dilepton channel. In 2006, Slovakia was officially accepted to the CDF collaboration. The physics focus of the Slovak group is a study of top quark properties, specifically top quark mass and charge. He is the author or co-author of more than 1200 scientific papers published in recognized scientific journals and cited over 45,000 times.

Juha Honkonen (Dr.) is a Senior Scientist at the National Defence University, Helsinki, Finland, and an Adjunct Professor in Theoretical Physics at the University of Helsinki, Finland. He received his M.Sc. degree in physics in 1980 and earned his Ph.D. in theoretical and mathematical physics in 1984 at the Leningrad State University under the direction of the late Alexandr Nikolaevich Vasilev. He has been working as a researcher and instructor at the Nordisk Institut for Teoretisk

Atomfysik (NORDITA) in Copenhagen, Denmark, and at the Helsinki University of Technology. Dr. Honkonen has authored and co-authored over 80 peer-reviewed scientific journal articles and numerous other publications in quantum statistical physics, theory of critical phenomena, stochastic dynamics, operations research, and military technology. He has been a keynote speaker and an invited speaker at several international and national conferences and meetings. He is the author of five university textbooks.

Preface to "Symmetry in Particle Physics"

This Special Issue of *Symmetry* contains seven articles dedicated to solving urgent problems of modern particle physics. In one way or another, they consider the problems associated with different symmetries and their violations in high energy physics.

Two of them are topical reviews and give a general idea of the experimental and theoretical status in the study of exotic quark states (Stanislav Dubnička et al.) and top quark physics (Roman Lysak).

The standard bound states predicted by QCD as colorless are composed of two (mesons) and three quarks (baryons). However, modern theoretical concepts do not exclude the possibility of the existence of more complex multiquark states. The authors of the review focused on the theoretical description of the decays of five exotic XYZ states within the framework of the closed covariate quark model.

The heaviest top quark, the last of the six quarks described by the Standard Model and acting in the three-generation scheme, was experimentally discovered relatively recently by the CDF and D0 collaborations (1995). One of the co-discoverers of the top quark is the editor of this Special Issue (Dr. Jaroslav Antoš). The author of the review, Roman Lysak, joined CDF after the discovery of the top quark and was very active in the experimental study of top quark properties. In the review, the main focus is on the current state of the experimental and theoretical studies of charge asymmetry in the formation of top anti-top pairs, since the explanation of such asymmetry may lead beyond the limits of the Standard Model.

A special branch in the physics of elementary particles is the study of the properties of neutrinos, primarily associated with the possibility of their mixing due to their nonzero masses. Many ongoing and planned experiments are aimed at measuring their masses and mixing angles (for example, Juno, Baikal). The authors of two articles on neutrinos (Fedor Simkovic et al. and Dimitrii Naumov et al.) are leading experts in neutrino physics and are members of the aforementioned collaborations on the experimental study of neutrinos. It is also known that the study of the fundamental properties of atmospheric, solar, and cosmic neutrinos can help to find the direction along which the expansion of the Standard Model should go.

To study the physical properties of neutrinos, Fedor Simkovic et al. proposed a scenario in which the Dirac–Majorana mass term in the corresponding Lagrangian is dominated by Dirac masses. By assuming a single small Majorana component of neutrino masses, oscillation probabilities and quantities measured in single and double beta decay experiments and in cosmology have been determined.

The article by Dimitrii Naumov et al. investigates a very important experimental aspect: propagation of high-energy neutrinos in a dense environment with accounting for neutrino masses, mixing, CP violation, refraction, and absorption.

Andrey Abruzov et al. presented results for plain QED, weak, and complete electroweak radiative corrections to various asymmetries in processes of electron–positron annihilation to be measured in future colliders. Asymmetries in electron–positron annihilation processes provide a powerful tool to verify the lepton universality hypothesis at a new level of precision. The results reveal an interplay between the weak and QED contributions to asymmetries, indicating the necessity of always considering these contributions in a combined way.

Vladimir Vechernin and Svetlana Belokurova investigated how fixing the number of quark–gluon strings affects long-range correlations between observables under the assumption of the existence of translational invariance in the rapidity space. Knowledge of this relationship is important when processing experimental data in modern collider experiments, such as those of the Relativistic Heavy Ion Collider (RHIC) or the Large Hadron Collider (LHC).

Oleg Teryaev studied the momentum energy tensor for hadrons described by QCD in the presence of classical gravity and inertia. He found the relation between the general space-time symmetry responsible for interactions of hadrons with gravity and the specific QCD dynamics.

We express our deep gratitude to all authors for their articles, which made up the content of the Special Issue. We are confident that the high professional level of their scientific research and the published results will raise the scientific standard of the *Symmetry* journal.

Michal Hnatič, Jaroslav Antoš, Juha Honkonen
Editors

Review

Dynamical Approach to Decays of XYZ States

Stanislav Dubnička [1], Anna Zuzana Dubničková [2], Mikhail A. Ivanov [3] and Andrej Liptaj [1,*]

1. Institute of Physics, Slovak Academy of Sciences, 845 11 Bratislava, Slovakia; stanislav.dubnicka@savba.sk
2. Department of Theoretical Physics, Comenius University, 842 48 Bratislava, Slovakia; Anna.Dubnickova@fmph.uniba.sk
3. Bogoliubov Laboratory of Theoretical Physics, Joint Institute for Nuclear Research, 141980 Dubna, Russia; ivanovm@theor.jinr.ru
* Correspondence: andrej.liptaj@savba.sk

Received: 6 May 2020; Accepted: 26 May 2020; Published: 29 May 2020

Abstract: We review the existing results on the exotic XYZ states and their decays obtained within the confined covariant quark model. This dynamical approach is based on a non-local Lagrangian of hadrons with quarks, has built-in quark confinement, and is suited well for the description of different multiquark states, including the four quark ones. We focus our analysis on the various decay modes of five exotic states, $X(3872)$, $Z_c(3900)$, $Y(4260)$, $Z_b(10610)$, and $Z_b'(10650)$, aiming to clarify their internal quark structures. By considering mostly branching fractions and decay widths using the molecular-type or the tetraquark-type interpolating currents, conclusions about the nature of these particles are drawn: the molecular structure is favored for $Z_c(3900)$, $Z_b(10610)$, and $Z_b'(10650)$ and the tetraquark for $X(3872)$ and $Y(4260)$.

Keywords: exotic states; confined covariant quark model; strong and radiative decays

1. Introduction

The concept of multiquark states composed of more then three quarks hypothesized decades ago [1] was for the first time confirmed in 2003 where multiquark state candidates were measured by the BES [2], BaBar [3], and Belle [4] experiments. The latter observation, seen in the $\pi^+\pi^- J/\psi$ invariant mass spectrum, was the first observation of a charmonium-like state $X(3872)$, which did not fit expectations of existing quark models for any conventional hadronic particle. The reason was mainly its measured mass 3872 MeV, not predicted by models, and also the difficulty in interpreting it as an excited charmonium ψ': its eventual decay into $\rho J/\psi$ is strongly suppressed because of isospin violation. In the following years, other heavy quarkonium-like states X, Y, Z were discovered, where Y usually denotes electrically neutral exotic (i.e., non-$c\bar{c}$) charmonia having quantum numbers $J^{PC} = 1^{--}$, Z is used for charged states, and X labels any non-Y and non-Z cases. With the aim to report on the results and achievements of the confined covariant quark model, we narrow our review of experimental outcomes to a relevant subset of the whole exotic meson family.

The first observation of the $X(3872)$ mentioned in the previous paragraph was later confirmed in the $p\bar{p}$ collisions by the CDF [5] and D0 [6] experiments in 2004, by the LHCb experiment [7] in 2011, and also by the BESIIIcollaboration [8] in 2014. Further experimental investigations [9–12] increased the mass measurement precision, established the quantum numbers, and put limits on several decay related observables. As of now [13], $X(3872)$ is a particle with the mass $m_{X(3872)} = 3871.69 \pm 0.17$ MeV, width $\Gamma_{X(3872)} < 1.2$ MeV, and quantum numbers $I^G(J^{PC}) = 0^+(1^{++})$, mostly decaying to $\overline{D}^{*0}(\to \overline{D}^0 \pi^0)D^0$.

Charmonium-like state $Y(4260)$ was for the first time observed by the BaBar experiment [14] in 2005 in the $J/\psi \pi^+ \pi^-$ mass distribution. Its existence was further confirmed by the CLEO [15] (2006),

Belle [16] (2007), and BESII [17] (2013) collaborations. Later investigations by BaBar [18] and BESIII [19] provided further updates on the mass and width parameters. With mass above the $D\bar{D}$ threshold, the $Y(4260)$ was also searched for in the open charm decay channels, however with negative results [20–24]. The $Y(4260)$ is [13] an $I^G(J^{PC}) = 0^-(1^{--})$ state with the mass and width $m_{Y(4260)} = 4230 \pm 8$ MeV, $\Gamma_{Y(4260)} = 55 \pm 19$ MeV.

The study of the $Y(4260)$ decay channel $J/\psi \pi^+ \pi^-$ by BESII [17] and Belle [25] in 2013 led to the discovery of the charged $Z_c(3900)$ resonance in the invariant mass distribution of $J/\psi \pi^\pm$. The $Z_c^\pm$ particle was in the same year observed also by the CLEO-c detector [26]. In addition, the latter experiment provided the first evidence of the neutral member of the Z_c isotriplet, the Z_c^0 state, discovered in the $\pi^0 J/\psi$ channel. A state $Z_c(3885)$ was seen in the $D\bar{D}^*$ spectrum of the $e^+e^- \to \pi^\pm(D\bar{D}^*)^\mp$ reaction at BESIII in 2014 [27]. Assuming it can be identified with the $Z_c(3900)$ particle, the measurement provided arguments in favor of $J^P = 1^+$ quantum numbers. The same experiment reaffirmed in 2015 the existence of the neutral Z_c state [28], in 2017 confirmed with high significance the $J^P = 1^+$ assignment [29], and in 2019 provided the evidence for the $\rho^\pm \eta_c$ decay channel [30]. The D0 collaboration published the observation of the $Z_c(3900)$ state in $p\bar{p}$ collision data in 2018 [31] and studied its mass and width in [32] (2019). The current Z_c parameters are [13] $m_{Z_c(3900)} = 3887.2 \pm 2.3$ MeV, $\Gamma_{Z_c(3900)} = 28.2 \pm 2.6$ MeV and $I^G(J^{PC}) = 1^+(1^{+-})$. $Z_c(3900)$ as a charmonium-like state with an electric charge is a prominent candidate for an exotic multiquark state and is largely discussed in the existing literature.

Two narrow bottomonium-like four quark state candidates were detected in the Belle detector [33] in 2012. They were labeled $Z_b(10610)$ and $Z_b'(10650)$ and were observed as peaks in the mass spectra of $\pi^\pm Y(ns)$, $(n = 1, 2, 3)$ and $\pi^\pm h_b(ms)$, $(m = 1, 2)$. The same experiment published two other papers dedicated to these exotics. In [34], the evidence was given for the quantum number assignment $I^G(J^P) = 1^+(1^+)$ for both of the states. In [35], they were observed in different decay channels $Z_b(10610) \to [B\bar{B}^* + cc]^\pm$ and $Z_b'(10650) \to [B^*\bar{B}^* + cc]^\pm$, where one can notice the proximity of the two states to the corresponding $B^{(*)}\bar{B}^*$ thresholds. These decays dominated the studied final states, which besides two bottom mesons, included also a pion and for which the Born cross-section was given. The decay into $B\bar{B}$ was found to be suppressed with respect to the two previous final states, and an upper limit was given. The masses and widths are $m_{Z_b^\pm(10610)} = 10,607.2 \pm 2.0$ MeV, $m_{Z_b'(10650)} = 10,652.2 \pm 1.5$ MeV, $\Gamma_{Z_b^\pm(10610)} = 18.4 \pm 2.4$ MeV, and $\Gamma_{Z_b'(10650)} = 11.5 \pm 2.2$ MeV.

Growing evidence suggests that the mentioned and also other, unmentioned exotic heavy quarkonium-like states observed since 2003 cannot be described as simple hadrons in the usual quark model. The effort to understand their nature combined with the non-applicability of the perturbative approach in the low energy domain of quantum chromodynamics (QCD) resulted in a large number of more or less model dependent strategies. In existing reviews [36–49], different ideas are analyzed. The proximity of the X, Y, Z masses to meson pair thresholds naturally leads to a popular concept of the hadronic molecule, more closely reviewed in different contexts. In [50], the authors studied the implications of the heavy quark flavor symmetry on molecular states. The authors of [51] argued in favor of a molecular picture using an isospin-exchange model, and a nice review of the molecular approach was given in [52]. A frequent treatment of four quark states is represented also by QCD sum rules [53–55] and different quark models. A dynamical approach based on a relativistic quark model with a diquark-antidiquark assumption was proposed in [56,57], where tetraquark masses were computed. A non-relativistic screened potential model, presented in [58], was used to compute the masses, electromagnetic decays, and E1 transitions of charmonium states. Treatment of tetraquarks as compact dynamical diquark-antidiquark systems in [59] had the ambition to explain why some of the exotic states preferred to decay into excited charmonia. Several hypotheses (molecular description, tetraquark description, hadro-charmonium picture) for different exotic states were investigated in [60] using tools based on the heavy quark spin symmetry: besides drawing conclusions for some XYZparticles, also possible discovery channels were given. The hybrid and tetraquark interpretation for several exotic states were discussed in paper [61] using the Born–Oppenheimer approximation.

A very complete review of exotic states with some emphasis on the chromomagnetic interaction was provided in a recent publication [62]. The ideas of coupled channels ([63]) and heavy quark limit ([64]) are also often seen in the context of the exotic quarkonia. Finally, one has to mention the possibility of peaks in invariant mass distributions being explained by the kinematic effect. This was investigated in detail in a recent text [65]. The arguments for X, Y, and Z states not being purely kinematic effects were given in [66].

In the present paper, we want to review the description of the exotic heavy quarkonia-like states by the confined covariant quark model (CCQM). The model [67–69] was proposed and developed as a practical and reliable tool for the theoretical description of exclusive reactions involving the mesons, baryons, and other multiquark states. It was based on a non-local interaction Lagrangian, which introduces a coupling between a hadron and its constituent quarks. The Lagrangian guarantees a full frame independence and the computations relay on standard quantum field theory techniques where matrix elements are given by the set of quark-loop Feynman diagrams according to the $1/N_c$ expansion. Earlier, a confinement was not implemented in the model, and thus, it was not suited for heavy particles (with baryon mass exceeding those of the constituent quarks summed). This was changed in [69], where a smart cutoff was introduced for integration over the space of Schwinger parameters. Since then, arbitrary heavy hadrons could be treated by the CCQM. The CCQM represents a framework where the hadron and the quarks coexist, which raises questions about the proper description of bound states and the double counting. They are solved using a so-called compositeness condition. It guarantees, by setting the hadron renormalization constant Z_H to zero, that the dressed state and the bare one have a vanishing overlap. In order to describe radiative decays, one also needs to introduce gauge fields properly in a non-local theory such as CCQM. This was done by the formalism developed in [70] where the path integral of gauge field appeared in the quark-field transformation exponential. One should also mention that the model had no gluons: their dynamics was effectively taken into account by the quark-hadron vertex functions, which depended on one hadron size related parameter. The model has a limited number of free parameters; besides the hadron related ones, it has six "global" parameters: five constituent quark masses and one universal cutoff. The model was applied to with success light and heavy mesons and baryons (e.g., [71–76]) and also to exotic four quark states [77–83]. The latter will be reviewed in the rest of this article.

All sketched features of the CCQM (interaction Lagrangian, confinement, compositeness condition, implementation of electromagnetic interaction) are addressed in more detail in Section 2. Section 3 is dedicated to the $X(3872)$ state and its decays to $J/\psi + \rho$ and $\bar{D} + D^*$. Its radiative decays are analyzed in Section 4. In Section 5, molecular and tetraquark hypotheses for the nature of $Z_c(3900)$ are put in place and the results compared with experimental data. The exotic to exotic reaction $Y(4260) \to Z_c(3900)^{\pm} + \pi^{\mp}$ and the decay of $Y(4260)$ to open charm are presented in Section 6. Decays of the bottomonium-like states $Z_b(10610)$ and $Z_b'(10650)$ to several different final states are studied within the molecular picture in Section 7. We close the text by a summary and conclusion given in Section 8.

2. Confined Covariant Quark Model

2.1. Interaction Lagrangian

The dynamical description of hadrons in the CCQM follows from the interaction Lagrangian:

$$\mathcal{L}_{\text{int}} = g_H \cdot H(x) \cdot J_H(x), \tag{1}$$

where the hadronic field is coupled to a non-local quark current. The latter takes different forms for different hadrons:

$$J_M(x) = \int dx_1 \int dx_2 \, F_M(x; x_1, x_2) \cdot \bar{q}_1^a(x_1) \, \Gamma_M \, q_2^a(x_2)$$

for the mesons,

$$J_B(x) = \int dx_1 \int dx_2 \int dx_3 \, F_B(x; x_1, x_2, x_3) \cdot \Gamma_1 q_1^{a_1}(x_1) \left(q_2^{a_2}(x_2) C \Gamma_2 q_3^{a_3}(x_3) \right) \cdot \varepsilon^{a_1 a_2 a_3}$$

for the baryons, and

$$J_T^\mu(x) = \int dx_1 \ldots \int dx_4 \, F_T(x; x_1, \ldots, x_4) \cdot \left(q_1^{a_1}(x_1) C \Gamma_1 q_2^{a_2}(x_2) \right) \cdot \left(\bar{q}_3^{a_3}(x_3) \Gamma_2 C \bar{q}_4^{a_4}(x_4) \right) \cdot \varepsilon^{a_1 a_2 c} \varepsilon^{a_3 a_4 c}$$

for the tetraquarks. Here, C stands for the charge conjugation matrix $C = \gamma^0 \gamma^2$ with $C = C^\dagger = C^{-1} = -C^T$ and Γ is an appropriate Dirac matrix (or string of Dirac matrices) to describe the spin quantum numbers of the hadron. One has $C \Gamma^T C^{-1} = \Gamma$ for the (pseudo)scalar and axial-vector fields and $C \Gamma^T C^{-1} = -\Gamma$ for vectors and tensors. The color indices are denoted by superscripts a_i, and $F_H(x; x_1, \ldots, x_n)$ represents a non-local vertex function, which characterizes the quark distribution inside the hadron. We assume it takes the form:

$$F_H(x; x_1, \ldots, x_n) = \delta^{(4)}\left(x - \sum_{i=1}^n w_i x_i \right) \Phi_H \left(\sum_{i<j}(x_i - x_j)^2 \right), \quad \text{where} \quad w_i = \frac{m_i}{\sum_{i=1}^n m_i}. \quad (2)$$

The first factor reflects the natural expectation that the barycenter of the quark system corresponds to the position of the hadron, and the second term has a general form dependent on the relative quark coordinates. Obviously, the vertex function is invariant under translations:

$$F_H(x + a; x_1 + a, \ldots, x_n + a) = F_H(x; x_1, \ldots, x_n)$$

for any four-vector a. In principle, any form of the function Φ_H is allowed as long as it has an appropriate fall-off behavior in the Euclidean momentum space to guarantee the ultraviolet convergence of the Feynman diagrams. Various alternatives of the vertex function for non-local quark currents were analyzed in [84], and it was found that the dependence of the observables on different choices was small. Because of convenience of performing calculations, the exponential form for the Fourier transform of the function Φ_H was adopted:

$$\tilde{\Phi}_H(-K^2) = \exp\left(\frac{K^2}{\Lambda_H^2} \right) \quad (3)$$

where K^2 is the combination of the loop and external momenta. The minus sign indicates that we are working in the Minkowski space, and the wicked-rotated argument $K^2 \to -K_E^2$ makes explicit the appropriate fall-off behavior in the Euclidean region. Λ_H is an adjustable parameter of the CCQM, which can be related to the hadron size. Additional free parameters are the constituent quark masses and a universal infrared cutoff (discussed in more detail later). Their values, summarized in Table 1, were determined by adjusting the model predictions to experimental data.

Table 1. Constituent quark masses and universal cutoff λ in GeV.

$m_{u,d}$	m_s	m_c	m_b	λ
0.241	0.428	1.67	5.04	0.181

2.2. Compositeness Condition

In the Lagrangian of the CCQM, quarks and hadrons are treated equally. However in nature, hadrons are made of quarks. Therefore, questions about an appropriate description of the bound states and double counting arise. The issue is resolved by imposing the so-called compositeness condition [85,86], which requires the renormalization constant of the hadron field to vanish. Since the renormalization constant $Z_H^{1/2}$ can be interpreted as the matrix element between the physical state and the corresponding bare state, $Z_H^{1/2} = 0$ implies that the physical state has no overlap with the bare state and is therefore described as a bound state. For a spin-one particle, the compositeness condition reads:

$$Z_H = 1 - g_H^2 \Pi_H'(m_H^2) = 0, \qquad (4)$$

where Π_H' is the derivative of the scalar part of the vector-meson mass operator:

$$\Pi_H^{\mu\nu}(p) = g^{\mu\nu} \Pi_H(p^2) + p^\mu p^\nu \Pi_H^{(1)}(p^2),$$

$$\Pi_H(p^2) = \frac{1}{3}\left(g_{\mu\nu} - \frac{p_\mu p_\nu}{p^2}\right)\Pi_H^{\mu\nu}(p).$$

The condition $Z_H^{1/2} = 0$ also effectively removes the constituent degrees of freedom from the space of physical states and so eliminates the double counting. A general tetraquark self-energy diagram to be used for the compositeness condition is show in Figure 1.

One should also notice that the application of the compositeness condition lowers the number of model parameters because its fulfillment is reached by tuning the coupling constant value. Equation (4) thus fixes the coupling and increases the predictive power of the CCQM over the wide range of hadronic data. The determination of g_H for all participating hadrons by means of the compositeness condition is the first step in the application of the CCQM. It should be remarked that the compositeness condition can be interpreted also in terms of the normalization of the electric form factor at $q^2 = 0$, as shown in [69].

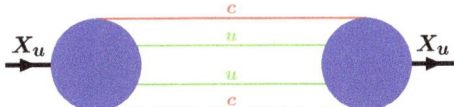

Figure 1. General confined covariant quark model (CCQM) tetraquark self-energy diagram.

2.3. Infrared Confinement

If the mass of a hadron reaches the limit defined by the sum of the masses of constituent quarks, then in a model without a confinement, the hadron becomes unstable and decays into its constituents. In order to correct this unphysical behavior and enlarge the applicability of the model also to the (increasing) experimental data on heavy hadrons, the confinement of quarks was introduced in [69]. Its implementation assumes the Schwinger representation of quark propagators:

$$S(k) = \frac{(m+\slashed{k})}{m^2 - k^2} = (m+\slashed{k}) \int_0^\infty d\beta \exp\{-\beta(m^2 - k^2)\} \qquad (5)$$

with the subsequent cutoff in the upper integration limit applied, in a clever way, to the whole structure of a Feynman diagram. The latter, containing l loops, m vertices, and n propagators, can be schematically written as:

$$\Pi(p_1,\ldots,p_m) = \int (d^4k)^l \prod_{i=1}^m \Phi_{i+n}\left\{-\sum_j (\kappa_{i+n}^{(j)} + v_{i+n}^{(j)})^2\right\} \cdot \prod_{k=1}^n S_k(\kappa_k + v_k)$$

$$= \int_0^\infty d^n\beta \, F(\beta_1,\ldots,\beta_n), \qquad (6)$$

where Φ symbolizes vertex functions, p denotes external momenta, v linear combinations of external momenta, k represents loop momenta, and κ linear combinations of the latter. The expression in curly brackets is the argument of the vertex function in the momentum representation. The second line makes explicit the integration over the space of the Schwinger parameters with the whole structure of the first line catch in the F symbol. The next step is to go from the integration over the Schwinger parameters to the integration over the $n-1$ simplex of the dimensionless Feynman parameters combined with a one-dimensional integral over a dimension variable t by using the simple insertion of unity in the above expression:

$$1 = \int_0^\infty dt\, \delta\left(t - \sum_{i=1}^n \beta_i\right).$$

One has:

$$\Pi = \int_0^\infty dt\, t^{n-1} \int d^n\alpha\, \delta\left(1 - \sum_{i=1}^n \alpha_i\right) F(t\alpha_1, \ldots, t\alpha_n). \tag{7}$$

Note that a dimension variable t is analogous of the Fock–Schwinger proper time. By performing an analytical continuation of the kinematical variables to the Minkowski space, one can encounter the branch points, which, in particular, correspond to the quark unitary thresholds. The appearance of the imaginary parts in Equation (7) is witnessed on the quark production in the physical spectrum, i.e., on the absence of the quark confinement. One possibility to resolve this problem is to cut the upper limit of the integration over the proper time t, i.e., $\infty \to 1/\lambda^2$ with λ being the "infrared" cutoff parameter. This allows one to remove all singularities of the diagram related to the local quark propagators. The integral becomes smooth and always convergent. For clarity, one can demonstrate the approach on a scalar one-loop two-point function. One starts with the loop integral in the Euclidean space:

$$\Pi(p^2) = \int \frac{d^4 k_E}{\pi^2} \frac{\exp(-s k_E^2)}{[m^2 + (k_E + p_E/2)^2][m^2 + (k_E - p_E/2)^2]}. \tag{8}$$

By using the above transformations, one gets:

$$\Pi(p^2) = \int_0^{\infty \to \frac{1}{\lambda^2}} dt\, \frac{t}{(s+t)^2} \int_0^1 d\alpha\, e^{-t[m^2 - \alpha(1-\alpha)p^2] + \frac{st}{s+t}(\alpha - 1/2)^2 p^2}, \tag{9}$$

where $p^2 = -p_E^2$. The expression has a branch point at $p^2 = 4m^2$ as follows from the vanishing of the first term inside the exponential at $\alpha = 1/2$. However, this singularity is removed by the cutoff.

We take the value of the cutoff parameter λ presented in Table 1 as universal for all processes we describe.

2.4. Electromagnetic Interactions

Inclusion of the electromagnetic (EM) interaction into the non-local CCQM in a gauge invariant way requires a dedicated approach. Our main interest will be in the radiative decays of neutral particles (i.e., the X(3872) tetraquark; see [79]), and so, we will focus on the EM interactions of quarks. The free part of the quark Lagrangian is gauged using the standard minimal coupling prescription:

$$\partial^\mu q \to (\partial^\mu - i e_q A^\mu) q, \qquad \partial^\mu \bar{q} \to (\partial^\mu + i e_q A^\mu) \bar{q}, \tag{10}$$

with $e_u = 2/3$, $e_d = -1/3$ in units of the proton charge. This defines the first part of the interaction Lagrangian of quarks with photons:

$$\mathcal{L}_{int}^{EM(1)} = \sum_q e_q A_\mu(x) J_q^\mu(x), \quad \text{with} \quad J_q^\mu(x) = \bar{q}(x) \gamma^\mu q(x). \tag{11}$$

A second term comes from gauging the non-local quark-hadron interaction Lagrangian (1). First, one multiplies each quark field by an exponential expression, which depends on the gauge field:

$$q(x_i) \to \exp\{-ie_q I(x_i, x, P)\} q(x_i), \qquad \bar{q}(x_i) \to \exp\{ie_q I(x_i, x, P)\} \bar{q}(x_i), \qquad (12)$$

with I being defined as the integral:

$$I(x_i, x, P) = \int_x^{x_i} A^\mu(z) dz_\mu$$

over the path that connects the hadron and quark positions. It can be readily seen that the local gauge transformations:

$$q(x_i) \to \exp\{ie_q f(x_i)\} q(x_i), \quad \bar{q}(x_i) \to \exp\{-ie_q f(x_i)\} \bar{q}(x_i),$$
$$A^\mu(z) \to A^\mu(z) + \partial^\mu f(z), \qquad (13)$$

leave the Lagrangian unchanged for any arbitrary function f. Indeed, the gauge field induced modification of the path integral $I(x_i, x, P) \to I(x_i, x, P) + f(x_i) - f(x)$ is canceled by the contribution coming from the quark transformations. The exact form of the gauged non-local Lagrangian $\mathcal{L}^{EM(2)}$ depends on the quark current structure (i.e., hadron quantum numbers), and we will write down an explicit form of it in the dedicated section. To use the gauged Lagrangian in perturbative calculations, one expands the gauge exponentials into the powers of the coupling constant (and thus, powers of A^μ) up to a desired order. The expansion contains only the derivatives of the path integral I, and using the approach proposed in [70,87], one can define them in a path independent manner:

$$\lim_{dx^\mu \to 0} dx^\mu \frac{\partial}{\partial x^\mu} I(x, y, P) = \lim_{dx^\mu \to 0} \left[I(x + dx, y, P') - I(x, y, P) \right]. \qquad (14)$$

Here, P' denotes a path derived from P by extending P from its endpoint by dx. This definition gives:

$$\frac{\partial}{\partial x^\mu} I(x, y, P) = A_\mu(x), \qquad (15)$$

where the independence of the derivative on the path P becomes explicit.

2.5. Selected Computational Aspects

To proceed with the calculations, it is convenient to use the following representation for the correlation function:

$$\Phi_X \left(\sum_{1 \le i < j \le 4} (x_i - x_j)^2 \right) = \left[\prod_{n=1}^{3} \int \frac{dq_n}{(2\pi)^4} \right] e^{-i \sum_{i=1}^{3} q_i (x_i - x_4)} \tilde{\Phi}_X \left(-\frac{1}{2} \sum_{1 \le i < j \le 3} q_i q_j \right). \qquad (16)$$

It may be easily obtained by using the Jacobi coordinates. The Gaussian function form of the correlation function $\tilde{\Phi}_X$ in Equation (3) can be joined with the exponents coming from the Schwinger representation of quark propagators given by Equation (5) into a single exponential function. Its argument takes a Gaussian form in loop momenta:

$$\exp(kak + 2kr + R), \qquad (17)$$

where a is a 3×3 matrix, $r = (r_1, r_2, r_3)$ is a vector constructed from external momenta, and the constant R behaves as a quadratic form of external momenta. As a result, one observes that, with respect to loop momenta, the general expression (6) is a product of a polynomial P (originating from evaluation of traces) with an exponential function. The tensorial loop momenta integration is then performed using the differential identity:

$$P\left(k_i^\mu\right) e^{2kr} = P\left(\frac{1}{2}\frac{\partial}{\partial r_{i\mu}}\right) e^{2kr}, \qquad (18)$$

which allows us the move the k independent differential operator in front of the integral. The action of the latter operator on the result of the integration is further simplified by applying a second operator identity:

$$P\left(\frac{1}{2}\frac{\partial}{\partial r_i}\right) e^{-ra^{-1}r} = e^{-ra^{-1}r} P\left(\frac{1}{2}\frac{\partial}{\partial r_i} - [a^{-1}r]_i\right), \qquad (19)$$

which permits commuting the differential operator with the exponential. The next steps are automated using a FORM program [88]. It repeatedly performs the differentiation using the chain rule, thus effectively commuting the differential operator to the right (and eventually making it vanish by acting on a constant).

At last, one is left with an integral over the space of the Schwinger parameters (see Section 2.3). The latter is computed numerically with the help of a FORTRAN code. Most of the time, one is interested in the q^2 dependent hadronic form factors: for the purposes of this text, the CCQM should be seen as a smart and effective tool that provides these form factors from the assumed quark currents as inputs.

3. Strong Decays of X(3872)

3.1. Decays $X \to D^{*0}(\to D^0\pi^0)\bar{D}^0$, $X \to \rho^0(\to \pi^+\pi^-)J/\psi$, and $X \to \omega(\to \pi^+\pi^-\pi^0)J/\psi$

The controversy raised by the discovery of the $X(3872)$ state can be best seen in the large number of publications it provoked (with many different interpretations). The proximity of the $D^{*0}\bar{D}^0$ threshold:

$$M_{X(3872)} - (M_{D^{*0}} + M_{D^0}) = -0.30 \pm 0.40 \text{ MeV} \qquad (20)$$

naturally suggests the idea of a loosely bound charm meson molecule. This idea was studied in several texts: implications of the molecular hypothesis for interference and binding effects were discussed in [89]; the authors of [90] found support for the molecular interpretation within a non-relativistic quark model; a published text [91] analyzed the molecular assumption in an effective field theory approach; and further works [92–94] based their analyses on an effective field theory with pion exchange, Monte Carlo simulations, and heavy quark spin symmetry. A rather strong support for the molecular picture was given in [95] (line shapes study) and [96] (potential model). The lattice study [97] found an explanation for $X(3872)$ in both the molecular and tetraquark scenario. An important group of analyses focused on charmonium [98–100] or mixed charmonium [101–104] explanations. Further arguments in favor of a charmonium structure followed from the Flatté analysis performed in [105], and both molecular and charmonium hypotheses were discussed in [106]. Several works [107–110] disfavored the molecular description. The authors of [107] based their conclusion on a non-relativistic quark model with the pion exchange, and the analysis presented in [108] favored the charmonium picture instead, while the conclusions in [109] were based on the pion and sigma exchange model. More rare were approaches based on the glueball picture [111] and chromomagnetic interaction [112]. The authors of [113] put in question the existence of a bound state at all. A hybrid hypothesis was considered in [114] and [115] (here, together with the molecular and charmonium one). Lattice computations in relation to X(3872) were used in [116,117], QCD sum rules in [118,119], and the coupled channel approach in [120–122]. One should also mention the studies based on quark models [56,123,124] and other strategies [125–127].

The description of the $X(3872)$ state by the CCQM was presented in [78]. There, one assumed a tetraquark structure, and within this assumption, decays $X \to J/\psi + 2\pi(3\pi)$ and $X \to \bar{D}^0 + D^0 +$

π^0, proceeding through the off-shell ρ/ω and D^* states respectively, were computed. In addition, possible implications of the $X(3872)$ dominance in the s-channel dissociation of J/ψ were discussed.

When describing the $X(3872)$ state, one follows the suggestions of [123,128], where a symmetric spin distribution of this $J^{PC} = 1^{++}$ state was proposed:

$$[cq]_{S=0}[\bar{c}\bar{q}]_{S=1} + [cq]_{S=1}[\bar{c}\bar{q}]_{S=0} \quad (q = u, d). \tag{21}$$

A non-local generalization of this diquark-antidiquark current is written as:

$$J^\mu_{X_q}(x) = \int dx_1 \ldots \int dx_4\, \delta\left(x - \sum_{i=1}^{4} w_i x_i\right) \Phi_X\left(\sum_{i<j}(x_i - x_j)^2\right) J^\mu_{4q}(x_1 \ldots, x_4),$$

$$J^\mu_{4q} = \tfrac{1}{\sqrt{2}} \varepsilon_{abc}\varepsilon_{dec}\left\{ [q_a(x_4) C \gamma^5 c_b(x_1)][\bar{q}_d(x_3)\gamma^\mu C \bar{c}_e(x_2)] + (\gamma^5 \leftrightarrow \gamma^\mu) \right\}, \tag{22}$$

with simplified weights resulting from only two quark flavors being present:

$$w_1 = w_2 = w_c = \frac{m_c}{2(m_q + m_c)}, \quad w_3 = w_4 = w_q = \frac{m_q}{2(m_q + m_c)}. \tag{23}$$

The strong isospin violation observed by comparing the ρ and ω vector meson mediated decays:

$$\frac{\mathcal{B}(X \to J/\psi \pi^+ \pi^- \pi^0)}{\mathcal{B}(X \to J/\psi \pi^+ \pi^-)} = 1.0 \pm 0.4(\text{stat}) \pm 0.3(\text{syst}) \tag{24}$$

experimentally established by Belle [129] suggested a mixed nature of the physical states X_l, X_h:

$$X_l \equiv X_{\text{low}} = X_u \cos\theta + X_d \sin\theta,$$
$$X_h \equiv X_{\text{high}} = -X_u \sin\theta + X_d \cos\theta,$$

where θ is the mixing angle. The state X_u breaks the isospin symmetry maximally:

$$X_u = \frac{1}{\sqrt{2}}\left\{ \underbrace{\frac{X_u + X_d}{\sqrt{2}}}_{I=0} + \underbrace{\frac{X_u - X_d}{\sqrt{2}}}_{I=1} \right\}.$$

The mixing angle is to be adjusted to fit the branching fraction ratio (24).

The first step in our calculation is to determine the coupling constant g_X by using the so-called compositeness condition discussed before. The derivative of the tetraquark mass operator needed for this can be written as:

$$\Pi'_X(p^2) = \frac{1}{2p^2} p^\alpha \frac{\partial}{\partial p^\alpha} \Pi_X(p^2) \tag{25}$$

$$= \frac{2 g_X^2}{3 p^2}\left(g_{\mu\nu} - \frac{p_\mu p_\nu}{p^2}\right) \prod_{i=1}^{3} \int \frac{d^4 k_i}{(2\pi)^4 i}\, \tilde{\Phi}_X^2(-K^2)$$

$$\times \Big\{ -w_c \text{tr}\left[S_c^{[12]} \not{p} S_c^{[2]} \gamma^5 S_q^{[2]} \gamma^5\right] \text{tr}\left[S_c^{[3]} \gamma^\mu S_q^{[13]} \gamma^\nu\right] + w_q \text{tr}\left[S_c^{[12]} \gamma^5 S_q^{[2]} \not{p} S_q^{[2]} \gamma^5\right] \text{tr}\left[S_c^{[3]} \gamma^\mu S_q^{[13]} \gamma^\nu\right]$$
$$- w_c \text{tr}\left[S_c^{[12]} \gamma^5 S_q^{[2]} \gamma^5\right] \text{tr}\left[S_c^{[3]} \not{p} S_c^{[3]} \gamma^\mu S_q^{[13]} \gamma^\nu\right] + w_q \text{tr}\left[S_c^{[12]} \gamma^5 S_q^{[2]} \gamma^5\right] \text{tr}\left[S_c^{[3]} \gamma^\mu S_q^{[13]} \not{p} S_q^{[13]} \gamma^\nu\right] \Big\},$$

where the short notations for the quark propagators and loop momenta are:

$$S_c^{[12]} = S_c(k_1 + k_2 - w_c p), \qquad S_c^{[3]} = S_c(k_3 - w_c p),$$
$$S_q^{[2]} = S_q(k_2 + w_q p), \qquad S_q^{[13]} = S_q(k_1 + k_3 + w_q p),$$
$$K^2 = \frac{1}{2} \sum_{i \le j} k_i k_j.$$

The evaluation of this expression is related to the determination of the size parameter Λ_X value and allows us to study the Λ_X dependence of the results.

Because the $X(3872)$ mass lies close to the studied thresholds:

$$m_X - (m_{J/\psi} + m_\rho) = -0.90 \pm 0.41 \, \text{MeV},$$
$$m_X - (m_{D^0} + m_{D^{*0}}) = -0.30 \pm 0.34 \, \text{MeV},$$

the off-mass-shell character of the ρ, ω, and D^* vector mesons has to be taken into account when evaluating the transition amplitudes $X \to J/\psi + \rho(\omega)$ and $X \to D^{*0}\bar{D}^0$. The Feynman diagrams to be considered within the CCQM are depicted in Figure 2.

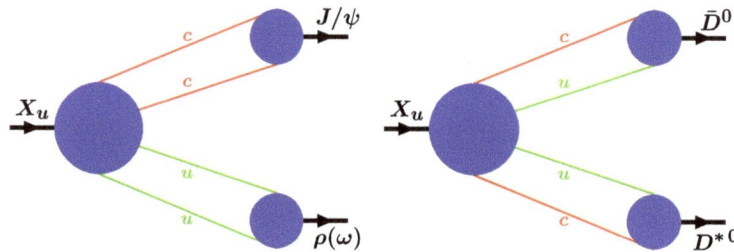

Figure 2. Feynman diagrams describing the decays $X \to J/\psi + \rho(\omega)$ and $X \to D + \bar{D}^*$.

In what follows, we use the notation for the light vector mesons $v^0 = \rho, \omega$. The amplitude of the decay $X_u \to \bar{D} + D^*$ is written as:

$$M^{\mu\nu}\left(X_u(p,\mu) \to \bar{D}(q_1) + D^*(q_2,\nu)\right) = 3\sqrt{2}\, g_X\, g_D\, g_{D^*} \int \frac{d^4 k_1}{(2\pi)^4 i} \int \frac{d^4 k_2}{(2\pi)^4 i} \widetilde{\Phi}_X\left(-K_2^2\right)$$
$$\times\ \widetilde{\Phi}_D\left(-(k_1+w_c q_1)^2\right) \widetilde{\Phi}_{D^*}\left(-(k_2+w_c q_2)^2\right)$$
$$\times\ \text{tr}\left[\gamma^5 S_c(k_1)\gamma^5 S_u(k_1+q_1)\gamma^\mu S_c(k_2)\gamma^\nu S_u(k_2+q_2)\right] + (m_u \leftrightarrow m_c, w_u \leftrightarrow w_c)$$
$$= g^{\mu\nu}\, M^{(1)}_{XDD^*} + q_1^\mu q_1^\nu M^{(2)}_{XDD^*} + q_1^\mu q_2^\nu M^{(3)}_{XDD^*} + q_2^\mu q_1^\nu M^{(4)}_{XDD^*} + q_2^\mu q_2^\nu M^{(5)}_{XDD^*} \qquad (26)$$

where the argument of the X-vertex function is equal to:

$$K_2^2 = \tfrac{1}{8}(k_1-k_2)^2 + \tfrac{1}{8}(k_1-k_2+q_1-q_2)^2 + \tfrac{1}{4}(k_1+k_2+w_c p)^2.$$

The amplitude of the decay $X_u \to J/\psi + v^0$ is written as:

$$M^{\mu\nu\rho}\left(X_u(p,\mu) \to J/\psi(q_1,\nu) + v^0(q_2,\rho)\right) = 6 g_X g_{J/\psi} g_{v^0} \int \frac{d^4 k_1}{(2\pi)^4 i} \int \frac{d^4 k_2}{(2\pi)^4 i} \widetilde{\Phi}_X\left(-K_1^2\right)$$
$$\times \widetilde{\Phi}_{J/\psi}\left(-(k_1 + \tfrac{1}{2}q_1)^2\right) \widetilde{\Phi}_{v^0}\left(-(k_2 + \tfrac{1}{2}q_2)^2\right)$$
$$\times \text{tr}\left[i\gamma^5 S_c(k_1) \gamma^\nu S_c(k_1 + q_1) \gamma^\mu S_u(k_2) \gamma^\rho S_u(k_2 + q_2)\right]$$
$$= \varepsilon^{q_1 q_2 \mu\nu} q_1^\rho M_{XJv}^{(1)} + \varepsilon^{q_1 q_2 \mu\nu} q_2^\rho M_{XJv}^{(2)} + \varepsilon^{q_1 q_2 \mu\rho} q_2^\nu M_{XJv}^{(3)} + \varepsilon^{q_1 q_2 \nu\rho} q_1^\mu M_{XJv}^{(4)}$$
$$+ \varepsilon^{q_1 \mu\nu\rho} M_{XJv}^{(5)} + \varepsilon^{q_2 \mu\nu\rho} M_{XJv}^{(6)} + \varepsilon^{q_1 q_2 \mu\rho} q_1^\nu M_{XJv}^{(7)} + \varepsilon^{q_1 q_2 \nu\rho} q_2^\mu M_{XJv}^{(8)} \quad (27)$$

where the argument of the X-vertex function is equal to:

$$K_1^2 = \tfrac{1}{2}(k_1 + \tfrac{1}{2}q_1)^2 + \tfrac{1}{2}(k_2 + \tfrac{1}{2}q_2)^2 + \tfrac{1}{4}(w_u q_1 - w_c q_2)^2.$$

In the latter expression, the number of Lorentz structures is reduced to six when X and J/ψ are on the mass-shell because, in that case, one has $\epsilon_\mu(q_1^\mu + q_2^\mu) = 0$ and $\epsilon_\nu q_1^\nu = 0$.

Obvious relations:

$$M(X_d \to J/\psi + \rho) = -M(X_u \to J/\psi + \rho), \qquad M(X_d \to J/\psi + \omega) = M(X_u \to J/\psi + \omega)$$

allow expressing all amplitudes of physical states transitions in terms of the X_u ones:

$$M(X_{\ell/h} \to J/\psi + \omega) = (\cos\theta \pm \sin\theta) M(X_u \to J/\psi + \omega),$$
$$M(X_{\ell/h} \to J/\psi + \rho) = (\pm\cos\theta - \sin\theta) M(X_u \to J/\psi + \rho).$$

The differential decay rate in the narrow-width approximation is written as [130]:

$$\frac{d\Gamma(X \to J/\psi + n\pi)}{dq^2} = \frac{1}{8 m_X^2 \pi} \cdot \frac{1}{3} |M_{XJv}|^2 \frac{\Gamma_{v^0} m_{v^0}}{\pi} \frac{p^*(q^2)}{(m_{v^0}^2 - q^2)^2 + \Gamma_{v^0}^2 m_{v^0}^2} B(v^0 \to n\pi), \quad (28)$$

$$\frac{1}{3}|M_{XJv}|^2 = \frac{1}{3}\sum_{\text{pol}} |\epsilon_X^\mu \epsilon_{J/\psi}^\nu \epsilon_{v^0}^\rho M_{\mu\nu\rho}|^2,$$

where $p^*(q^2) = \lambda^{1/2}(m_X^2, m_{J/\psi}^2, q^2)/2m_X$ is the momentum of the v^0 in the X rest frame. The allowed kinematic range is given by:

$$(n\, m_\pi)^2 \le q^2 \le (m_X - m_{J/\psi})^2,$$

where $n = 2$ for the ρ meson and $n = 3$ for the ω meson. The masses, decay widths, and branching fractions appearing in (28) were taken from PDG [13]. In addition to the model parameter values presented in Table 1, further model parameters are needed, namely the size parameters of the appearing mesons. Their values have been settled earlier and are presented in Table 2.

Table 2. Size parameters for selected mesons in GeV.

Λ_π	$\Lambda_{\rho/\omega}$	Λ_D	Λ_{D*}	$\Lambda_{J/\psi}$	Λ_{η_c}
0.711	0.295	1.4	2.3	3.3	3.0

Two adjustable parameters remain, the size parameter Λ_X and the mixing angle θ. It was found out that the dependence of the branching fraction:

$$\frac{\Gamma(X_u \to J/\psi + 3\pi)}{\Gamma(X_u \to J/\psi + 2\pi)} \approx 0.25 \quad (29)$$

on the size parameter Λ_X is in the CCQM small and close to $1/4$. Using this observation and the central value of the experimental ratio in Equation (24), one can deduce the mixing angle from:

$$\frac{\Gamma(X_{l,h} \to J/\psi + 3\pi)}{\Gamma(X_{l,h} \to J/\psi + 2\pi)} \approx 0.25 \cdot \left(\frac{1 \pm \tan\theta}{1 \mp \tan\theta}\right)^2 \approx 1. \tag{30}$$

The latter equation yields $\theta \approx 18.4°$ for X_l and $\theta \approx -18.4°$ for X_h. When not considering the ratio, the sensitivity of the decay widths on the size parameter is more important. One may expect the size parameter value to be close to those of the charmonia $\Lambda_{J/\psi}$ and Λ_{η_c}, i.e., to be in the range $3\,\text{GeV} < \Lambda_X < 4\,\text{GeV}$. This range was scanned, and the behavior of the decay width is depicted in Figure 3.

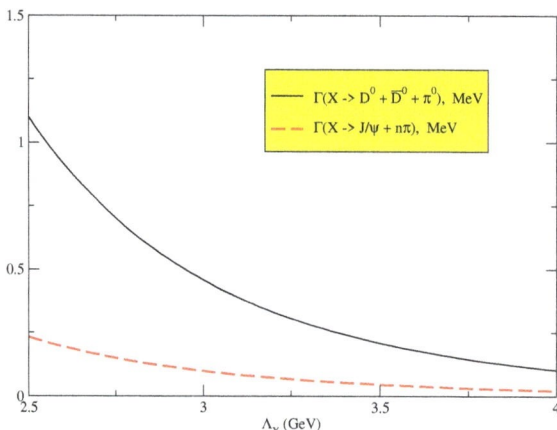

Figure 3. The dependence of the decay widths $\Gamma(X_l \to \bar{D}^0 D^0 \pi^0)$ and $\Gamma(X \to J/\psi + n\pi)$ on the size parameter Λ_X.

One can conclude that the predicted values in the interval $2.5 \le q^2 \le 3.5$ GeV lie in the range $0.05\,\text{MeV} < \Gamma_{X(3872)} < 0.23\,\text{MeV}$, which is in agreement with the upper limit of 1.2 MeV.

The differential rate of the decay $X(3872) \to \bar{D}^0 D^0 \pi^0$ in the narrow-width approximation is written as:

$$\frac{d\Gamma(X_u \to \bar{D}^0 D^0 \pi^0)}{dq^2} = \frac{1}{2m_X^2 \pi} \cdot \frac{1}{3}|M_{XDD^*}|^2 \cdot \frac{\Gamma_{D^{*0}} m_{D^{*0}}}{\pi} \frac{p^*(q^2) \mathcal{B}(D^{*0} \to D^0 \pi^0)}{(m_{D^{*0}}^2 - q^2)^2 + \Gamma_{D^{*0}}^2 m_{D^{*0}}^2}, \tag{31}$$

$$\frac{1}{3}|M_{XDD^*}|^2 = \frac{1}{3}\sum_{pol}|\varepsilon_X^\mu \varepsilon_{D^{*0}}^\nu M_{\mu\nu}|^2,$$

where $p^*(q^2) = \lambda^{1/2}(m_X^2, m_{D^0}^2, q^2)/2m_X$ is the momentum of D^{*0} in the X rest frame. The matrix element $M_{\mu\nu}$ was defined above by Equation (26). One has to note that the allowed kinematic range:

$$3.99928\,\text{GeV}^2 \approx (m_{D^0} + m_{\pi^0})^2 \le q^2 \le (m_X - m_{D^0})^2 \approx 4.02672\,\text{GeV}^2$$

is very narrow. Taking the masses, widths, and branching fractions of appearing D^* mesons from [13,89,131–134], we can calculate the decay width:

$$\Gamma(X_l \to \bar{D}^0 D^0 \pi^0) = \cos^2\theta\, \Gamma(X_u \to \bar{D}^0 D^0 \pi^0)$$

and study its dependence on the size parameter Λ_X. This is shown in Figure 3. By using the experimental data from PDG [13] for the ratio:

$$10^5 \mathcal{B}(B^\pm \to K^\pm X) \cdot \mathcal{B}(X \to J/\psi \pi^+ \pi^-) = 0.95 \pm 0.19,$$

$$10^5 \mathcal{B}(B^\pm \to K^\pm X) \cdot \mathcal{B}(X \to D^0 \bar{D}^0 \pi^0) = 10.0 \pm 4.0, \quad (32)$$

one finds:

$$\frac{\Gamma(X \to D^0 \bar{D}^0 \pi^0)}{\Gamma(X \to J/\psi \pi^+ \pi^-)} = 10.5 \pm 4.7. \quad (33)$$

The latter is to be compared to the CCQM prediction:

$$\left. \frac{\Gamma(X \to D^0 \bar{D}^0 \pi^0)}{\Gamma(X \to J/\psi \pi^+ \pi^-)} \right|_{\text{CCQM}} = 6.0 \pm 0.2, \quad (34)$$

where the uncertainty of the result reflects the uncertainty on Λ_X. One can see that the two numbers agree within errors.

3.2. Implications of X(3872) in the Charm Dissociation Process by Light Mesons

It is interesting to check the significance of $X(3872)$ in the reaction of the charm dissociation process $J/\psi + \rho(\omega) \to X(3872) \to \bar{D}D^*$, which plays an important role in heavy ion physics. This state will contribute to the s channel of the process. The X-addition to the full cross-section is written as:

$$\sigma(J/\psi + v^0 \to D(\bar{D}) + \bar{D}^*(D^*)) = 2(\cos\theta \mp \sin\theta)^2 \, \sigma(J/\psi + v^0 \to X_u \to \bar{D} + D^*), \quad (35)$$

$$\sigma(J/\psi + v^0 \to X_u \to \bar{D} + D^*) = \frac{1}{16 \pi s} \frac{\lambda^{1/2}(s, m_D^2, m_{D^*}^2)}{\lambda^{1/2}(s, m_{J/\psi}^2, m_{v^0}^2)} \cdot \frac{1}{9} \sum_{\text{pol}} \frac{|A|^2}{(s - m_X^2)^2 + \Gamma_X^2 m_X^2},$$

$$A = \varepsilon_{J/\psi}^\nu \varepsilon_{v^0}^\rho M_{\mu\nu\rho} \left(-g^{\mu\alpha} + \frac{p^\mu p^\alpha}{m_X^2} \right) \varepsilon_{D^*}^\beta M_{\alpha\beta},$$

where $p = p_1 + p_2 = q_1 + q_2$. The $\mp$ sign in the first equation is negative for the ρ meson and positive for ω. A Breit–Wigner propagator is used with $\Gamma_X = 1$ MeV, and the size parameter value is fixed to $\Lambda_X = 3.5$ GeV. With this setting, the dependence of the cross-section on the energy $E = \sqrt{s}$ is shown in Figure 4.

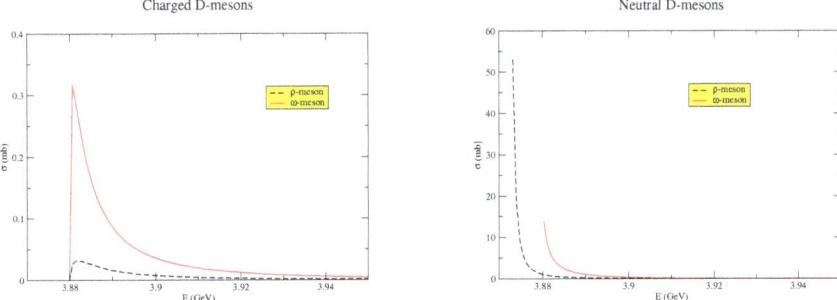

Figure 4. The cross-sections of the processes $J/\psi + v^0 \to X \to D + D^*$. Charged D-mesons in left panel; neutral D-mesons in the right panel.

One can compare the predicted behavior to available results for the charged D-mesons: At $E = 4.0$ GeV, a theoretical evaluation [135] predicts $\sigma(J/\psi + \pi \to D + \bar{D}^*) = 0.9$ mb, and the work in [136] predicted $\sigma(J/\psi + \rho \to D + \bar{D}^*) = 2.9$ mb at $E = 3.9$ GeV. In the case of $X(3872)$, the cross-section reaches the maximum of approximately 0.32 mb at $E = 3.88$ GeV, and one can conclude that the expected contribution of $X(3872)$ in the charm dissociation is non-negligible.

4. Radiative Decays of $X(3872)$

The first experimental evidence for the radiative decay of the $X(3872)$ particle was given in [129] by the Belle experiment. From the measured branching fraction product:

$$\mathcal{B}(B \to XK) \cdot \mathcal{B}(X \to \gamma + J/\psi) = (1.8 \pm 0.6 \, (\text{stat}) \pm 0.1 \, (\text{syst})) \times 10^{-6} \quad (36)$$

the partial width ratio was deduced:

$$\frac{\Gamma(X \to \gamma + J/\psi)}{\Gamma(X \to \pi^+\pi^- J/\psi)} = 0.14 \pm 0.05. \quad (37)$$

This finding was supported by the BaBar observation [137]:

$$\mathcal{B}(B^+ \to XK^+) \cdot \mathcal{B}(X \to \gamma + J/\psi) = (3.3 \pm 1.0 \, (\text{stat}) \pm 0.3 \, (\text{syst})) \times 10^{-6} \quad (38)$$

which had a limited significance of 3.4 σ. The same experiment reaffirmed the observation in 2009 [138] with smaller errors:

$$\mathcal{B}(B^\pm \to XK^\pm) \cdot \mathcal{B}(X \to \gamma + J/\psi) = (2.8 \pm 0.8 \, (\text{stat}) \pm 0.1 \, (\text{syst})) \times 10^{-6} \quad (39)$$

from which one can deduce [36]:

$$\frac{\Gamma(X \to \gamma + J/\psi)}{\Gamma(X \to \pi^+\pi^- J/\psi)} = 0.22 \pm 0.06. \quad (40)$$

BaBar also presented a result related to $\psi(2s)$:

$$\mathcal{B}(B^\pm \to XK^\pm) \cdot \mathcal{B}(X \to \gamma + \psi(2S)) = (9.5 \pm 2.7 \, (\text{stat}) \pm 0.6 \, (\text{syst})) \times 10^{-6}. \quad (41)$$

In 2011, the Belle collaboration published measurements with J/ψ and $\psi(2s)$ in the final state [139]:

$$\begin{aligned}
\mathcal{B}(B^\pm \to XK^\pm) \cdot \mathcal{B}(X \to \gamma + J/\psi) &= (1.78^{+0.48}_{-0.44} \, (\text{stat}) \pm 0.12 \, (\text{syst})) \times 10^{-6}, \\
\mathcal{B}(B^\pm \to XK^\pm) \cdot \mathcal{B}(X \to \gamma + \psi(2S)) &< 3.45 \times 10^{-6}.
\end{aligned} \quad (42)$$

The first result was in good agreement with the previous one from the same experiment (36); however, the second number brought some tension when compared to BaBar and a later LHCb measurement [140]:

$$\frac{\Gamma(X \to \psi(2s) + \gamma)}{\Gamma(X \to J/\psi + \gamma)} = \begin{cases} 3.4 \pm 1.4 & \text{BaBar} \\ < 2.0 \, (90\% \, \text{CL}) & \text{Belle} \\ 2.46 \pm 0.64 \, (\text{stat}) \pm 0.29 \, (\text{sys}) & \text{LHCb} \end{cases}. \quad (43)$$

The theoretical study of radiative $X(3872)$ decays includes several different approaches. Such decays were analyzed in [98] in the charmonium picture. The authors studied excited 1D and 2P states and their decays in relation with the electric dipole radiation and provided implications for quantum number assignments. The molecular hypothesis was considered in [106]. There, the authors argued that the validity of the molecular picture could be determined from the study of several $X(3872)$ decay channels (including some with the photon emission). The work in [141] was dedicated to radiative decays with two D mesons in the final state. It was claimed that the discrimination between the molecular and charmonium picture could be obtained via analysis of the photon spectrum. Several decay modes, which also included $J/\psi + \gamma$, were examined in [142] within a phenomenological Lagrangian approach. The predicted value of the radiative decay width depended on the model parameters and varied from 125 KeV to 250 KeV. In [143], $X(3872)$ was described as a mixture of charmonium and exotic molecular states and treated using QCD sum rules. The predicted radiative decay width ratio $\Gamma_X(J/\psi\gamma)/\Gamma_X(J/\psi\pi^+\pi^-) = 0.19 \pm 0.13$ was in agreement with experimental measurements. The excited charmonium hypothesis and study of E1 decay widths within the relativistic Salpeter method was presented in [144]. A description based on a charmonium-like picture with high spin 2^{-+} using a light front quark model was proposed in [145]. Later works [146–149] were mostly interested in the puzzling $\Gamma_X(\psi(2s)\gamma)/\Gamma_X(J/\psi\gamma)$ ratio (43) and analyzed it with different approaches (quark potential model, single-channel approximation, coupled-channel approach, charmonium-molecule hybrid model, and an effective theory framework).

Here, we focus on the J/ψ decay channel, which was studied using the CCQM in [79]. The non-local quark current for the $X(3872)$ hadron was given in the previous section; see Equation (22). The J/ψ quark current is written as:

$$J^\mu_{J/\psi}(y) = \int dy_1 \int dy_2 \, \delta\left(y - \frac{1}{2}(y_1 + y_2)\right) \times \Phi_{J/\psi}\left((y_1 - y_2)^2\right) \bar{c}_a(y_1) \gamma^\mu c_a(y_2). \quad (44)$$

The related size parameter was established in earlier works and has the value of $\Lambda_{J/\psi} = 1.738\,\text{GeV}$. The knowledge of the quark currents enables us to give more details concerning the interaction with photons, addressed before in Section 2.4. The second part of the electromagnetic interaction Lagrangian stands:

$$\mathcal{L}^{\text{EM}(2)}_{\text{int}}(x) = g_X X_{q\mu}(x) \cdot J^\mu_{X_q-\text{em}}(x) + g_{J/\psi} J/\psi_\mu(x) \cdot J^\mu_{J/\psi-\text{em}}(x), \quad (q = u, d)$$

$$J^\mu_{X_q-\text{em}} = \int d\vec{p}\, \Phi_X(\vec{p}^{\,2})\, J^\mu_{4q}(x_1,\ldots,x_4) \left\{ ie_q \left[I^{x_3}_x - I^{x_4}_x\right] + ie_c \left[I^{x_2}_x - I^{x_1}_x\right]\right\},$$

$$J^\mu_{J/\psi-\text{em}} = \int d\rho\, \Phi_{J/\psi}(\rho^2)\, J^\mu_{2q}(x_1,x_2)\, ie_c \left[I^{x_1}_x - I^{x_2}_x\right], \quad I^{x_i}_x \equiv I(x_i, x, P).$$

where J^μ_{4q} and J^μ_{2q} correspond to the parts of usual currents (22), (44) not containing the vertex function. In order to make use of the definition (15), it is convenient to switch to the Fourier transforms of the vertex functions and quark fields:

$$\Phi_X(\vec{p}^{\,2}) = \int \frac{d^4\vec{\omega}}{(2\pi)^4} \widetilde{\Phi}_X(-\vec{\omega}^{\,2}) e^{-i\vec{p}\vec{\omega}} = \widetilde{\Phi}_X(\vec{\partial}^2_\rho)\, \delta^{(4)}(\vec{\rho}),$$

$$\Phi_{J/\psi}(\rho^2) = \int \frac{d^4\omega}{(2\pi)^4} \widetilde{\Phi}_{J/\psi}(-\omega^2) e^{-i\rho\omega} = \widetilde{\Phi}_{J/\psi}(\partial^2_\rho)\, \delta^{(4)}(\rho),$$

$$q(x_i) = \int \frac{d^4 p_i}{(2\pi)^4} e^{-ip_i x_i} \tilde{q}(p_i), \quad \bar{q}(x_i) = \int \frac{d^4 p_i}{(2\pi)^4} e^{ip_i x_i} \tilde{\bar{q}}(p_i),$$

so that the differential operator can be placed in front of the path integrals:

$$J^\mu_{X_q-\text{em}} = \prod_{i=1}^{4} \int \frac{d^4 p_i}{(2\pi)^4} \widetilde{J}^\mu_{4q}(p_1,\ldots,p_4) \int d\vec{\rho}\, \delta^{(4)}(\vec{\rho}) \widetilde{\Phi}_X(\vec{\partial}^2_\rho) e^{-i(p_1 x_1 - p_2 x_2 - p_3 x_3 + p_4 x_4)} \cdot Q_X$$

$$= \prod_{i=1}^{4} \int \frac{d^4 p_i}{(2\pi)^4} \widetilde{J}^\mu_{4q}(p_1,\ldots,p_4) e^{-i(p_1-p_2-p_3+p_4)x} \int d\vec{\rho}\, \delta^{(4)}(\vec{\rho}) e^{-i\vec{\rho}\vec{\omega}} \widetilde{\Phi}_X(\vec{D}^2_\rho) \cdot Q_X$$

$$Q_X = ie_q \left[I^{x_3}_x - I^{x_4}_x\right] + ie_c \left[I^{x_2}_x - I^{x_1}_x\right],$$

15

$$J^\mu_{J/\psi-em} = \prod_{i=1}^{2}\int \frac{d^4p_i}{(2\pi)^4} \tilde{J}^\mu_{2q}(p_1,p_2)\int d\rho\, \delta^{(4)}(\rho)\tilde{\Phi}_{J/\psi}(\partial^2_\rho)\, e^{i(p_1x_1-p_2x_2)} \cdot Q_{J/\psi}$$

$$= \prod_{i=1}^{2}\int \frac{d^4p_i}{(2\pi)^4} \tilde{J}^\mu_{2q}(p_1,p_2)e^{i(p_1-p_2)x}\int d\rho\, \delta^{(4)}(\rho)e^{i p\rho}\tilde{\Phi}_{J/\psi}(D^2_\rho)\cdot Q_{J/\psi}$$

$$Q_{J/\psi} = ie_c\,[I^{x_1}_x - I^{x_2}_x],$$

where the long derivatives are defined as $D^\mu_{\rho_i} = \partial^\mu_{\rho_i} - i\omega^\mu_i$ and $D^\mu_\rho = \partial^\mu_\rho + ip^\mu$, $p=\frac{1}{2}(p_1+p_2)$ with ω_i being combinations of the integration four-vectors p_i and mass parameters w_q and w_c. Next, the identity involving the operator function action on the path integral [150] is applied:

$$F(D^2_{\rho_i})I^{x_i}_x = \int_0^1 d\tau F'(\tau D^2_{\rho_i} - (1-\tau)\omega^2_j)\, w_{ij}\cdot\left(\partial^\nu_{\rho_i}A_\nu(x_i) - 2i\omega^\nu_j A_\nu(x_i)\right) + F(-\omega^2_j)I^{x_i}_x. \tag{45}$$

Its validity extends to all functions F analytic at zero. The result for $X(3872)$ reads:

$$J^\mu_{X_q-em}(x) = \prod_{i=1}^{4}\int d^4x_i \int d^4y\, J^\mu_{4q}(x_1,\ldots,x_4)\, A_\rho(y)\cdot E^\rho_X(x;x_1,\ldots,x_4,y), \tag{46}$$

$$E^\rho_X(x;x_1,\ldots,x_4,y) = \prod_{i=1}^{4}\int \frac{d^4p_i}{(2\pi)^4}\int \frac{d^4r}{(2\pi)^4} e^{-ip_1(x-x_1)+ip_2(x-x_2)+ip_3(x-x_3)-ip_4(x-x_4)-ir(x-y)}\tilde{E}^\rho_X(p_1,\ldots,p_4,r),$$

$$\tilde{E}^\rho_X(p_1,\ldots,p_4,r) = \int_0^1 d\tau \sum_{j=1}^{3}\left\{e_c\left[-\tilde{\Phi}'_X(-z_{1j})\,l^\rho_{1j} + \tilde{\Phi}'_X(-z_{2j})\,l^\rho_{2j}\right]\right.$$
$$\left. + e_q\left[-\tilde{\Phi}'_X(-z_{4j})\,l^\rho_{4j} + \tilde{\Phi}'_X(-z_{3j})\,l^\rho_{3j}\right]\right\},$$

$$l_{ij} = w_{ij}(w_{ij}r + 2\omega_j), \qquad (i=1,\ldots,4;\ j=1,\ldots,3),$$
$$z_{i1} = \tau(w_{i1}r+\omega_1)^2 + (1-\tau)\,\omega^2_1 + \omega^2_2 + \omega^2_3,$$
$$z_{i2} = (w_{i1}r+\omega_1)^2 + \tau(w_{i2}r+\omega_2)^2 + (1-\tau)\,\omega^2_2 + \omega^2_3,$$
$$z_{i3} = (w_{i1}r+\omega_1)^2 + (w_{i2}r+\omega_2)^2 + \tau(w_{i3}r+\omega_3)^2 + (1-\tau)\,\omega^2_3.$$

For J/ψ, one obtains:

$$J^\nu_{J/\psi-em}(y) = \int d^4y_1\int d^4y_2\int d^4z\, J^\nu_{2q}(y_1,y_2)\,A_\rho(z)\,E^\rho_{J/\psi}(y;y_1,y_2,z), \tag{47}$$

$$E^\rho_{J/\psi}(y;y_1,y_2,z) = \int \frac{d^4p_1}{(2\pi)^4}\int \frac{d^4p_2}{(2\pi)^4}\int \frac{d^4q}{(2\pi)^4} e^{-ip_1(y_1-y)+ip_2(y_2-y)+iq(z-y)}\tilde{E}^\rho_{J/\psi}(p_1,p_2,q),$$

$$\tilde{E}^\rho_{J/\psi}(p_1,p_2,q) = e_c\int_0^1 d\tau\left\{-\tilde{\Phi}'_{J/\psi}(-z_-)\,l^\rho_- - \tilde{\Phi}'_{J/\psi}(-z_+)\,l^\rho_+\right\},$$

$$z_\mp = \tau(p\mp\tfrac{1}{2}q)-(1-\tau)p^2,\qquad l_\mp = p\mp\tfrac{1}{2}q,\qquad p=\tfrac{1}{2}(p_1+p_2).$$

The amplitude evaluation requires evaluation of four Feynman diagrams displayed in Figure 5. The corresponding expression stands:

$$M(X_q(p)\to J/\psi(q_1)\gamma(q_2)) = i(2\pi)^4\delta^{(4)}(p-q_1-q_2)\,\varepsilon^\mu_X\,\varepsilon^\rho_\gamma\,\varepsilon^\nu_{J/\psi}\,T_{\mu\rho\nu}(q_1,q_2), \tag{48}$$

where $T_{\mu\rho\nu}(q_1,q_2)$ can be expanded in terms of appropriate Lorentz structures. Using the on-mass shell condition, gauge invariance, and Schouten identities [151], one can show that only two independent structures remain:

$$T_{\mu\rho\nu} = W_A\,\varepsilon_{q_1 q_2 \mu\rho}q_{2\nu} + W_B\,\varepsilon_{q_1 q_2 \nu\rho}q_{1\mu}. \tag{49}$$

The functions $W_{A/B}$ are to be extracted from the expression following from the CCQM computation:

$$T_{\mu\rho\nu}(q_1, q_2) = \sum_{i=a,b,c,d} T^{(i)}_{\mu\rho\nu}(q_1, q_2), \qquad (50)$$

where the separate contributions are written down:

$$T^{(a)}_{\mu\rho\nu} = 6\sqrt{2} g_X g_{J/\psi} e_q \int \frac{d^4k_1}{(2\pi)^4 i} \int \frac{d^4k_2}{(2\pi)^4 i} \widetilde{\Phi}_X\left(-K_a^2\right) \widetilde{\Phi}_{J/\psi}\left(-(k_1 + \tfrac{1}{2}q_1)^2\right)$$
$$\times \tfrac{1}{2} \mathrm{tr}\left[\gamma_5 S_c(k_1)\gamma_\nu S_c(k_1+q_1)\gamma_\mu S_q(k_2)\gamma_\rho S_q(k_2+q_2) - (\gamma_5 \leftrightarrow \gamma_\mu)\right],$$
$$K_a^2 = \tfrac{1}{2}(k_1+\tfrac{1}{2}q_1)^2 + \tfrac{1}{2}(k_2+\tfrac{1}{2}q_2)^2 + \tfrac{1}{4}(w_q q_1 - w_c q_2)^2,$$

$$T^{(b)}_{\mu\rho\nu} = 6\sqrt{2} g_X g_{J/\psi} \int \frac{d^4k_1}{(2\pi)^4 i} \int \frac{d^4k_2}{(2\pi)^4 i} \widetilde{\Phi}_{J/\psi}\left(-(k_2+\tfrac{1}{2}q_1)^2\right) \widetilde{E}_{X\rho}(p_1,\ldots,p_4,r)$$
$$\times \tfrac{1}{2}\mathrm{tr}\left[\gamma_5 S_q(k_1)\gamma_\mu S_c(k_2)\gamma_\nu S_c(k_2+q_1) - (\gamma_5 \leftrightarrow \gamma_\mu)\right],$$
$$p_1 = k_2, \quad p_2 = k_2 + q_1, \quad p_3 = p_4 = -k_1, \quad r = -q_2,$$

$$T^{(c)}_{\mu\rho\nu} = 6\sqrt{2} g_X g_{J/\psi} e_c \int \frac{d^4k_1}{(2\pi)^4 i} \int \frac{d^4k_2}{(2\pi)^4 i} \widetilde{\Phi}_X\left(-K_c^2\right) \widetilde{\Phi}_{J/\psi}\left(-(k_2+q_2+\tfrac{1}{2}q_1)^2\right)$$
$$\times \tfrac{1}{2}\mathrm{tr}\left[\gamma_5 S_q(k_1)\gamma_\mu S_c(k_2)\gamma_\rho S_c(k_2+q_2)\gamma_\nu S_c(k_2+p) - (\gamma_5 \leftrightarrow \gamma_\mu)\right],$$
$$K_c^2 = \tfrac{1}{2}k_1^2 + \tfrac{1}{2}(k_2+\tfrac{1}{2}p)^2 + \tfrac{1}{4}w_q^2 p^2,$$

$$T^{(d)}_{\mu\rho\nu} = 6\sqrt{2} g_X g_{J/\psi} e_c \int \frac{d^4k_1}{(2\pi)^4 i} \int \frac{d^4k_2}{(2\pi)^4 i} \widetilde{\Phi}_X\left(-K_c^2\right) \widetilde{E}_{J/\psi\rho}(p_1,p_2,q)$$
$$\times \tfrac{1}{2}\mathrm{tr}\left[\gamma_\mu S_q(k_1)\gamma_5 S_c(k_2)\gamma_\nu S_c(k_2+p) - (\gamma_5 \leftrightarrow \gamma_\mu)\right],$$
$$p_1 = -k_2 - p, \quad p_2 = -k_2, \quad q = -q_2.$$

One evaluates the traces and the loop momenta integrals, and the expression is re-arranged in two terms following the mentioned Lorentz structure. The behavior of coefficient functions $W_{A/B}$ is predicted using a numerical integration over the Schwinger parameters:

$$W_{A,B} = \int_0^\infty dt \int_0^1 d^3\beta\, F_{A,B}(t,\beta_1,\beta_2,\beta_3). \qquad (51)$$

The decay width is expressed as:

$$\Gamma(X \to \gamma J/\psi) = \frac{1}{12\pi} \frac{|\vec{q}_2|}{m_X^2}\left(|H_L|^2 + |H_T|^2\right), \qquad (52)$$

where H_i denote the helicity amplitudes:

$$H_L = i\frac{m_X^2}{m_{J/\psi}}|\vec{q}_2|^2 W_A, \qquad H_T = -i m_X |\vec{q}_2|^2 W_B \qquad (53)$$

with $|\vec{q}_2| = \left(m_X^2 - m_{J/\psi}^2\right)/(2m_X)$. The dependence of the predicted decay width on the size parameter Λ_X is shown in Figure 6.

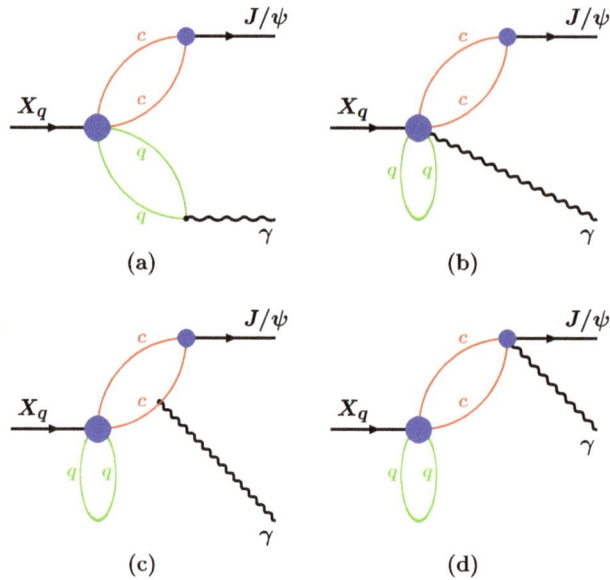

Figure 5. Four Feynman diagrams describing the decay $X \to \gamma + J/\psi$. One with the photon emission form the light quark line (**a**) and three bubble graphs (**b**–**d**).

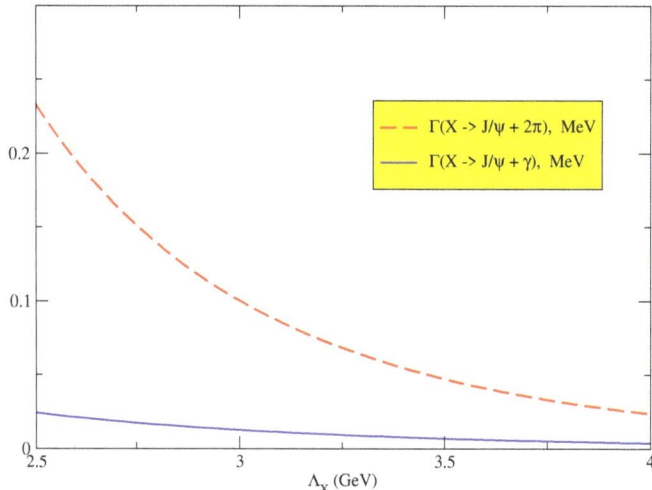

Figure 6. The dependence of the decay widths $\Gamma(X_l \to \gamma + J/\psi)$ and $\Gamma(X_l \to J/\psi\, 2\pi)$ on the size parameter Λ_X.

If we follow the approach from the previous section and take $\Lambda_X = 3.0 \pm 0.5$ GeV, then the model predicts:

$$\frac{\Gamma(X_l \to \gamma + J/\psi)}{\Gamma(X_l \to J/\psi + 2\pi)}\bigg|_{\text{CCQM}} = 0.15 \pm 0.03, \tag{54}$$

which is to be compared with the experimental results Equation (37) and Equation (40). One may conclude that the bound-tetraquark description of the $X(3872)$ state by the CCQM is in an agreement with the experimental observations.

5. Nature of $Z_c(3900)$

As stated in the Introduction, the detected $Z_c(3900)$ decays include both the $\pi^\pm J/\psi$ and D^*D final states (assuming $Z_c(3885)$ and $Z_c(3900)$ are the same particle). The ratio of the decay widths of these two channels was measured by BESIII [27]:

$$\frac{\Gamma(Z_c \to D\bar{D}^*)}{\Gamma(Z_c \to \pi J/\psi)} = 6.2 \pm 1.1(\text{stat}) \pm 2.7(\text{syst}) \tag{55}$$

and represents a quantitative observation to be explained by the theorists. There are many different theoretical approaches that are trying to understand the nature of this state.

The tetraquark interpretation was intensively discussed within QCD sum rules [152–154] and also in the color flux-tube model [155]. The molecular scenario seems to be more abundant in the literature and is discussed or preferred in several theoretical frameworks. A light front theory description was presented in [156]; an effective field theory description was proposed in [157]; and QCD sum rules were used in [158,159]. The molecular interpretation was also supported by the quark model developed in [160]. The authors of [161] made a proposal for BESIII and forthcoming Belle II measurements by using also the molecular scenario. Further molecular picture oriented works can be found in [162] (constituent quark model, coupled channels) and in [163] (quark interchange model). It is interesting to note that most of the lattice QCD based studies obtained different results from previous ones: some did not see (within the approach they used) a bound state at all [164–167], invoked a threshold cusp explanation [168,169], or indicated that the understanding of Z_c within the lattice QCD was only approaching [170]. For completeness, one can mention the charmonium hybrid interpretation studied in [171], the hadro charmonium picture presented in [172] with the tetraquark and molecular interpretation and the color magnetic interaction [173]. Further ideas can be found in [174–184].

The description of $Z_c(3900)$ in the framework of the CCQM was presented in [80]. Two options were tested: the molecular interpretation and the tetraquark hypothesis. For each option, the strong decays into $J/\psi\pi^+$, $\eta_c\rho^+$, $\bar{D}^0D^{*+}$, and $\bar{D}^{*0}D^+$ were computed and compared to available experimental data. First, we investigate the tetraquark hypothesis. In this scenario, the non-local Z_c current is written as:

$$J_{Z_c}^\mu(x) = \int dx_1 \ldots \int dx_4 \, \delta\left(x - \sum_{i=1}^4 w_i x_i\right) \cdot \Phi_{Z_c}\left(\sum_{i<j}(x_i - x_j)^2\right) J_{4q}^\mu(x_1, \ldots, x_4), \tag{56}$$

$$J_{4q}^\mu = \frac{i}{\sqrt{2}} \varepsilon_{abc}\varepsilon_{dec} \left\{ [u_a(x_4)C\gamma_5 c_b(x_1)][\bar{d}_d(x_3)\gamma^\mu C\bar{c}_e(x_2)] - (\gamma_5 \leftrightarrow \gamma^\mu) \right\}.$$

The tetraquark mass operator looks like:

$$\Pi_{Z_c}^{\mu\nu}(p) = 6 \prod_{i=1}^3 \int \frac{d^4k_i}{(2\pi)^4 i} \tilde{\Phi}_{Z_c}^2\left(-\tilde{\omega}^2\right) \tag{57}$$

$$\times \left\{ \text{tr}\left(S_4(\hat{k}_4)\gamma_5 S_1(\hat{k}_1)\gamma_5\right) \text{tr}\left(S_3(\hat{k}_3)\gamma^\mu S_2(\hat{k}_2)\gamma^\nu\right) \right.$$

$$\left. + \text{tr}\left(S_4(\hat{k}_4)\gamma^\nu S_2(\hat{k}_2)\gamma^\mu\right) \text{tr}\left(S_3(\hat{k}_3)\gamma_5 S_1(\hat{k}_1)\gamma_5\right) \right\},$$

where the momenta are defined by:

$$\hat{k}_1 = k_1 - w_1 p, \quad \hat{k}_2 = k_2 - w_2 p, \quad \hat{k}_3 = k_3 + w_3 p, \quad \hat{k}_4 = k_1 + k_2 - k_3 + w_4 p,$$
$$\tilde{\omega}^2 = 1/2\,(k_1^2 + k_2^2 + k_3^2 + k_1 k_2 - k_1 k_3 - k_2 k_3).$$

The matrix elements of the decays $Z_c^+ \to J/\psi + \pi^+$ and $Z_c^+ \to \eta_c + \rho^+$ are written down:

$$M^{\mu\nu}\left(Z_c(p,\epsilon_p^\mu) \to J/\psi(q_1,\epsilon_{q_1}^\nu) + \pi^+(q_2)\right) = \frac{6}{\sqrt{2}} g_{Z_c} g_{J/\psi} g_\pi$$

$$\times \int \frac{d^4k_1}{(2\pi)^4 i} \int \frac{d^4k_2}{(2\pi)^4 i} \tilde{\Phi}_{Z_c}\left(-\tilde{\eta}^2\right) \tilde{\Phi}_{J/\psi}\left(-(k_1+v_2q_1)^2\right) \tilde{\Phi}_\pi\left(-(k_2+u_4q_2)^2\right)$$

$$\times \left\{ \text{tr}\left(\gamma_5 S_4(k_2)\gamma_5 S_3(k_2+q_2)\gamma^\mu S_2(k_1)\gamma^\nu S_1(k_1+q_1)\right) + (\gamma_5 \leftrightarrow \gamma^\mu) \right\}$$

$$= A_{J/\psi\pi} g^{\mu\nu} + B_{J/\psi\pi} q_1^\mu q_2^\nu, \tag{58}$$

$$M^{\mu\alpha}\left(Z_c(p,\epsilon_p^\mu) \to \eta_c(q_1) + \rho(q_2, \epsilon_{q_2}^\alpha)\right) = \frac{6}{\sqrt{2}} g_{Z_c} g_{\eta_c} g_\rho$$

$$\times \int \frac{d^4k_1}{(2\pi)^4 i} \int \frac{d^4k_2}{(2\pi)^4 i} \tilde{\Phi}_{Z_c}\left(-\tilde{\eta}^2\right) \tilde{\Phi}_{\eta_c}\left(-(k_1+v_2q_1)^2\right) \tilde{\Phi}_\rho\left(-(k_2+u_4q_2)^2\right)$$

$$\times \left\{ \text{tr}\left[\gamma_5 S_4(k_2)\gamma^\alpha S_3(k_2+q_2)\gamma^\mu S_2(k_1)\gamma_5 S_1(k_1+q_1)\right] + (\gamma_5 \leftrightarrow \gamma^\mu) \right\}$$

$$= A_{\eta_c\rho} g^{\mu\alpha} - B_{\eta_c\rho} q_2^\mu q_1^\alpha, \tag{59}$$

where the argument of the Z_c-vertex function is given by:

$$\eta_1 = \frac{1}{2\sqrt{2}}(2k_1 + (1-w_1+w_2)q_1 - (w_1-w_2)q_2),$$

$$\eta_2 = \frac{1}{2\sqrt{2}}(2k_2 - (w_3-w_4)q_1 + (1-w_3+w_4)q_2),$$

$$\eta_3 = \frac{1}{2}((w_3+w_4)q_1 - (w_1+w_2)q_2), \qquad \tilde{\eta}^2 = \eta_1^2 + \eta_2^2 + \eta_3^2.$$

The notations used are as follows: $m_1 = m_2 = m_c$, $m_3 = m_4 = m_d = m_u$, $v_1 = m_1/(m_1+m_2)$, $v_2 = m_2/(m_1+m_2)$, $u_3 = m_3/(m_3+m_4)$, and $u_4 = m_4/(m_3+m_4)$.

The amplitudes of the $Z_c^+ \to \bar{D}^0 + D^{*+}$ and $Z_c^+ \to \bar{D}^{*0} + D^+$ decays are:

$$M^{\mu\nu}\left(Z_c(p,\epsilon_p^\mu) \to \bar{D}^0(q_1) + D^{*+}(q_2, \epsilon_{q_2}^\nu)\right) = \frac{6}{\sqrt{2}} g_{Z_c} g_D g_{D^*}$$

$$\times \int \frac{d^4k_1}{(2\pi)^4 i} \int \frac{d^4k_2}{(2\pi)^4 i} \tilde{\Phi}_{Z_c}\left(-\tilde{\delta}^2\right) \tilde{\Phi}_D\left(-(k_2+v_2q_2)^2\right) \tilde{\Phi}_{D^*}\left(-(k_1+u_1q_2)^2\right)$$

$$\times \left\{ \text{tr}\left(\gamma_5 S_4(k_2+q_1)\gamma_5 S_1(k_1)\gamma^\nu S_3(k_1+q_2)\gamma^\mu S_2(k_2)\right) - (\gamma_5 \leftrightarrow \gamma^\mu) \right\}$$

$$= A_{\bar{D}D^*} g^{\mu\nu} - B_{\bar{D}D^*} q_2^\mu q_1^\nu, \tag{60}$$

$$M^{\mu\alpha}\left(Z_c(p,\epsilon_p^\mu) \to \bar{D}^{*0}(q_1, \epsilon_{q_1}^\alpha) + D^+(q_2,)\right) = \frac{6}{\sqrt{2}} g_{Z_c} g_{D^*} g_D$$

$$\times \int \frac{d^4k_1}{(2\pi)^4 i} \int \frac{d^4k_2}{(2\pi)^4 i} \tilde{\Phi}_{Z_c}\left(-\tilde{\delta}^2\right) \tilde{\Phi}_{D^*}\left(-(k_1+\hat{v}_1q_1)^2\right) \tilde{\Phi}_D\left(-(k_2+\hat{u}_4q_2)^2\right)$$

$$\times \left\{ \text{tr}\left(S_4(k_2+q_1)\gamma_5 S_1(k_1)\gamma_5 S_3(k_1+q_2)\gamma^\mu S_2(k_2)\gamma^\alpha\right) - (\gamma_5 \leftrightarrow \gamma^\mu) \right\}$$

$$= A_{D^*D} g^{\mu\alpha} + B_{D^*D} q_1^\mu q_2^\alpha, \tag{61}$$

with the argument of the Z_c-vertex function being:

$$\begin{aligned}
\delta_1 &= -\tfrac{1}{2\sqrt{2}}(k_1 - k_2 + (w_1 - w_2)(q_1 + q_2)), \\
\delta_2 &= +\tfrac{1}{2\sqrt{2}}(k_1 - k_2 - (1 + w_3 - w_4)q_1 + (1 - w_3 + w_4)q_2), \\
\delta_3 &= -\tfrac{1}{2}(k_1 + k_2 + (w_1 + w_2)(q_1 + q_2)), \qquad \vec{\delta}^2 = \delta_1^2 + \delta_2^2 + \delta_3^2.
\end{aligned} \tag{62}$$

Now, the notation used is $m_1 = m_2 = m_c$, $m_3 = m_4 = m_d = m_u$, $\hat{v}_2 = m_2/(m_2 + m_4)$, $\hat{v}_4 = m_4/(m_2 + m_4)$, $\hat{u}_1 = m_1/(m_1 + m_3)$, and $\hat{u}_3 = m_3/(m_1 + m_3)$.

The decay width for the $1^+(p) \to 1^-(q_v) + 0^-(q_s)$ transition is given by:

$$\Gamma = \frac{1}{8\pi} \frac{1}{2s+1} \frac{|\mathbf{q}_v|}{m^2} (|H_{+1+1}|^2 + |H_{-1-1}|^2 + |H_{00}|^2), \tag{63}$$

where H denotes the helicity amplitudes and $\mathbf{q}_v$ is the three-momentum of the final state vector particle $q_v^\mu = (E_v, 0, 0, |\mathbf{q}_v|)$. The helicity amplitudes can be related to the invariant amplitudes $\mathcal{A}_1$ and $\mathcal{A}_2$, which parametrize the matrix element in terms of the Lorentz structures:

$$M = \mathcal{A}_1 \, m \, g^{\mu\rho} + \mathcal{A}_2 \, \frac{1}{m} q_1^\mu q_2^\rho \tag{64}$$

by means of the relations:

$$H_{00} = -\frac{m}{m_1} E_v \mathcal{A}_1 - \frac{1}{m_1}|\mathbf{q}_v|^2 \mathcal{A}_2, \qquad H_{+1+1} = H_{-1-1} = -m \mathcal{A}_1.$$

From the comparison of Equation (64) with Equations (58)–(61), one can express $\mathcal{A}_{1,2}$ as a function of A_{xy}, B_{xy}. The results are importantly influenced by the fact that the amplitudes $A_{\bar{D}D^*}$ and A_{D^*D} (Formulas (60) and (61)) vanish exactly within the CCQM description $A_{\bar{D}D^*} = A_{D^*D} = 0$, and the contributions from the non-zero B amplitudes are strongly suppressed by the $|\mathbf{q}_v|^5$ factor. Before arriving at the numerical predictions, the size parameters need to be specified, and a strategy with respect to the choice of Λ_{Z_c} value has to be settled. The numerical values of the size parameters were in [80] (i.e., the herein presented Z_c analysis) re-adjusted with respect to those in [78] and are shown in Table 3.

Table 3. The size parameters for selected mesons in GeV used in the $Z_c(3900)$ analysis.

Λ_π	$\Lambda_{\rho/\omega}$	Λ_D	Λ_{D^*}	$\Lambda_{J/\psi}$	Λ_{η_c}
0.711	0.295	1.4	2.3	3.3	3.0

As concerns the Λ_{Z_c} parameter, first, it is taken as $\Lambda_{Z_c} = 2.24 \pm 0.10$ GeV to make the predicted value of the decay width $\Gamma(Z_c^+ \to J/\psi + \pi^+)$ close to the one from [152,176]. One obtains:

$$\begin{aligned}
\Gamma(Z_c^+ \to J/\psi + \pi^+) &= (27.9^{+6.3}_{-5.0}) \text{ MeV}, & \Gamma(Z_c^+ \to \bar{D}^0 + D^{*+}) &\propto 10^{-8} \text{ MeV}, \\
\Gamma(Z_c^+ \to \eta_c + \rho^+) &= (35.7^{+6.3}_{-5.2}) \text{ MeV}, & \Gamma(Z_c^+ \to \bar{D}^{*0} + D^+) &\propto 10^{-8} \text{ MeV}.
\end{aligned} \tag{65}$$

These outputs contradict the experimental number (see Equation (55)), which indicates a larger coupling to DD^* than to the $J/\psi\pi$ mode. If trying to adjust the Λ_{Z_c} parameter to a more realistic value, the results do not become any better. Assuming $\Lambda_{Z_c} = 3.3 \pm 1.1$ GeV, one gets:

$$\begin{aligned}
\Gamma(Z_c^+ \to J/\psi + \pi^+) &= (4.3^{+0.7}_{-0.6}) \text{ MeV}, & \Gamma(Z_c^+ \to \bar{D}^0 + D^{*+}) &\propto 10^{-9} \text{ MeV}, \\
\Gamma(Z_c^+ \to \eta_c + \rho^+) &= (8.0^{+1.2}_{-1.0}) \text{ MeV}, & \Gamma(Z_c^+ \to \bar{D}^{*0} + D^+) &\propto 10^{-9} \text{ MeV}.
\end{aligned} \tag{66}$$

These predictions suggest that the tetraquark picture is not appropriate for the $Z_c(3900)$ state.

The molecular description of $Z_c(3900)$ appears as a natural alternative. In such a scenario, the non-local interpolation quark current is written as [53]:

$$J^\mu_{4q} = \frac{1}{\sqrt{2}}\left\{(\bar{d}(x_3)\gamma_5 c(x_1))(\bar{c}(x_2)\gamma^\mu u(x_4)) + (\bar{d}(x_3)\gamma^\mu c(x_1))(\bar{c}(x_2)\gamma_5 u(x_4))\right\}. \tag{67}$$

By using similar steps as in the tetraquark analysis, one writes down the Fourier transformed Z_c mass operator in the form:

$$\Pi^{\mu\nu}_{Z_c}(p) = \frac{9}{2}\prod_{i=1}^{3}\int\frac{d^4 k_i}{(2\pi)^4 i}\tilde{\Phi}^2_{Z_c}\left(-\tilde{\omega}^2\right) \tag{68}$$

$$\times \left\{\text{tr}\left[\gamma_5 S_1(\hat{k}_1)\gamma_5 S_3(\hat{k}_3)\right]\cdot\text{tr}\left[\gamma^\mu S_4(\hat{k}_4)\gamma^\nu S_2(\hat{k}_2)\right]\right.$$

$$\left. +\text{tr}\left[\gamma^\mu S_1(\hat{k}_1)\gamma^\nu S_3(\hat{k}_3)\right]\cdot\text{tr}\left[\gamma_5 S_4(\hat{k}_4)\gamma_5 S_2(\hat{k}_2)\right]\right\}$$

in order to pin down the Λ_{Z_c} dependence of the coupling g_{Z_c}. Next, the transition amplitudes are constructed:

$$M^{\mu\nu}\left(Z_c(p,\epsilon^\mu_p)\to J/\psi(q_1,\epsilon^\nu_{q_1})+\pi^+(q_2)\right) = \frac{3}{\sqrt{2}}g_{Z_c}g_{J/\psi}g_\pi$$

$$\times \int\frac{d^4k_1}{(2\pi)^4 i}\int\frac{d^4k_2}{(2\pi)^4 i}\tilde{\Phi}_{Z_c}\left(-\tilde{\eta}^2\right)\tilde{\Phi}_{J/\psi}\left(-(k_1+v_1 q_1)^2\right)\tilde{\Phi}_\pi\left(-(k_2+u_4 q_2)^2\right)$$

$$\times \left\{\text{tr}\left(\gamma_5 S_1(k_1)\gamma^\nu S_2(k_1+q_1)\gamma^\mu S_4(k_2)\gamma_5 S_3(k_2+q_2)\right)+(\gamma_5\leftrightarrow\gamma^\mu)\right\}$$

$$= A_{J/\psi\pi}g^{\mu\nu}+B_{J/\psi\pi}q_1^\mu q_2^\nu. \tag{69}$$

$$M^{\mu\alpha}\left(Z_c(p,\epsilon^\mu_p)\to\eta_c(q_1)+\rho(q_2,\epsilon^\alpha_{q_2})\right) = \frac{3}{\sqrt{2}}g_{Z_c}g_{\eta_c}g_\rho$$

$$\times \int\frac{d^4k_1}{(2\pi)^4 i}\int\frac{d^4k_2}{(2\pi)^4 i}\tilde{\Phi}_{Z_c}\left(-\tilde{\eta}^2\right)\tilde{\Phi}_{\eta_c}\left(-(k_1+v_1 q_1)^2\right)\tilde{\Phi}_\rho\left(-(k_2+u_4 q_2)^2\right)$$

$$\times \left\{\text{tr}\left(\gamma_5 S_1(k_1)\gamma_5 S_2(k_1+q_1)\gamma^\mu S_4(k_2)\gamma^\alpha S_3(k_2+q_2)\right)+(\gamma_5\leftrightarrow\gamma^\mu)\right\}$$

$$= A_{\eta_c\rho}g^{\mu\alpha}-B_{\eta_c\rho}q_2^\mu q_1^\alpha, \tag{70}$$

$$M^{\mu\nu}\left(Z_c(p,\epsilon^\mu_p)\to\bar{D}^0(q_1)+D^{*+}(q_2,\epsilon^\nu_{q_2})\right) = \frac{9}{\sqrt{2}}g_{Z_c}g_D g_{D^*}$$

$$\times \int\frac{d^4k_1}{(2\pi)^4 i}\int\frac{d^4k_2}{(2\pi)^4 i}\tilde{\Phi}_{Z_c}\left(-\tilde{\delta}^2\right)\tilde{\Phi}_D\left(-(k_2+v_4 q_1)^2\right)\tilde{\Phi}_{D^*}\left(-(k_1+u_1 q_2)^2\right)$$

$$\times \text{tr}\left(\gamma^\mu S_1(k_1)\gamma^\nu S_3(k_1+q_2)\right)\cdot\text{tr}\left(\gamma_5 S_4(k_2)\gamma_5 S_2(k_2+q_1)\right)$$

$$= A_{\bar{D}D^*}g^{\mu\nu}-B_{\bar{D}D^*}q_2^\mu q_1^\nu, \tag{71}$$

$$M^{\mu\alpha}\left(Z_c(p,\epsilon_p^\mu) \to \bar{D}^{*0}(q_1,\epsilon_{q_1}^\alpha) + D^+(q_2)\right) = \frac{9}{\sqrt{2}} g_{Z_c} g_{D^*} g_D$$

$$\times \int \frac{d^4k_1}{(2\pi)^4 i} \int \frac{d^4k_2}{(2\pi)^4 i} \tilde{\Phi}_{Z_c}\left(-\vec{\delta}^2\right) \tilde{\Phi}_{D^*}\left(-(k_1+\hat{v}_1 q_1)^2\right) \tilde{\Phi}_D\left(-(k_2+\hat{u}_4 q_2)^2\right)$$

$$\times \operatorname{tr}\left(\gamma_5 S_1(k_1) \gamma_5 S_3(k_1+q_2)\right) \cdot \operatorname{tr}\left(\gamma^\mu S_4(k_2) \gamma^\alpha S_2(k_2+q_1)\right)$$

$$= A_{D^*D}\, g^{\mu\alpha} + B_{D^*D}\, q_1^\mu q_2^\alpha, \tag{72}$$

where the argument of the function $\tilde{\Phi}_{Z_c}$ is given by:

$$\begin{aligned}
\delta_1 &= -\tfrac{1}{2\sqrt{2}}\left(k_1+k_2+(1+w_1-w_2)q_1+(w_1-w_2)q_2\right), \\
\delta_2 &= +\tfrac{1}{2\sqrt{2}}\left(k_1+k_2-(w_3-w_4)q_1+(1-w_3+w_4)q_2\right), \\
\delta_3 &= +\tfrac{1}{2}\left(-k_1+k_2+(1-w_1-w_2)q_1-(w_1+w_2)q_2\right), \qquad \vec{\delta}^2=\delta_1^2+\delta_2^2+\delta_3^2. \tag{73}
\end{aligned}$$

The meaning of all other letters and symbols is the same as was in the previous paragraph dedicated to the tetraquark description. The decay widths are also evaluated in a fully analogous way. However, the parameter Λ_{Z_c} needs to be adjusted independently. Tuning its value in such a way so as to provide the best description of the BESIII measurement [27], one gets $\Lambda_{Z_c}=3.3\pm 1.1$ GeV with the following values for the decay widths:

$$\begin{aligned}
\Gamma(Z_c^+ \to J/\psi + \pi^+) &= (1.8\pm 0.3)\,\text{MeV}, & \Gamma(Z_c^+ \to \bar{D}^0 + D^{*+}) &= (10.0^{+1.7}_{-1.4})\,\text{MeV}, \\
\Gamma(Z_c^+ \to \eta_c + \rho^+) &= (3.2^{+0.5}_{-0.4})\,\text{MeV}, & \Gamma(Z_c^+ \to \bar{D}^{*0} + D^+) &= (9.0^{+1.6}_{-1.3})\,\text{MeV}. \tag{74}
\end{aligned}$$

One can see that the obtained results at this time are in agreement with the experimental observations by showing an enhancement of the DD^* sector and are in agreement with the observed branching fraction ratio in Equation (55) within the errors. One can conclude that the CCQM supports the molecular picture of the $Z_c(3900)$ state.

6. The Nature of $Y(4260)$

The distinctive characteristics of the $Y(4260)$ are its mass, which does not fit any charmonium in the same mass region, the suppression of open charm decays with respect to the $J/\psi \pi^+\pi^-$ final state, and the appearance of the exotic charmonium $Z_c(3900)$ among its decay products. This interesting mix of properties is addressed in quite a few theoretical works, and like in other cases, the molecular, tetraquark, and several other explanations are invoked.

A support for the molecular picture was provided by the QCD lattice computations in [185], by QCD sum rules in [186], by a meson exchange model in [187], and also by the authors of [188], which favored it over the hadro-charmonium interpretation. Further arguments for $Y(4260)$ being a molecule were based on the line shape study in [189], and the authors of [190] proposed an unconventional state with a large, but not completely dominant molecular component. An interesting paper [191] came up with a baryonic molecule concept, and the molecular hypothesis was also analyzed in [192–195].

On the contrary, the molecular scenario is strongly disfavored in [196] because of reasons related to the heavy quark spin symmetry and the molecular scenario was rejected in [197] in favor of a charmonium hybrid one. Here, the crux of the argument lies in an important separation between $Y(4260)$ mass and its decay threshold. Further arguments to support the charmonium or hybrid-charmonium picture were given in the publications [198–200].

One should also mention different quark models [201–204] with some of them favoring the tetraquark description of $Y(4260)$. The tetraquark hypothesis was also analyzed in the QCD sum rules study [205], and the coupled channels approach combined with the three-particle Faddeev equations was used to describe $Y(4260)$ in [206].

The analysis of Y(4260) is within the CCQM [207] done in a similar way to the Z_c case: its decay modes are analyzed in both the molecular and tetraquark scenario. With quantitative measurements related to Y(4260) not being very numerous, one can analyze the partial decay widths to $J/\psi \pi^+ \pi^-$ and open charm final states and see whether the latter ones are suppressed. The Feynman diagrams describing the studied transitions are drawn in Figure 7. The considered open charm final states include $D\bar{D}$, $D\bar{D}^*$, $D^*\bar{D}$, and $D^*\bar{D}^*$. As follows from the previous section, $Z_c(3900)$ is described as a molecular state (67).

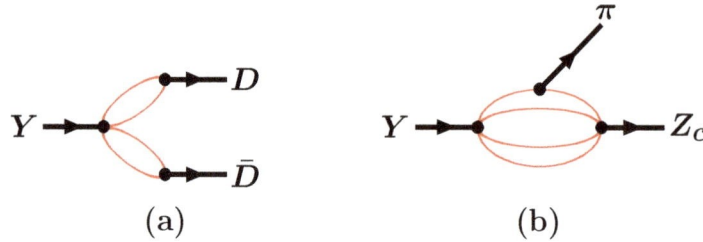

Figure 7. Feynman diagrams of the Y(4260) decay to open charm (a) and $Z_c \pi$ (b).

The molecular-type non-local interpolating current for Y(4260) is written as:

$$J^\mu_{Ymol}(x) = \int dx_1 \ldots \int dx_4 \, \delta\left(x - \sum_{i=1}^{4} w_i x_i\right) \Phi_Y\left(\sum_{i<j}(x_i - x_j)^2\right) J^\mu_{Ymol;4q}(x_1, \ldots, x_4), \quad (75)$$

$$J^\mu_{Ymol;4q} = \frac{1}{\sqrt{2}}\left\{(\bar{q}(x_3)\gamma_5 c(x_1)) \cdot (\bar{c}(x_2)\gamma^\mu \gamma_5 q(x_4)) - (\gamma_5 \leftrightarrow \gamma^\mu \gamma_5)\right\}, \quad (q = u, d)$$

with:

$$w_1 = w_2 = \frac{m_c}{2(m_q + m_c)}, \quad w_3 = w_4 = \frac{m_q}{2(m_q + m_c)}.$$

The matrix element corresponding to the open charm production is given by:

$$M\left(Y_u(p, \epsilon^\mu_p) \to D^0_1(p_1) + \bar{D}^0_2(p_2)\right) = \frac{9}{\sqrt{2}} g_Y g_{D_1} g_{D_2} \quad (76)$$

$$\times \int \frac{d^4 k_1}{(2\pi)^4 i} \int \frac{d^4 k_2}{(2\pi)^4 i} \tilde{\Phi}_Y\left(-\Omega_q^2\right) \tilde{\Phi}_{D_1}\left(-\ell_1^2\right) \tilde{\Phi}_{D_2}\left(-\ell_2^2\right)$$

$$\times \left\{\text{tr}\left(\gamma_5 S_c(k_1) \Gamma_2 S_u(k_3)\right) \text{tr}\left(\gamma^\mu \gamma_5 S_u(k_2) \Gamma_1 S_c(k_4)\right) - (\gamma_5 \leftrightarrow \gamma^\mu \gamma_5)\right\},$$

where:

$$\Gamma_1 \otimes \Gamma_2 = \begin{cases} \gamma_5 \otimes \gamma_5 & \text{for } D\bar{D} \\ \epsilon^*_{\nu_1} \gamma^{\nu_1} \otimes \gamma_5 & \text{for } D^*\bar{D} \\ \epsilon^*_{\nu_1} \gamma^{\nu_1} \otimes \epsilon^*_{\nu_2} \gamma^{\nu_2} & \text{for } D^*\bar{D}^* \end{cases} \quad (77)$$

and the momenta are defined as:

$$\Omega_q^2 = \frac{1}{2}\sum_{i\leq j} q_i q_j, \quad q_1 = -k_1 - w^Y_1 p, \quad q_2 = k_4 - w^Y_2 p, \quad q_3 = k_3 - w^Y_3 p,$$

$$\ell_1 = k_2 + w^D_u p_1, \quad \ell_2 = -k_1 - w^D_c p_2, \quad k_3 = k_1 + p_2, \quad k_4 = k_2 + p_1.$$

The decay into $Z_c + \pi$ involves a three-loop diagram, and the corresponding matrix element is:

$$M\left(Y_u(p,\epsilon^\mu) \to Z_c^+(p_1,\epsilon^\nu) + \pi^-\right) = \frac{9}{2} g_Y g_{Z_c} g_\pi \tag{78}$$

$$\times \prod_{j=1}^3 \left[\int \frac{d^4k_j}{(2\pi)^4 i}\right] \tilde\Phi_Y\left(-\Omega_q^2\right) \tilde\Phi_{Z_c}\left(-\Omega_r^2\right) \tilde\Phi_\pi\left(-\ell^2\right) \epsilon_\mu(p)\epsilon_\nu^*(p_1)$$

$$\times \sum_\Gamma \text{tr}\left(\Gamma_1 S_c(k_1)\Gamma_2 S_u(k_2)\right) \text{tr}\left(\Gamma_3 S_u(k_3)\Gamma_4 S_d(k_4)\Gamma_5 S_c(k_5)\right),$$

where:

$$\sum_\Gamma [\Gamma_1 \otimes \Gamma_2]\cdot[\Gamma_3 \otimes \Gamma_4 \otimes \Gamma_5] = [\gamma_5 \otimes \gamma_5]\cdot[\gamma^\mu\gamma_5 \otimes \gamma_5 \otimes \gamma^\nu]$$

$$- [\gamma^\mu\gamma_5 \otimes \gamma^\nu]\cdot[\gamma_5 \otimes \gamma_5 \otimes \gamma_5] - [\gamma^\mu\gamma_5 \otimes \gamma_5]\cdot[\gamma_5 \otimes \gamma_5 \otimes \gamma^\nu]$$

and the momenta are defined as:

$$\Omega_q^2 = \frac{1}{2}\sum_{i\le j} q_i q_j; \quad q_1 = -k_1 - w_1^Y p, \quad q_2 = k_5 - w_2^Y p, \quad q_3 = k_2 - w_3^Y p,$$

$$\Omega_r^2 = \frac{1}{2}\sum_{i\le j} r_i r_j; \quad r_1 = -k_5 + w_1^Z p_1, \quad r_2 = k_1 + w_2^Z p_1, \quad r_3 = k_4 - w_3^Z p_1,$$

$$\ell = k_3 + w_u^\pi p_2, \quad k_4 = k_3 + p_2, \quad k_5 = k_1 - k_2 + k_3 + p.$$

In the tetraquark scenario, the non-local $Y(4260)$ current takes the form:

$$J_{\text{Ytet}}^\mu(x) = \int dx_1 \ldots \int dx_4 \, \delta\left(x - \sum_{i=1}^4 w_i^Y x_i\right)\Phi_Y\left(\sum_{i<j}(x_i - x_j)^2\right) J_{\text{Ytet;4}q}^\mu(x_1,\ldots,x_4), \tag{79}$$

$$J_{\text{Ytet;4}q}^\mu = \frac{1}{\sqrt{2}}\epsilon_{abc}\epsilon_{dec}\left\{(q_a(x_4)C\gamma_5 c_b(x_1))(\bar q_d(x_3)\gamma^\mu\gamma_5 C\bar c_e(x_2)) - (\gamma_5 \leftrightarrow \gamma^\mu\gamma_5)\right\}. \tag{80}$$

The matrix element of the decay into $D\bar D$ is expressed as:

$$M\left(Y_u^{\text{tet}}(p,\epsilon_p^\mu) \to D_1^0(p_1) + \bar D_2^0(p_2)\right) = \frac{6}{\sqrt{2}} g_Y g_{D_1} g_{D_2} \tag{81}$$

$$\times \int \frac{d^4k_1}{(2\pi)^4 i}\int \frac{d^4k_2}{(2\pi)^4 i} \tilde\Phi_Y\left(-\Omega_q^2\right) \tilde\Phi_{D_1}\left(-\ell_1^2\right) \tilde\Phi_{D_2}\left(-\ell_2^2\right)$$

$$\times \left\{\text{tr}\left(\gamma_5 S_c(k_1)\Gamma_2^D S_u(k_3)\gamma^\mu\gamma_5 S_c(k_2)\Gamma_1^D S_u(k_4)\right) - (\gamma_5 \leftrightarrow \gamma^\mu\gamma_5)\right\},$$

with the momenta:

$$\Omega_q^2 = \frac{1}{2}\sum_{i\le j} q_i q_j; \quad q_1 = -k_1 - w_1^Y p, \quad q_2 = -k_2 - w_2^Y p, \quad q_3 = k_3 - w_3^Y p,$$

$$\ell_1 = -k_2 - w_c^D p_1, \quad \ell_2 = -k_1 - w_c^D p_2, \quad k_3 = k_1 + p_2, \quad k_4 = k_2 + p_1.$$

The matrix element of the decay into $Z_c\pi$ is given by:

$$M\left(Y_u^{\text{tet}}(p,\epsilon^\mu) \to Z_c^+(p_1,\epsilon^\nu) + \pi^-(p_2)\right) = 3 g_Y g_{Z_c} g_\pi \tag{82}$$

$$\times \prod_{j=1}^3 \left[\int \frac{d^4k_j}{(2\pi)^4 i}\right] \tilde\Phi_Y\left(-\Omega_q^2\right)\tilde\Phi_{Z_c}\left(-\Omega_r^2\right)\tilde\Phi_\pi\left(-\ell^2\right)$$

$$\times \epsilon_\mu(p)\epsilon_\nu^*(p_1)\sum_\Gamma \text{tr}\left[\Gamma_1^Y S_c(k_1)\Gamma_2^Z S_u(k_2)\Gamma_2^Y S_c(k_3)\tilde\Gamma_1^Z S_d(k_4)\gamma_5 S_u(k_5)\right],$$

with the momenta:

$$\Omega_q^2 = \frac{1}{2}\sum_{i\leq j} q_i q_j; \quad q_1 = -k_1 - w_1^Y p, \quad q_2 = -k_3 - w_2^Y p, \quad q_3 = k_2 - w_3^Y p,$$

$$\Omega_r^2 = \frac{1}{2}\sum_{i\leq j} r_i r_j; \quad r_1 = k_3 + w_1^Z p_1, \quad r_2 = k_1 + w_2^Z p_1, \quad r_3 = -k_4 + w_3^Z p_1,$$

$$\ell = -k_4 - w_d^\pi p_2, \quad k_4 = k_1 - k_2 + k_3 + p_1, \quad k_5 = k_1 - k_2 + k_3 + p.$$

Here, the summation over Γ is defined by:

$$\sum_\Gamma = [\gamma_5 \otimes \gamma^\mu \gamma_5 - \gamma^\mu \gamma_5 \otimes \gamma_5]^Y \otimes [\gamma_5 \otimes \gamma^\nu - \gamma^\nu \otimes \gamma_5]^Z.$$

The considered decays comprise different combinations of pseudoscalar, vector, and axial-vector particles in the final state. The relevant expressions for the matrix elements and decay widths are written down:

$$M(V(p) \to P(p_1) + P(p_2)) = \epsilon_V^\mu q_\mu G_{VPP}, \quad q = p_1 - p_2,$$

$$\Gamma(V \to PP) = \frac{|\mathbf{p_1}|^3}{6\pi m^2} G_{VPP}^2,$$

$$M(V(p) \to A(p_1) + P(p_2)) = \epsilon_V^\mu \epsilon_A^{*\nu} (g_{\mu\nu} A + p_{1\mu} p_\nu B),$$

$$\Gamma(V \to AP) = \frac{|\mathbf{p_1}|}{24\pi m^2}\left\{\left(3 + \frac{|\mathbf{p_1}|^2}{m_1^2}\right) A^2 + \frac{m^2}{m_1^2}|\mathbf{p_1}|^4 B^2 + \frac{m^2 + m_1^2 - m_2^2}{m_1^2}|\mathbf{p_1}|^2 AB\right\},$$

$$M(V(p) \to V(p_1) + P(p_2)) = \epsilon_V^\mu \epsilon_V^{*\nu_1} \varepsilon_{\mu\nu_1\alpha\beta} p^\alpha p_1^\beta G_{VVP},$$

$$\Gamma(V \to VP) = \frac{|\mathbf{p_1}|^3}{12\pi} G_{VVP}^2,$$

$$M(V(p) \to V(p_1) + V(p_2)) = \epsilon_V^\mu \epsilon_V^{*\nu_1} \epsilon_V^{*\nu_2}\{p_{1\mu} p_{1\nu_2} p_{2\nu_1} A + g_{\mu\nu_1} p_{1\nu_2} B + g_{\mu\nu_2} p_{2\nu_1} C + g_{\nu_1\nu_2} p_{1\mu} D\},$$

$$\Gamma(V \to V_1 V_2) = \frac{|\mathbf{p_1}|^3}{24\pi m_1^2 m_2^2}\Big\{m^2|\mathbf{p_1}|^4 A^2 + [|\mathbf{p_1}|^2 - 3m_1^2]B^2 + [|\mathbf{p_1}|^2 + 3m_2^2]C^2$$

$$+ [|\mathbf{p_1}|^2 + 3\frac{m_1^2 m_2^2}{m^2}]D^2 + |\mathbf{p_1}|^2[m^2 + m_1^2 - m_2^2]AB + |\mathbf{p_1}|^2[-m^2 + m_1^2 - m_2^2]AC$$

$$+ |\mathbf{p_1}|^2[m^2 - m_1^2 - m_2^2]AD + [2|\mathbf{p_1}|^2 - m^2 + m_1^2 + m_2^2]BC + [2|\mathbf{p_1}|^2 + m_1^2 + \frac{m_1^2}{m^2}(m_2^2 - m_1^2)]BD$$

$$+ [-2|\mathbf{p_1}|^2 - m_2^2 + \frac{m_2^2}{m^2}(m_2^2 - m_1^2)]CD\Big\}.$$

The value of Λ_{Z_c} is set to 3.3 GeV, and guided by our experience, we assume that $\Lambda_{Y(4260)} = 3.3 \pm 0.1$ GeV. The numerical evaluation leads to the results presented in Table 4.

In both scenarios, the open charm decays are suppressed with respect to the $J/\psi\pi$ decay channel. The discrimination between them is provided by the total decay width $\Gamma[Y(4260)] = 55 \pm 19$ MeV, which is in contradiction with the molecular description. Thus, one can conclude that the CCQM approach favors the tetraquark structure of $Y(4260)$.

Table 4. Decay widths of the selected Y(4260) transition in MeV.

Mode	Molecular-Type Current	Tetraquark Current
$Y \to Z_c^+ + \pi^-$	146 ± 13	5.77 ± 0.39
$Y \to D^0 + \bar{D}^0$	11 ± 2	$(0.42 \pm 0.16) \cdot 10^{-3}$
$Y \to D^{*0} + \bar{D}^0$	$(0.39 \pm 0.14) \cdot 10^{-2}$	0.32 ± 0.09
$Y \to D^{*0} + \bar{D}^{*0}$	0	$(0.19 \pm 0.08) \cdot 10^{-3}$

7. Bottomonium-Like States $Z_b(10610)$ and $Z_b'(10650)$

Exotic quarkonia states appear also in the bottomonium sector: $Z_b(10610)$ and $Z_b'(10650)$ are two examples. Even though the exotic bottomonia masses tend to be significantly higher than the charmonia ones, the underlying dynamics is similar, and one finds the molecular, tetraquark, and other hypotheses in theoretical approaches that describe them.

$Z_b(10610)$ and $Z_b'(10650)$ were seen as molecules in the boson exchange model of [208], and the molecular picture was also favored in [209], where the spin structure of these two particles was analyzed. Further support of the molecular scenario came from the quark model based on a phenomenological Lagrangian used by the authors of [210] and also from other analyses preformed in [211] (QCD multipole expansion), [212] (effective field theory), [213] (pion exchange model), [214] (QCD sum rules, only $Z_b'(10650)$ included), [215] (heavy quark spin symmetry and coupled channels analysis),and [216] (coupled channels approach with pion exchange model). A different set of works supports, with various intensity, the tetraquark structure of the two bottomonia states. In [217], the conclusion followed from an effective diquark-antidiquark Hamiltonian combined with meson-loop induced effects. The authors of [218] based their analysis on the QCD sum rules and interpreted Z_b and Z_b' as axial-vector tetraquarks. The two works [219,220] also drew their conclusions from the QCD sum rules and allowed the tetraquark and molecular scenario. The former work suggested that Z_b and Z_b' could have both the diquark-antidiquark and molecular components (following from a mixed interpolating current). The latter one excluded neither the tetraquark nor molecular the interpretation of $Z_b(10610)$, and the idea of a mixed current appeared also. The mentioned analyses could be supplemented by numerous other works [221–241] where further ideas and approaches were exploited.

The theoretical analysis of the $Z_b(10610)$ and $Z_b'(10650)$ states by the CCQM was performed in [81]. The work assumed a molecular-type interpolating current, which is favored by most theoretical approaches when interpreting the experimental results. It is a natural choice reflecting the proximity of the particle masses to the corresponding thresholds:

$$m(Z_b^+) = 10607.2 \pm 2.0 \, \text{MeV}, \quad m(B^*\bar{B}) = 10604 \, \text{MeV},$$
$$m(Z_b'^+) = 10652.2 \pm 1.5 \, \text{MeV}, \quad m(B^*\bar{B}^*) = 10649 \, \text{MeV}.$$

The quantum numbers of the two states $I^G(J^{PC}) = 1^+(1^{+-})$ lead to the choice of (local) interpolating currents:

$$J^\mu_{Z_b^+} = \tfrac{1}{\sqrt{2}} \left[(\bar{d}\gamma_5 b)(\bar{b}\gamma^\mu u) + (\bar{d}\gamma^\mu b)(\bar{b}\gamma_5 u) \right], \tag{83}$$

$$J^{\mu\nu}_{Z_b'^+} = \varepsilon^{\mu\nu\alpha\beta}(\bar{d}\gamma_\alpha b)(\bar{b}\gamma_\beta u), \tag{84}$$

which guarantees that, when considering the transitions into $B^{(*)}\bar{B}^{(*)}$, the Z_b state can decay only to the $[\bar{B}^*B + c.c.]$ pair, while the Z_b' state can decay only to a $\bar{B}^*B^*$ pair. Decays into the BB channels are not allowed.

Further decay channels include a bottomonium particle accompanied with a charged light meson. Taking into account the G parity, which is conserved in strong interactions and kinematic considerations, only three possible bottomonium-meson decay channels are available: $Z_b^+ \to Y + \pi^+$,

$Z_b^+ \to h_b + \pi^+$ and $Z_b^+ \to \eta_b + \rho^+$. All mentioned Z_b^+ transition can be arranged into three groups with respect to the spin kinematics:

$$1^+ \to 1^- + 0^- : \quad Z_b^+ \to Y + \pi^+, \quad Z_b^+ \to [\bar{B}^{*0}B^+ + c.c.], \quad Z_b^+ \to \eta_b + \rho^+,$$
$$1^+ \to 1^+ + 0^- : \quad Z_b^+ \to h_b + \pi^+,$$
$$1^+ \to 1^- + 1^- : \quad Z_b^+ \to \bar{B}^{*0}B^{*+}.$$

The classification of the bottomonia particles based on their quantum numbers is shown in Table 5.

Table 5. The bottomonium states $^{2S+1}L_J$. We use the notation $\overleftrightarrow{\partial} = \overrightarrow{\partial} - \overleftarrow{\partial}$.

Quantum Number $I^G(J^{PC})$	Name	Quark Current	Mass (MeV)
$0^+(0^{-+})$ ($S=0, L=0$)	$^1S_0 = \eta_b(1S)$	$\bar{b}i\gamma^5 b$	9399.00 ± 2.30
$0^-(1^{--})$ ($S=1, L=0$)	$^3S_1 = Y$	$\bar{b}\gamma^\mu b$	9460.30 ± 0.26
$0^+(0^{++})$ ($S=1, L=1$)	$^3P_0 = \chi_{b0}$	$\bar{b}b$	9859.44 ± 0.52
$0^+(1^{++})$ ($S=1, L=1$)	$^3P_1 = \chi_{b1}$	$\bar{b}\gamma^\mu\gamma^5 b$	9892.72 ± 0.40
$0^-(1^{+-})$ ($S=0, L=1$)	$^1P_1 = h_b(1P)$	$\bar{b}\overleftrightarrow{\partial}^\mu \gamma^5 b$	9899.30 ± 0.80

The expressions for matrix elements and decay widths depend on the spin structure and are for the three cases as follows.

- For $1^+ \to 1^- + 0^-$ transitions, the matrix element can be parameterized with two Lorentz structures:

$$\langle 1^-(q_1;\delta), 0^-(q_2)|\, T\, |1^+(p;\mu)\rangle = \left(A\, g^{\mu\delta} + B\, q_1^\mu q_2^\delta\right) \varepsilon_\mu \varepsilon_{1\delta}^*. \tag{85}$$

The invariant amplitudes A and B can be combined into the helicity amplitudes:

$$H_{00} = -\frac{E_1}{M_1} A - \frac{M}{M_1}|\mathbf{q}_1|^2 B, \quad H_{+1+1} = H_{-1-1} = -A,$$

which are practical to express the decay width. For the derivation of the latter, it is useful to work in the rest frame of the initial particle, where $|\mathbf{q}_1| = \lambda^{1/2}(M^2, M_1^2, M_2^2)/2M$ is the three-momentum and $E_1 = (M^2 + M_1^2 - M_2^2)/2M$ is the energy of the final state vector. Furthermore, the on-mass-shell character of the initial and final state particles is taken into account by $p^2 = M^2$, $q_1^2 = M_1^2$, $q_2^2 = M_2^2$, and $p^\mu \varepsilon_\mu = 0$. One arrives at:

$$\Gamma = \frac{|\mathbf{q}_1|}{24\pi M^2}\left\{|H_{+1+1}|^2 + |H_{-1-1}|^2 + |H_{00}|^2\right\}. \tag{86}$$

- The matrix element for the $1^+ \to 1^+ + 0^-$ transitions is expressed through one covariant term only:

$$\langle 1^+(q_1;\delta), 0^-(q_2)|\, T\, |1^+(p;\mu)\rangle = C\, q_{1\alpha} q_{2\beta}\, \varepsilon^{\alpha\beta\mu\delta}\, \varepsilon_\mu \varepsilon_{1\delta}^*. \tag{87}$$

The decay with the formula can be written as:

$$\Gamma = \frac{|\mathbf{q}_1|^3}{12\pi M^2}\, C^2, \tag{88}$$

where one can note the p-wave suppression factor $|\mathbf{q}_1|^3$.

- As shown in [81], the matrix element for $1^+ \to 1^- + 1^-$ decay can be parameterized using three amplitudes:

$$\langle 1^-(q_1;\delta), 1^-(q_2;\rho)|\, T\, |1^+(p;\mu)\rangle = \left(B_1\, \varepsilon^{q_1 q_2 \rho \delta}\, q_1^\mu + B_2\, \varepsilon^{q_1 \mu \rho \delta} + B_3\, \varepsilon^{q_2 \mu \rho \delta}\right) \varepsilon_\mu \varepsilon_\delta \varepsilon_\rho. \tag{89}$$

The relation between the helicity amplitudes $H_{\lambda;\lambda_1\lambda_2}$ ($\lambda = \lambda_1 - \lambda_2$) and the invariant amplitudes can be shown to be:

$$H_{0;+1+1} = -H_{0;-1-1} = -E_1 A_1 - E_2 A_2 - M|\mathbf{q}_1|^2 A_5,$$

$$H_{+1;+10} = -H_{-1;-10} = \frac{(E_1 M - M_1^2)}{M_2} A_1 + M_2 A_2 - \frac{M^2}{M_2}|\mathbf{q}_1|^2 A_4,$$

$$H_{-1;0+1} = -H_{+1;0-1} = M_1 A_1 + \frac{(E_1 M - M_1^2)}{M_1} A_2 - \frac{M^2}{M_1}|\mathbf{q}_1|^2 A_3. \tag{90}$$

The rate of the decay $1^+(p) \to 1^-(q_1) + 1^-(q_2)$, finally, reads:

$$\Gamma = \frac{|\mathbf{q}_1|}{24\pi M^2} \cdot 2 \left\{ |H_{0;+1+1}|^2 + |H_{+1;+1}|^2 + |H_{-1;0+1}|^2 \right\}. \tag{91}$$

Coming back to the CCQM description, one can write the non-local versions of Equations (83) and (84) as follows:

$$J^\mu_{Z_b^+}(x) = \int dx_1 \ldots \int dx_4\, \delta\left(x - \sum_{i=1}^4 w_i x_i\right) \Phi_{Z_b;Z_b;Z_b}\left(\sum_{i<j}(x_i - x_j)^2\right) J^\mu_{4q}(x_1,\ldots,x_4), \tag{92}$$

$$J^\mu_{Z_b;4q} = \tfrac{1}{\sqrt{2}}\left\{ (\bar{d}(x_3)\gamma_5 b(x_1))(\bar{b}(x_2)\gamma^\mu u(x_4)) + (\bar{d}(x_3)\gamma^\mu b(x_1))(\bar{b}(x_2)\gamma_5 u(x_4)) \right\},$$

$$J^{\mu\nu}_{Z_b'^+}(x) = \int dx_1 \ldots \int dx_4\, \delta\left(x - \sum_{i=1}^4 w_i x_i\right) \Phi_{Z_b'}\left(\sum_{i<j}(x_i - x_j)^2\right) J^{\mu\nu}_{Z_b';4q}(x_1,\ldots,x_4), \tag{93}$$

$$J^{\mu\nu}_{Z_b';4q} = \varepsilon^{\mu\nu\alpha\beta}\, (\bar{d}(x_3)\gamma_\alpha b(x_1))(\bar{b}(x_2)\gamma_\beta u(x_4)),$$

The interaction Lagrangian is constructed in the usual way for Z_b; in the case of Z_b', the stress tensor of the field is introduced $Z'_{b,\mu\nu} = \partial_\mu Z'_{b,\nu} - \partial_\nu Z'_{b,\mu}$:

$$\mathcal{L}_{\text{int},Z_b} = g_{Z_b} Z_{b,\mu}(x) \cdot J^\mu_{Z_b}(x) + \text{H.c.}, \tag{94}$$

$$\mathcal{L}_{\text{int},Z_b'} = \frac{g_{Z_b'}}{2M_{Z_b'}} Z'_{b,\mu\nu}(x) \cdot J^{\mu\nu}_{Z_b'}(x) + \text{H.c.}. \tag{95}$$

The factor $2M_{Z_b'}$ is put into the denominator in order to preserve the same physical dimensions of the g_{Z_b} and $g_{Z_b'}$ couplings. The link between these couplings and the size parameters is done via the compositeness condition, which is based on the evaluation of hadronic mass operators. The latter are written in the momentum space as:

$$\tilde{\Pi}^{\mu\nu}_{Z_b}(p) = \frac{9}{2} \prod_{i=1}^3 \int \frac{d^4 k_i}{(2\pi)^4 i} \tilde{\Phi}^2_{Z_b}(-\tilde{\omega}^2) \tag{96}$$

$$\times \left\{ \operatorname{tr}\left[\gamma_5 S_1(\hat{k}_1)\gamma_5 S_3(\hat{k}_3)\right] \operatorname{tr}\left[\gamma^\mu S_4(\hat{k}_4)\gamma^\nu S_2(\hat{k}_2)\right] \right.$$

$$\left. + \operatorname{tr}\left[\gamma^\mu S_1(\hat{k}_1)\gamma^\nu S_3(\hat{k}_3)\right] \operatorname{tr}\left[\gamma_5 S_4(\hat{k}_4)\gamma_5 S_2(\hat{k}_2)\right] \right\},$$

$$\tilde{\Pi}^{\mu\nu}_{Z_b'}(p) = -9\frac{\varepsilon^{\mu\rho\alpha\beta}\varepsilon^{\nu\rho\rho\sigma}}{M^2_{Z_b'}} \prod_{i=1}^3 \int \frac{d^4 k_i}{(2\pi)^4 i} \tilde{\Phi}^2_{Z_b'}(-\tilde{\omega}^2) \tag{97}$$

$$\times \operatorname{tr}\left[\gamma_\rho S_1(\hat{k}_1)\gamma_\alpha S_3(\hat{k}_3)\right] \operatorname{tr}\left[\gamma_\beta S_4(\hat{k}_4)\gamma_\sigma S_2(\hat{k}_2)\right],$$

where $\bar{\omega}^2 = 1/2\,(k_1^2 + k_2^2 + k_3^2 + k_1 k_2 - k_1 k_3 - k_2 k_3)$ and:

$$\hat{k}_1 = k_1 - w_1 p, \quad \hat{k}_2 = k_2 - w_2 p, \quad \hat{k}_3 = k_3 + w_3 p,$$
$$\hat{k}_4 = k_1 + k_2 - k_3 + w_4 p, \quad \varepsilon^{\mu\rho\alpha\beta} = p_\nu \varepsilon^{\mu\nu\alpha\beta}.$$

A list of matrix elements for different decay reactions as predicted by the CCQM is given in what follows. For each element, we provide, in the last line of the corresponding expression, the form factor parametrization of the matrix element to be compared with the appropriate expression from Equations (85), (87), and (89). Beforehand, let us also define the argument of $\tilde{\Phi}_{Z_b}(\vec{\eta}^2)$. One has:

$$\vec{\eta}^2 = \sum_{i=1}^3 \eta_i^2 \qquad \eta_1 = +\frac{1}{2\sqrt{2}}(2k_1 + (1 + w_1 - w_2)q_1 + (w_1 - w_2)q_2),$$

$$\eta_2 = +\frac{1}{2\sqrt{2}}(2k_2 - (w_3 - w_4)q_1 + (1 - w_3 + w_4)q_2),$$

$$\eta_3 = +\frac{1}{2}((1 - w_1 - w_2)q_1 - (w_1 + w_2)q_2),$$

where w_i denotes four body reduced masses $w_i = m_i / \sum_{j=1}^4 m_j$ and quarks are indexed as $q_1 = q_2 = b$, $q_3 = q_4 = d = u$.

- $1^+ \to 1^- + 0^-$ matrix elements parametrized as in Equation (85):

$$M^{\mu\delta}\left(Z_b(p,\mu) \to Y(q_1,\delta) + \pi^+(q_2)\right) = \frac{3}{\sqrt{2}} 8 z_b g_Y g_\pi \qquad (98)$$

$$\int \frac{d^4 k_1}{(2\pi)^4 i} \int \frac{d^4 k_2}{(2\pi)^4 i} \tilde{\Phi}_{Z_b}\left(-\vec{\eta}^2\right) \tilde{\Phi}_Y\left(-(k_1 + v_1 q_1)^2\right) \tilde{\Phi}_\pi\left(-(k_2 + u_4 q_2)^2\right)$$

$$\times \left\{ \mathrm{tr}\left[\gamma_5 S_1(k_1) \gamma^\delta S_2(k_1 + q_1) \gamma^\mu S_4(k_2) \gamma_5 S_3(k_2 + q_2)\right] \right.$$
$$\left. + \mathrm{tr}\left[\gamma^\mu S_1(k_1) \gamma^\delta S_2(k_1 + q_1) \gamma_5 S_4(k_2) \gamma_5 S_3(k_2 + q_2)\right] \right\}$$

$$= A_{Z_b Y \pi}\, g^{\mu\delta} + B_{Z_b Y \pi}\, q_1^\mu q_2^\delta,$$

$$M^{\mu\delta}\left(Z_b'(p,\mu) \to Y(q_1,\delta) + \pi^+(q_2)\right) = 3\, g_{Z_b'} g_Y g_\pi \frac{i\varepsilon^{\mu\rho\alpha\beta}}{M_{Z_b'}} \qquad (99)$$

$$\int \frac{d^4 k_1}{(2\pi)^4 i} \int \frac{d^4 k_2}{(2\pi)^4 i} \tilde{\Phi}_{Z_b'}\left(-\vec{\eta}^2\right) \tilde{\Phi}_Y\left(-(k_1 + v_1 q_1)^2\right) \tilde{\Phi}_\pi\left(-(k_2 + u_4 q_2)^2\right)$$

$$\times\ \mathrm{tr}\left[\gamma_\alpha S_1(k_1) \gamma^\delta S_2(k_1 + q_1) \gamma_\beta S_4(k_2) \gamma_5 S_3(k_2 + q_2)\right]$$

$$= A_{Z_b' Y \pi}\, g^{\mu\delta} + B_{Z_b' Y \pi}\, q_1^\mu q_2^\delta,$$

$$M^{\mu\rho}\left(Z_b(p,\mu) \to \eta_b(q_1) + \rho(q_2,\rho)\right) = \frac{3}{\sqrt{2}} g_{Z_b} g_{\eta_b} g_\rho \qquad (100)$$

$$\int \frac{d^4 k_1}{(2\pi)^4 i} \int \frac{d^4 k_2}{(2\pi)^4 i} \widetilde{\Phi}_{Z_b}\left(-\vec{\eta}^2\right) \widetilde{\Phi}_{\eta_b}\left(-(k_1+v_1 q_1)^2\right) \widetilde{\Phi}_\rho\left(-(k_2+u_4 q_2)^2\right)$$

$$\times \left\{ \mathrm{tr}\left[\gamma_5 S_1(k_1)\gamma_5 S_2(k_1+q_1)\gamma^\mu S_4(k_2)\gamma^\rho S_3(k_2+q_2)\right]\right.$$
$$\left. + \mathrm{tr}\left[\gamma^\mu S_1(k_1)\gamma_5 S_2(k_1+q_1)\gamma_5 S_4(k_2)\gamma^\rho S_3(k_2+q_2)\right] \right\}$$

$$= A_{Z_b \eta_b \rho}\, g^{\mu\rho} - B_{Z_b \eta_b \rho}\, q_2^\mu q_1^\rho,$$

$$M^{\mu\rho}\left(Z_b'(p,\mu) \to \eta_b(q_1) + \rho(q_2,\rho)\right) = 3 g_{Z_b'} g_{\eta_b} g_\rho \frac{i\varepsilon^{\mu\rho\alpha\beta}}{M_{Z_b'}} \qquad (101)$$

$$\int \frac{d^4 k_1}{(2\pi)^4 i} \int \frac{d^4 k_2}{(2\pi)^4 i} \widetilde{\Phi}_{Z_b'}\left(-\vec{\eta}^2\right) \widetilde{\Phi}_{\eta_b}\left(-(k_1+v_1 q_1)^2\right) \widetilde{\Phi}_\rho\left(-(k_2+u_4 q_2)^2\right)$$

$$\times \mathrm{tr}\left[\gamma_\alpha S_1(k_1)\gamma_5 S_2(k_1+q_1)\gamma_\beta S_4(k_2)\gamma^\rho S_3(k_2+q_2)\right]$$

$$= A_{Z_b' \eta_b \rho}\, g^{\mu\rho} - B_{Z_b' \eta_b \rho}\, q_2^\mu q_1^\rho.$$

- $1^+ \to 1^+ + 0^-$ matrix elements parametrized as in Equation (87):

$$M^{\mu\delta}\left(Z_b^+(p,\mu) \to h_b(q_1,\delta) + \pi^+(q_2)\right) = \frac{3}{\sqrt{2}} g_{Z_b} g_{h_b} g_\pi \qquad (102)$$

$$\int \frac{d^4 k_1}{(2\pi)^4 i} \int \frac{d^4 k_2}{(2\pi)^4 i} \widetilde{\Phi}_{Z_b}\left(-\vec{\eta}^2\right) \widetilde{\Phi}_{h_b}\left(-(k_1+v_1 q_1)^2\right) \widetilde{\Phi}_\pi\left(-(k_2+u_4 q_2)^2\right)$$

$$\times \left\{ \mathrm{tr}\left[\gamma_5 S_1(k_1)\gamma_5 \cdot (2k_1^\delta) S_2(k_1+q_1)\gamma^\mu S_4(k_2)\gamma_5 S_3(k_2+q_2)\right]\right.$$
$$\left. + \mathrm{tr}\left[\gamma^\mu S_1(k_1)\gamma_5 \cdot (2k_1^\delta) S_2(k_1+q_1)\gamma_5 S_4(k_2)\gamma_5 S_3(k_2+q_2)\right] \right\}$$

$$= \varepsilon^{\mu\delta q_1 q_2} A_{Z_b h_b \pi},$$

$$M^{\mu\delta}\left(Z_b'(p,\mu) \to h_b(q_1,\delta) + \pi^+(q_2)\right) = 3 g_{Z_b'} g_{h_b} g_\pi \frac{i\varepsilon^{\mu\rho\alpha\beta}}{M_{Z_b'}} \qquad (103)$$

$$\int \frac{d^4 k_1}{(2\pi)^4 i} \int \frac{d^4 k_2}{(2\pi)^4 i} \widetilde{\Phi}_{Z_b'}\left(-\vec{\eta}^2\right) \widetilde{\Phi}_{h_b}\left(-(k_1+v_1 q_1)^2\right) \widetilde{\Phi}_\pi\left(-(k_2+u_4 q_2)^2\right)$$

$$\times \mathrm{tr}\left[\gamma_\alpha S_1(k_1)\gamma_5 \cdot (2k_1^\delta) S_2(k_1+q_1)\gamma_\beta S_4(k_2)\gamma_5 S_3(k_2+q_2)\right]$$

$$= \varepsilon^{\mu\delta q_1 q_2} A_{Z_b' h_b \pi}.$$

The matrix elements describing decays to a pair of B mesons can be also listed within two groups depending on the quantum numbers. The argument of the Z_b-vertex function $\vec{\delta}^2$ is defined as:

$$\vec{\delta}^2 = \sum_{i=1}^{3} \delta_i^2; \quad \delta_1 = -\frac{1}{2\sqrt{2}}\left(k_1 + k_2 + (w_1 - w_2)q_1 + (1 + w_1 - w_2)q_2\right),$$

$$\delta_2 = +\frac{1}{2\sqrt{2}}\left(k_1 + k_2 + (1 - w_3 + w_4)q_1 - (w_3 - w_4)q_2\right),$$

$$\delta_3 = +\frac{1}{2}\left(k_1 - k_2 + (w_1 + w_2)q_1 - (1 - w_1 - w_2)q_2\right). \qquad (104)$$

The quark indices are similar to the previous case $q_1 = q_2 = b$, $q_3 = q_4 = d = u$, $\hat{v}_2 = m_2/(m_2 + m_4)$, $\hat{v}_4 = m_4/(m_2 + m_4)$, $\hat{u}_1 = m_1/(m_1 + m_3)$, and $\hat{u}_3 = m_3/(m_1 + m_3)$.

- $1^+ \to 1^- + 0^-$ matrix elements parametrized as in Equation (85):

$$M^{\mu\rho}\left(Z_b^+(p,\mu) \to \bar{B}^0(q_1) + B^{*+}(q_2,\rho)\right) = \frac{9}{\sqrt{2}} g_{Z_b} g_B g_{B^*} \qquad (105)$$

$$\int \frac{d^4 k_1}{(2\pi)^4 i} \int \frac{d^4 k_2}{(2\pi)^4 i} \tilde{\Phi}_{Z_b}\left(-\bar{\delta}^2\right) \tilde{\Phi}_B\left(-(k_2 + v_4 q_1)^2\right) \tilde{\Phi}_{B^*}\left(-(k_1 + u_1 q_2)^2\right)$$

$$\times \mathrm{tr}\left[\gamma^\mu S_1(k_1) \gamma^\rho S_3(k_1 + q_2)\right] \mathrm{tr}\left[\gamma_5 S_4(k_2) \gamma_5 S_2(k_2 + q_1)\right]$$

$$= A_{Z_b BB^*} g^{\mu\rho} - B_{Z_b BB^*} q_2^\mu q_1^\rho ,$$

$$M^{\mu\alpha}\left(Z_b^+(p,\mu) \to \bar{B}^{*0}(q_1,\delta) + B^+(q_2)\right) = \frac{9}{\sqrt{2}} g_{Z_b} g_{B^*} g_B \qquad (106)$$

$$\int \frac{d^4 k_1}{(2\pi)^4 i} \int \frac{d^4 k_2}{(2\pi)^4 i} \tilde{\Phi}_{Z_b}\left(-\bar{\delta}^2\right) \tilde{\Phi}_{B^*}\left(-(k_1 + \hat{v}_1 q_1)^2\right) \tilde{\Phi}_B\left(-(k_2 + \hat{u}_4 q_2)^2\right)$$

$$\times \mathrm{tr}\left[\gamma_5 S_1(k_1) \gamma_5 S_3(k_1 + q_2)\right] \mathrm{tr}\left[\gamma^\mu S_4(k_2) \gamma^\delta S_2(k_2 + q_1)\right]$$

$$= A_{Z_b B^* B} g^{\mu\delta} + B_{Z_b B^* B} q_1^\mu q_2^\delta .$$

- $1^+ \to 1^- + 1^-$ matrix elements parametrized as in Equation (89):

$$M^{\mu\delta\rho}(Z_b^{\prime +}(p,\mu) \to B^{*0}(q_1,\delta) + \bar{B}^{*+}(q_2,\rho)) = 9 g_{Z_b'} g_{B^*} g_{B^*} \frac{\epsilon^{\mu\rho\alpha\delta}}{M_{Z_b'}} \qquad (107)$$

$$\int \frac{d^4 k_1}{(2\pi)^4 i} \int \frac{d^4 k_2}{(2\pi)^4 i} \tilde{\Phi}_{Z_b'}\left(-\bar{\delta}^2\right) \tilde{\Phi}_{B^*}\left(-(k_1 + \hat{v}_1 q_1)^2\right) \tilde{\Phi}_{B^*}\left(-(k_2 + \hat{u}_4 q_2)^2\right)$$

$$\times \mathrm{tr}\left[\gamma_\alpha S_1(k_1) \gamma^\delta S_3(k_1 + q_1)\right] \mathrm{tr}\left[\gamma_\beta S_4(k_2) \gamma^\rho S_2(k_2 + q_2)\right]$$

$$= B_1 q_1^\mu \epsilon^{q_1 q_2 \rho \delta} + B_2 \epsilon^{q_1 \mu \rho \delta} + B_3 \epsilon^{q_2 \mu \rho \delta} .$$

With all the above theoretical expressions, one can proceed to the numerical evaluation of the decay widths. The first step is the adjustment of the size parameters Λ_Z and Λ_Z'. They are tuned so as to respect the observables measured by the Belle collaboration [35]:

$$\Gamma_{Z_b}(BB^*\pi) = (25 \pm 7)\,\mathrm{MeV}, \qquad \mathcal{B}(Z_b^+ \to [B^+\bar{B}^{*0} + \bar{B}^0 B^{*+}]) = 85.6^{+1.5+1.5}_{-2.0-2.1}\%,$$

$$\Gamma_{Z_b'}(B^*B^*\pi) = (23 \pm 8)\,\mathrm{MeV}, \qquad \mathcal{B}(Z_b'^+ \to \bar{B}^{*+} B^{*0}) = 73.7^{+3.4+2.7}_{-4.4-3.5}\%, \qquad (108)$$

leading to:

$$\Lambda_{Z_b} = 3.45 \pm 0.05\,\mathrm{GeV} \qquad \Lambda_{Z_b'} = 3.00 \pm 0.05\,\mathrm{GeV}. \qquad (109)$$

With the decays into B pairs dominating all other decay channels, we approximate the total decay width as the sum of all herein evaluated channels. The CCQM gives:

$$\Gamma_{Z_b} = 30.9^{+2.3}_{-2.1}\,\mathrm{MeV}, \qquad \Gamma_{Z_b'} = 34.1^{+2.8}_{-2.5}\,\mathrm{MeV}, \qquad (110)$$

which is in fair agreement with (108). The predicted partial decay widths of $Z_b(10610)$ and $Z_b'(10650)$ particles are summarized in Table 6.

Table 6. Particle decay widths for the $Z_b^+(10610)$ and $Z_b^+(10650)$.

Channel	Widths, MeV	
	$Z_b(10610)$	$Z_b(10650)$
$Y(1S)\pi^+$	5.9 ± 0.4	$9.5^{+0.7}_{-0.6}$
$h_b(1P)\pi^+$	$(0.14 \pm 0.01) \cdot 10^{-1}$	$0.74^{+0.05}_{-0.04} \cdot 10^{-3}$
$\eta_b \rho^+$	4.4 ± 0.3	$7.5^{+0.6}_{-0.5}$
$B^+\bar{B}^{*0} + \bar{B}^0 B^{*+}$	$20.7^{+1.6}_{-1.5}$	–
$B^{*+}\bar{B}^{*0}$	–	$17.1^{+1.5}_{-1.4}$

The Z_b and Z_b' decays are dominated [13] by $\Gamma_{Z_b}(B^+\bar{B}^{*0}+B^{*+}\bar{B}^0) = (85.6^{+2.1}_{-2.9})\%$ and $\Gamma_{Z_b'}(B^{*+}\bar{B}^{*0}) = (74^{+4}_{-6})\%$, respectively, meaning that the bottomonia modes should not exceed 15 and 25 percent. This is observed for the $h_b(1P)\pi^+$ final state; the other bottomonia channels are suppressed, but not so much as seen in the data:

$$\frac{\Gamma(Z_b \to Y(1S)\pi)}{\Gamma(Z_b \to B\bar{B}^* + c.c.)} \approx 0.29, \qquad \frac{\Gamma(Z_b \to \eta_b\rho)}{\Gamma(Z_b \to B\bar{B}^* + c.c.)} \approx 0.21,$$

$$\frac{\Gamma(Z_b' \to Y(1S)\pi)}{\Gamma(Z_b' \to B^*\bar{B}^*)} \approx 0.56, \qquad \frac{\Gamma(Z_b' \to \eta_b\rho)}{\Gamma(Z_b' \to B^*\bar{B}^*)} \approx 0.44.$$

The model also allows us to make predictions:

$$R_{Y(1S)\pi} = \frac{\Gamma(Z_b \to Y(1S)\pi)}{\Gamma(Z_b' \to Y(1S)\pi)} = 0.62 \pm 0.06, \qquad R_{\eta_b\rho} = \frac{\Gamma(Z_b \to \eta_b\rho)}{\Gamma(Z_b' \to \eta_b\rho)} = 0.59 \pm 0.06. \tag{111}$$

One can conclude that the CCQM provides, within a molecular picture, a fair description of $Z_b(10610)/Z_b'(10650)$ states and related decay observables and catches the tendencies seen in experimental data. Some deviations are observed when the fraction of bottomonium in final states is considered.

8. Summary and Conclusions

The confined covariant quark model is an approach based on a non-local interaction Lagrangian of quarks and hadrons. It has many appealing features: a full Lorentz invariance, confinement, large applicability range (from mesons to exotic hadrons), inclusion of the electromagnetic interaction, and a limited number of free parameters. As a practical tool, it allows overcoming the difficulties related to the non-applicability of the perturbative approach for bound states in QCD. In this text, we used it to describe four quark exotic states $X(3872)$, $Z_c(3900)$, $Y(4260)$, $Z_b(10610)$, and $Z_b'(10650)$. We demonstrated that the CCQM had enough predictive power to make the distinctions between various hypothesis, with respect to the exotic quarkonia mostly related to their structure (molecular versus tetraquark one). At the same time, the model provides a good description of experimental data without large deviations and predictions for future measurements. Concerning the structure of the studied particles, the molecular picture is favored for $Z_c(3900)$, $Z_b(10610)$, and $Z_b'(10650)$ and the tetraquark one for $X(3872)$ and $Y(4260)$. These conclusions follow from the measured decay characteristics of the considered exotic states and the related model description: with the expected increase in the number and quality of experimental data, one may hope the quarkonia-structure puzzle will be solved in the years to come.

Author Contributions: Conceptualization, M.A.I.; methodology, M.A.I., A.Z.D. and S.D.; software, M.A.I.; validation, A.Z.D., S.D., M.A.I. and A.L.; formal analysis, M.A.I. and A.L.; investigation, A.Z.D., S.D., M.A.I. and A.L.; resources, A.Z.D., S.D., M.A.I. and A.L.; data curation, M.A.I. and A.L.; writing—original draft preparation, A.L.; writing—review and editing, A.Z.D., S.D., M.A.I. and A.L.; visualization, M.A.I.; supervision, M.A.I.; project administration, M.A.I. and A.L.; funding acquisition, A.Z.D, S.D., M.A.I. and A.L. All authors have read and agreed to the published version of the manuscript.

Funding: This research was funded by the Joint Research Project of the Institute of Physics, Slovak Academy of Sciences (SAS), and Bogoliubov Laboratory of Theoretical Physics, Joint Institute for Nuclear Research (JINR), Grant No. 01-3-1135-2019/2023. A.Z.D., S.D., and A.L. acknowledge the funding from the Slovak Grant Agency for Sciences (VEGA), Grant No. 2/0153/17.

Acknowledgments: We would like to thank Thomas Gutsche, Jürgen Körner, Valery Lyubovitskij, Pietro Santorelli, A. Issadykov, F. Goerke, K. Xu, and G. G. Saidullaeva for their collaboration, by which the results discussed in this review have been obtained.

Conflicts of Interest: The authors declare no conflict of interest.

References

1. Gell–Mann, M. A Schematic Model of Baryons and Mesons. *Phys. Lett.* **1964**, *8*, 214–215. [CrossRef]
2. Bai, J.Z.; Ban, Y.; Bian, J.G.; Cai, X.; Chang, J.F.; Chen, H.F.; Chen, H.S.; Chen, J.; Chen, J.C. Observation of a near threshold enhancement in the $p\bar{p}$ mass spectrum from radiative $J/\psi \to \gamma p\bar{p}$ decays. *Phys. Rev. Lett.* **2003**, *91*, 022001.
3. Aubert, B.; Barate, R.; Boutigny, D.; Gaillard, J.-M.; Hicheur, A.; Karyotakis, Y.; Lees, J.P.; Robbe, P.; Tisserand, V.; Zghiche, A.; et al. Observation of a narrow meson decaying to $D_s^+ \pi^0$ at a mass of 2.32-GeV/c^2. *Phys. Rev. Lett.* **2003**, *90*, 242001. [CrossRef] [PubMed]
4. Choi, S.K.; Olsen, S.L.; Abe, K.; Abe, T.; Adachi, I.; Ahn, B.S.; Aihara, H.; Akai, K.; Akatsu, M.; Akemoto, M.; et al. Observation of a narrow charmonium-like state in exclusive $B^{\pm} \to K^{\pm}\pi^+\pi^- J/\psi$ decays. *Phys. Rev. Lett.* **2003**, *91*, 262001. [CrossRef]
5. Acosta, D.; Affolder, T.; Ahn, M.H.; Akimoto, T.; Albrow, M.G.; Ambrose, D.; Amerio, S.; Amidei, D.; Anastassov, A.; Anikeev, K.; et al. Observation of the narrow state $X(3872) \to J/\psi\pi^+\pi^-$ in $\bar{p}p$ collisions at $\sqrt{s} = 1.96$ TeV. *Phys. Rev. Lett.* **2004**, *93*, 072001. [CrossRef]
6. Abazov, V.M.; Abbott, B.; Abolins, M.; Acharya, B.S; Adams, D.L.; Adams, M.; Adams, T.; Agelou, M.; Agram, J.L.; Ahmed, S.N.; et al. Observation and properties of the $X(3872)$ decaying to $J/\psi\pi^+\pi^-$ in $p\bar{p}$ collisions at $\sqrt{s} = 1.96$ TeV. *Phys. Rev. Lett.* **2004**, *93*, 162002. [CrossRef]
7. Aaij, R.; Abellan Beteta, C.; Adeva, B.; Adinolfi, M.; Adrover, C.; Affolder, A.; Ajaltouni, Z.; Albrecht, J.; Alessio, F. Observation of $X(3872)$ production in pp collisions at $\sqrt{s} = 7$ TeV. *Eur. Phys. J. C* **2012**, *72*, 1972. [CrossRef]
8. Ablikim, M.; Achasov, M.N.; Ai, X.C.; Albayrak, O.; Ambrose, D.J.; An, F.F.; An, Q.; Bai, J.Z.; Baldini Ferroli, R.; Ban, Y.; et al. Observation of $e^+e^- \to \gamma X(3872)$ at BESIII. *Phys. Rev. Lett.* **2014**, *112*, 092001. [CrossRef]
9. Abulencia, A.; Adelman, J.; Affolder, T.; Akimoto, T.; Albrow, M.G.; Ambrose, D.; Amerio, S.; Amidei, D.; Anastassov, A.; Anikeev, K.; et al. Analysis of the quantum numbers J^{PC} of the $X(3872)$. *Phys. Rev. Lett.* **2007**, *98*, 132002. [CrossRef]
10. Aaltonen, T.; Adelman, J.; Akimoto, T.; Alvarez Gonzalezt, B.; Ameriozs, S.; Amidei, D.; Anastassov, A.; Annovi, A.; Antos, J.; Apollinari, G.; et al. Precision Measurement of the $X(3872)$ Mass in $J/\psi\pi^+\pi^-$ Decays. *Phys. Rev. Lett.* **2009**, *103*, 152001. [CrossRef]
11. Choi, S.K.; Olsen, L.; Trabelsi, K.; Adachi, I.; Aihara, H.; Arinstein, K.; Asner, D.M.; Aushev, T.; Bakich, A.M.; Barberio, E.; et al. Bounds on the width, mass difference and other properties of $X(3872) \to \pi^+\pi^- J/\psi$ decays. *Phys. Rev. D* **2011**, *84*, 052004. [CrossRef]
12. Aaij, R.; Abellan Beteta, C.; Adeva, B.; Adinolfi, M.; Adrover, C.; Affolder, A.; Ajaltouni, Z.; Albrecht, J.; Alessio, F.; Alexander, M.; et al. Determination of the $X(3872)$ meson quantum numbers. *Phys. Rev. Lett.* **2013**, *110*, 222001. [CrossRef] [PubMed]
13. Tanabashi, M.; Hagiwara, K.; Hikasa, K.; Nakamura, K.; Sumino, Y.; Takahashi, F.; Tanaka, J.; Agashe, K.; Aielli, G.; Allanach, B.C.; et al. Review of Particle Physics. *Phys. Rev. D* **2018**, *98*, 030001 [CrossRef]
14. Aubert, B.; Barate, R.; Boutigny, D.; Couderc, F.; Karyotakis, Y.; Lees, J.P.; Poireau, V.; Tisserand, V.; Zghiche, A.; Grauges, E.; et al. Observation of a broad structure in the $\pi^+\pi^- J/\psi$ mass spectrum around 4.26-GeV/c^2. *Phys. Rev. Lett.* **2005**, *95*, 142001. [CrossRef] [PubMed]
15. He, Q.; Insler, J.; Muramatsu, H.; Park, C.S.; Thorndike, E.H.; Yang, F.; Coan, T. E.; Gao, Y. S.; Artuso, M.; Blusk, S.; et al. Confirmation of the $Y(4260)$ resonance production in ISR. *Phys. Rev. D* **2006**, *74*, 091104. [CrossRef]

16. Yuan, C.Z.; Shen, C.P.; Wang, P.; McOnie, S.; Adachi, I.; Aihara, H.; Aulchenko, V.; Aushev, T.; Bahinipati, S.; Balagura, V.; et al. Measurement of $e^+e^- \to \pi^+\pi^- J/\psi$ cross-section via initial state radiation at Belle. *Phys. Rev. Lett.* **2007**, *99*, 182004. [CrossRef]
17. Ablikim, M.; Achasov, M.N.; Ai, X.C.; Albayrak, O.; Ambrose, D.J.; An, F.F.; An, Q.; Bai, J.Z.; Baldini Ferroli, R.; Ban Y.; et al. Observation of a Charged Charmoniumlike Structure in $e^+e^- \to \pi^+\pi^- J/\psi$ at $\sqrt{s} = 4.26$ GeV. *Phys. Rev. Lett.* **2013**, *110*, 252001. [CrossRef]
18. Lees, J.P.; Poireau, V.; Tisserand, V.; Garra Tico, J.; Grauges, E.; Palanoab, A.; Eigen, G.; Stugu, B.; Brown, D.N.; Kerth, L.T.; et al. Study of the reaction $e^+e^- \to J/\psi \pi^+\pi^-$ via initial-state radiation at BaBar. *Phys. Rev. D* **2012**, *86*, 051102.
19. Ablikim, M.; Achasov, M.N.; Ahmed, S.; Ai, X.C.; Albayrak, O.; Albrecht, M.; Ambrose, D.J.; Amoroso, A.; An, F.F.; An, Q.; et al. Precise measurement of the $e^+e^- \to \pi^+\pi^- J/\psi$ cross-section at center-of-mass energies from 3.77 to 4.60 GeV. *Phys. Rev. Lett.* **2017**, *118*, 092001.
20. Cronin-Hennessy, D.; Gao, K.Y.; Hietala, J.; Kubota, Y.; Klein, T.; Lang, B.W.; Poling, R.; Scott, A.W.; Zweber, P.; Dobbs, S.; et al. Measurement of Charm Production Cross Sections in e^+e^- Annihilation at Energies between 3.97 and 4.26-GeV. *Phys. Rev. D* **2009**, *80*, 072001. [CrossRef]
21. Pakhlova, G.; Abe, K.; Adachi, I.; Aihara, H.; Anipko, D.; Aulchenko, V.; Aushev, T.; Bakich, A.M.; Balagura, V.; Barberio E.; et al. Measurement of the near-threshold $e^+e^- \to D^{(*)\pm}D^{*\mp}$ cross-section using initial-state radiation. *Phys. Rev. Lett.* **2007**, *98*, 092001. [CrossRef] [PubMed]
22. Aubert, B.; Barate, R.; Bona, M.; Boutigny, D.; Couderc, F.; Karyotakis, Y.; Lees, J.P.; Poireau, V.; Tisserand, V.; Zghiche A.; et al. Study of the Exclusive Initial-State Radiation Production of the $D\bar{D}$ System. *Phys. Rev. D* **2007**, *76*, 111105. [CrossRef]
23. Aubert, B.; Karyotakis, Y.; Lees, J. P.; Poireau, V.; Prencipe, E.; Prudent, X.; Tisserand, V.; Garra Tico, J.; Grauges, E.; Lopez, L.; et al. Exclusive Initial-State-Radiation Production of the $D\bar{D}$, $D^*\bar{D}^*$, and $D^*\bar{D}^*$ Systems. *Phys. Rev. D* **2009**, *79*, 092001. [CrossRef]
24. del Amo Sanchez, P.; Lees, J.P.; Poireau, V.; Prencipe, E.; Tisserand, V.; Garra Tico, J.; Grauges, E.; Martinelli, M.; Palano, A.; Pappagallo, M.; et al. Exclusive Prod. $D_s^+D_s^-$, $D_s^{*+}D_s^-$, $D_s^{*+}D_s^{*-}$ via e^+E^- Annihil. *Initial State Phys. Rev. D* **2010**, *82*, 052004.
25. Liu, Z.Q.; Shen, C.P.; Yuan, C.Z.; Adachi, I.; Aihara, H.; Asner, D.M.; Aulchenko, V.; Aushev, T.; Aziz, T.; Bakich, A.M.; et al. Study of $e^+e^- \to \pi^+\pi^- J/\psi$ and Observation of a Charged Charmoniumlike State at Belle. *Phys. Rev. Lett.* **2013**, *110* 252002. [CrossRef]
26. Xiao, T.; Dobbs, S.; Tomaradze, A.; Seth, K.K. Observation of the Charged Hadron $Z_c^\pm(3900)$ and Evidence for the Neutral $Z_c^0(3900)$ in $e^+e^- \to \pi\pi J/\psi$ at $\sqrt{s} = 4170$ MeV. *Phys. Lett. B* **2013**, *727*, 366. [CrossRef]
27. Ablikim, M.; Achasov, M.N.; Albayrak, O.; Ambrose, D.J.; An, F.F.; An, Q.; Bai, J.Z.; Baldini Ferroli, R.; Ban, Y.; Becker, J.; et al. Observation of a charged $(D\bar{D}^*)^\pm$ mass peak in $e^+e^- \to \pi D\bar{D}^*$ at $\sqrt{s} = 4.26$ GeV. *Phys. Rev. Lett.* **2014**, *112*, 022001. [CrossRef]
28. Ablikim, M.; Achasov, M.N.; Ai, X.C.; Albayrak, O.; Albrecht, M.; Ambrose, D.J.; Amoroso, A.; An, F.F.; An, Q.; Bai, J.Z.; et al. Observation of $Z_c(3900)^0$ in $e^+e^- \to \pi^0\pi^0 J/\psi$. *Phys. Rev. Lett.* **2015**, *115*, 112003. [CrossRef]
29. Ablikim, M.; Achasov, M.N.; Ai, X.C.; Albayrak, O.; Albrecht, M.; Ambrose, D.J.; Amoroso, A.; An, F.; An, Q.; Bai, J.Z.; et al. Determination of the Spin and Parity of the $Z_c(3900)$. *Phys. Rev. Lett.* **2017**, *119*, 072001. [CrossRef]
30. Ablikim, M.; Achasov, M.N.; Ahmed, S.; Albrecht, M.; Alekseev, M.; Amoroso, A.; An, F.F.; An, Q.; Bai, Y.; Bakina, O.; et al. Study of $e^+e^- \to \pi^+\pi^-\pi^0\eta_c$ and evidence for $Z_c(3900)^\pm$ decaying into $\rho^\pm\eta_c$. *Phys. Rev. D* **2019**, *100*, 111102. [CrossRef]
31. Abazov, V.M.; Abbott, B.; Acharya, B.S.; Adams, M.; Adams, T.; Agnew, J.P.; Alexeev, G.D.; Alkhazov, G.; Altona, A.; Askew, A.; et al. Evidence for $Z_c^\pm(3900)$ in semi-inclusive decays of b-flavored hadrons. *Phys. Rev. D* **2018**, *98*, 052010.
32. Abazov, V.M.; Abbott, B.; Acharya, B.S.; Adams, M.; Adams, T.; Agnew, J.P.; Alexeev, G.D.; Alkhazov, G.; Altona, A.; Askew, A.; et al. Properties of $Z_c^\pm(3900)$ Produced in $p\bar{p}$ Collision. *Phys. Rev. D* **2019**, *100*, 012005. [CrossRef]
33. Bondar, A.; Garmash, A.; Mizuk, R.; Santel, D.; Kinoshita, K.; Adachi, I.; Aihara, H.; Arinstein, K.; Asner, D.M.; Aushev, T.; et al. Observation of two charged bottomonium-like resonances in $Y(5S)$ decays. *Phys. Rev. Lett.* **2012**, *108*, 122001. [CrossRef]

34. Garmash, A.; Bondar, A.; Kuzmin, A.; Abdesselam, A.; Adachi, I.; Aihara, H.; Al Said, S.; Asner, D.M.; Aulchenko, V.; Aushev, T.; et al. Amplitude analysis of $e^+e^- \to \Upsilon(nS)\pi^+\pi^-$ at $\sqrt{s} = 10.865$ GeV. *Phys. Rev. D* **2015**, *91*, 072003. [CrossRef]
35. Garmash, A.; Abdesselam, A.; Adachi, I.; Aihara, H.; Asner, D.M.; Aushev, T.; Ayad, R.; Aziz, T.; Babu, V.; Badhrees, I.; et al. Observation of $Z_b(10610)$ and $Z_b(10650)$ Decaying to *B* Mesons. *Phys. Rev. Lett.* **2016**, *116*, 212001. [CrossRef]
36. Klempt, E.; Zaitsev, A. Glueballs, Hybrids, Multiquarks. Experimental facts versus QCD inspired concepts. *Phys. Rep.* **2007**, *454*, 1. [CrossRef]
37. Godfrey, S.; Olsen, S.L. The Exotic XYZ Charmonium-like Mesons. *Ann. Rev. Nucl. Part. Sci.* **2008**, *58*, 51. [CrossRef]
38. Olsen, S.L. A New Hadron Spectroscopy. *Front. Phys.* **2015**, *10*, 121. [CrossRef]
39. Hosaka, A.; Iijima, T.; Miyabayashi, K.; Sakai, Y.; Yasui, S. Exotic hadrons with heavy flavors: *X*, *Y*, *Z*, and related states. *PTEP* **2016**, *2016*, 062C01.
40. Richard, J.M. Exotic hadrons: Review and perspectives. *Few Body Syst.* **2016**, *57*, 1185.
41. Chen, H.X.; Chen, W.; Liu, X.; Zhu, S.L. The hidden-charm pentaquark and tetraquark states. *Phys. Rep.* **2016**, *639*, 1. [CrossRef]
42. Esposito, A.; Pilloni, A.; Polosa, A.D. Multiquark Resonances. *Phys. Rep.* **2017**, *668*, 1. [CrossRef]
43. Lebed, R.F.; Mitchell, R.E.; Swanson, E.S. Heavy-Quark QCD Exotica. *Prog. Part. Nucl. Phys.* **2017**, *93*, 143. [CrossRef]
44. Ali, A.; Lange, J.S.; Stone, S. Exotics: Heavy Pentaquarks and Tetraquarks. *Prog. Part. Nucl. Phys.* **2017**, *97*, 123. [CrossRef]
45. Olsen, S.L.; Skwarnicki, T.; Zieminska, D. Nonstandard heavy mesons and baryons: Experimental evidence. *Rev. Mod. Phys.* **2018**, *90*, 015003. [CrossRef]
46. Karliner, M.; Rosner, J.L.; Skwarnicki, T. Multiquark States. *Ann. Rev. Nucl. Part. Sci.* **2018**, *68*, 17. [CrossRef]
47. Yuan, C.Z. The *XYZ* states revisited. *Int. J. Mod. Phys. A* **2018**, *33*, 1830018. [CrossRef]
48. Brambilla, N.; Eidelman, S.; Hanhart, C.; Nefediev, A.; Shen, C.; Thomas, C.E.; Vairo, A.; Yuan, C.Z. The *XYZ* states: Experimental and theoretical status and perspectives. *arXiv* **2019**, arXiv:1907.07583.
49. Agaev, S.; Azizi, K.; Sundu, H. Four-quark exotic mesons. *Turk. J. Phys.* **2020**, *44*, 95. [CrossRef]
50. Guo, F.K.; Hidalgo-Duque, C.; Nieves, J.; Valderrama, M.P. Consequences of Heavy Quark Symmetries for Hadronic Molecules. *Phys. Rev. D* **2013**, *88*, 054007. [CrossRef]
51. Karliner, M.; Rosner, J.L. New Exotic Meson and Baryon Resonances from Doubly-Heavy Hadronic Molecules. *Phys. Rev. Lett.* **2015**, *115*, 122001. [CrossRef] [PubMed]
52. Guo, F.K.; Hanhart, C.; Meißner, U.G.; Wang, Q.; Zhao, Q.; Zou, B.S. Hadronic molecules. *Rev. Mod. Phys.* **2018**, *90*, 015004. [CrossRef]
53. Nielsen, M.; Navarra, F.S.; Lee, S.H. New Charmonium States in QCD Sum Rules: A Concise Review. *Phys. Rep.* **2010**, *497*, 41. [CrossRef]
54. Kleiv, R.T.; Steele, T.G.; Zhang, A.; Blokland, I. Heavy-light diquark masses from QCD sum rules and constituent diquark models of tetraquarks. *Phys. Rev. D* **2013**, *87*, 125018. [CrossRef]
55. Albuquerque, R.M.; Dias, J.M.; Khemchandani, K.P.; Martínez Torres, A.; Navarra, F.S.; Nielsen, M.; Zanetti, C.M. QCD sum rules approach to the *X*, *Y* and *Z* states. *J. Phys. G* **2019**, *46*, 093002. [CrossRef]
56. Ebert, D.; Faustov, R.N.; Galkin, V.O. Masses of heavy tetraquarks in the relativistic quark model. *Phys. Lett. B* **2006**, *634*, 214. [CrossRef]
57. Ebert, D.; Faustov, R.N.; Galkin, V.O.; Lucha, W. Masses of tetraquarks with two heavy quarks in the relativistic quark model. *Phys. Rev. D* **2007**, *76*, 114015. [CrossRef]
58. Li, B.Q.; Chao, K.T. Higher Charmonia and *X*, *Y*, *Z* states with Screened Potential. *Phys. Rev. D* **2009**, *79*, 094004. [CrossRef]
59. Brodsky, S.J.; Hwang, D.S.; Lebed, R.F. Dynamical Picture for the Formation and Decay of the Exotic XYZ Mesons. *Phys. Rev. Lett.* **2014**, *113*, 112001. [CrossRef]
60. Cleven, M.; Guo, F.K.; Hanhart, C.; Wang, Q.; Zhao, Q. Employing spin symmetry to disentangle different models for the XYZ states. *Phys. Rev. D* **2015**, *92*, 014005. [CrossRef]
61. Braaten, E.; Langmack, C.; Smith, D.H. Born-Oppenheimer Approximation for the XYZ Mesons. *Phys. Rev. D* **2014**, *90*, 014044. [CrossRef]

62. Liu, Y.R.; Chen, H.X.; Chen, W.; Liu, X.; Zhu, S.L. Pentaquark and Tetraquark states. *Prog. Part. Nucl. Phys.* **2019**, *107*, 237. [CrossRef]
63. Eichten, E.J.; Lane, K.; Quigg, C. New states above charm threshold. *Phys. Rev. D* **2006**, *73*, 014014. [CrossRef]
64. Ma, L.; Liu, X.H.; Liu, X.; Zhu, S.L. Strong decays of the XYZ states. *Phys. Rev. D* **2015**, *91*, 034032. [CrossRef]
65. Guo, F.K.; Liu, X.H.; Sakai, S. Threshold cusps and triangle singularities in hadronic reactions. *Prog. Part. Nucl. Phys.* **2020**, *112*, 103757. [CrossRef]
66. Guo, F.K.; Hanhart, C.; Wang, Q.; Zhao, Q. Could the near-threshold XYZ states be simply kinematic effects? *Phys. Rev. D* **2015**, *91*, 051504. [CrossRef]
67. Efimov, G.V.; Ivanov, M.A. Confinement and quark structure of light hadrons. *Int. J. Mod. Phys. A* **1989**, *4*, 2031. [CrossRef]
68. Efimov, G.V.; Ivanov, M.A. *The Quark Confinement Model of Hadrons*; CRC Press: Boca Raton, FL, USA, 1993.
69. Branz, T.; Faessler, A.; Gutsche, T.; Ivanov, M.A.; Körner, J.G.; Lyubovitskij, V.E. Relativistic constituent quark model with infrared confinement. *Phys. Rev. D* **2010**, *81*, 034010. [CrossRef]
70. Terning, J. Gauging nonlocal Lagrangians. *Phys. Rev. D* **1991**, *44*, 887–897. [CrossRef]
71. Gutsche, T.; Ivanov, M.A.; Körner, J.G.; Lyubovitskij, V.E.; Santorelli, P. Light baryons and their electromagnetic interactions in the covariant constituent quark model. *Phys. Rev. D* **2012**, *86*, 074013. [CrossRef]
72. Gutsche, T.; Ivanov, M.A.; Körner, J.G.; Kovalenko, S.; Lyubovitskij, V.E. Nucleon tensor form factors in a relativistic confined quark model. *Phys. Rev. D* **2016**, *94*, 114030.
73. Lyubovitskij, V.E.; Faessler, A.; Gutsche, T.; Ivanov, M.A.; Körner, J.G. Heavy baryons in the relativistic quark model. *Prog. Part. Nucl. Phys.* **2003**, *50*, 329.
74. Faessler, A.; Gutsche, T.; Ivanov, M.A.; Körner, J.G.; Lyubovitskij, V.E. Semileptonic decays of double heavy baryons in a relativistic constituent three-quark model. *Phys. Rev. D* **2009**, *80*, 034025. [CrossRef]
75. Faessler, A.; Gutsche, T.; Ivanov, M.A.; Körner, J.G.; Lyubovitskij, V.E. Decay properties of double heavy baryons. *AIP Conf. Proc.* **2010**, *1257*, 311.
76. Branz, T.; Faessler, A.; Gutsche, T.; Ivanov, M.A.; Körner, J.G.; Lyubovitskij, V.E.; Oexl, B. Radiative decays of double heavy baryons in a relativistic constituent three–quark model including hyperfine mixing. *Phys. Rev. D* **2010**, *81*, 114036.
77. Ivanov, M. Dynamical picture for the exotic XYZ states. *EPJ Web Conf.* **2018**, *192*, 00042. [CrossRef]
78. Dubnicka, S.; Dubnickova, A.Z.; Ivanov, M.A.; Körner, J.G. Quark model description of the tetraquark state X(3872) in a relativistic constituent quark model with infrared confinement. *Phys. Rev. D* **2010**, *81*, 114007. [CrossRef]
79. Dubnicka, S.; Dubnickova, A.Z.; Ivanov, M.A.; Körner, J.G.; Santorelli, P.; Saidullaeva, G.G. One-photon decay of the tetraquark state $X(3872) \to \gamma + J/\psi$ in a relativistic constituent quark model with infrared confinement. *Phys. Rev. D* **2011**, *84*, 014006. [CrossRef]
80. Goerke, F.; Gutsche, T.; Ivanov, M.A.; Körner, J.G.; Lyubovitskij, V.E.; Santorelli, P. Four-quark structure of Zc(3900), Z(4430) and Xb(5568) states. *Phys. Rev. D* **2016**, *94*, 094017. [CrossRef]
81. Goerke, F.; Gutsche, T.; Ivanov, M.A.; Körner, J.G.; Lyubovitskij, V.E. $Z_b(10610)$ and $Z'_b(10650)$ decays in a covariant quark model. *Phys. Rev. D* **2017**, *96*, 054028.
82. Gutsche, T.; Ivanov, M.A.; Körner, J.G.; Lyubovitskij, V.E.; Xu, K. Test of the multiquark structure of $a_1(1420)$ in strong two-body decays. *Phys. Rev. D* **2017**, *96*, 114004.
83. Gutsche, T.; Ivanov, M.A.; Körner, J.G.; Lyubovitskij, V.E. Isospin-violating strong decays of scalar single-heavy tetraquarks. *Phys. Rev. D* **2016**, *94*, 094012.
84. Anikin, I.V.; Ivanov, M.A.; Kulimanova, N.B.; Lyubovitskij, V.E. The Extended Nambu-Jona-Lasinio model with separable interaction: Low-energy pion physics and pion nucleon form-factor. *Z. Phys. C* **1995**, *65*, 681. [CrossRef]
85. Salam, A. Lagrangian theory of composite particles. *Nuovo Cim.* **1962**, *25*, 224–227. [CrossRef]
86. Weinberg, S. Elementary particle theory of composite particles. *Phys. Rev.* **1963**, *130*, 776–783.
87. Mandelstam, S. Quantum electrodynamics without potentials. *Ann. Phys.* **1962**, *19*, 1–24. [CrossRef]
88. Vermaseren, J.A.M. New features of FORM. *arXiv* **2000**, arXiv: math-ph/0010025.
89. Voloshin, M.B. Interference and binding effects in decays of possible molecular component of X(3872). *Phys. Lett. B* **2004**, *579*, 316.
90. Wong, C.Y. Molecular states of heavy quark mesons. *Phys. Rev. C* **2004**, *69*, 055202.

91. AlFiky, M.T.; Gabbiani, F.; Petrov, A.A. X(3872): Hadronic molecules in effective field theory. *Phys. Lett. B* **2006**, *640*, 238. [CrossRef]
92. Fleming, S.; Kusunoki, M.; Mehen, T.; van Kolck, U. Pion interactions in the X(3872). *Phys. Rev. D* **2007**, *76*, 034006. [CrossRef]
93. Bignamini, C.; Grinstein, B.; Piccinini, F.; Polosa, A.D.; Sabelli, C. Is the X(3872) Production Cross Section at Tevatron Compatible with a Hadron Molecule Interpretation? *Phys. Rev. Lett.* **2009**, *103*, 162001. [CrossRef] [PubMed]
94. Hidalgo-Duque, C.; Nieves, J.; Valderrama, M.P. Light flavor and heavy quark spin symmetry in heavy meson molecules. *Phys. Rev. D* **2013**, *87*, 076006.
95. Braaten, E.; Lu, M. Line shapes of the X(3872). *Phys. Rev. D* **2007**, *76*, 094028. [CrossRef]
96. Lee, I.W.; Faessler, A.; Gutsche, T.; Lyubovitskij, V.E. X(3872) as a molecular DD* state in a potential model. *Phys. Rev. D* **2009**, *80*, 094005. [CrossRef]
97. Chiu, T.-W.; Hsieh, T.-H. X(3872) in lattice QCD with exact chiral symmetry. *Phys. Lett. B* **2007**, *646*, 95. [CrossRef]
98. Barnes, T.; Godfrey, S. Charmonium options for the X(3872). *Phys. Rev. D* **2004**, *69*, 054008. [CrossRef]
99. Eichten, E.J.; Lane, K.; Quigg, C. Charmonium levels near threshold and the narrow state $X(3872) \to \pi^+\pi^- J/\psi$. *Phys. Rev. D* **2004**, *69*, 094019. [CrossRef]
100. Kalashnikova, Y.S. Coupled-channel model for charmonium levels and an option for X(3872). *Phys. Rev. D* **2005**, *72*, 034010. [CrossRef]
101. Matheus, R.D.; Navarra, F.S.; Nielsen, M.; Zanetti, C.M. QCD Sum Rules for the X(3872) as a mixed molecule-charmoniun state. *Phys. Rev. D* **2009**, *80*, 056002.
102. Ferretti, J.; Galata, G.; Santopinto, E. Interpretation of the X(3872) as a charmonium state plus an extra component due to the coupling to the meson-meson continuum. *Phys. Rev. C* **2013**, *88*, 015207. [CrossRef]
103. Takizawa, M.; Takeuchi, S. X(3872) as a hybrid state of charmonium and the hadronic molecule. *PTEP* **2013**, *2013*, 093D01. [CrossRef]
104. Ferretti, J.; Galata, G.; Santopinto, E. Quark structure of the X(3872) and $\chi_b(3P)$ resonances. *Phys. Rev. D* **2014**, *90*, 054010. [CrossRef]
105. Zhang, O.; Meng, C.; Zheng, H.Q. Ambiversion of X(3872). *Phys. Lett. B* **2009**, *680*, 453. [CrossRef]
106. Swanson, E.S. Diagnostic decays of the X(3872). *Phys. Lett. B* **2004**, *598*, 197.
107. Swanson, E.S. Short range structure in the X(3872). *Phys. Lett. B* **2004**, *588*, 189. [CrossRef]
108. Suzuki, M. The X(3872) boson: Molecule or charmonium. *Phys. Rev. D* **2005**, *72*, 114013. [CrossRef]
109. Liu, Y.R.; Liu, X.; Deng, W.Z.; Zhu, S.L. Is X(3872) Really a Molecular State? *Eur. Phys. J. C* **2008**, *56*, 63. [CrossRef]
110. Coito, S.; Rupp, G.; van Beveren, E. X(3872) is not a true molecule. *Eur. Phys. J. C* **2013**, *73*, 2351.
111. Seth, K.K. An Alternative Interpretation of X(3872). *Phys. Lett. B* **2005**, *612*, 1. [CrossRef]
112. Hogaasen, H.; Richard, J.M.; Sorba, P. A Chromomagnetic mechanism for the X(3872) resonance. *Phys. Rev. D* **2006**, *73*, 054013. [CrossRef]
113. Ortega, P.G.; Ruiz Arriola, E. Is X(3872) a bound state? *Chin. Phys. C* **2019**, *43*, 124107. [CrossRef]
114. Li, B.A. Is X(3872) a possible candidate of hybrid meson. *Phys. Lett. B* **2005**, *605*, 306. [CrossRef]
115. Close, F.E.; Page, P.R. The D*0 anti-D0 threshold resonance. *Phys. Lett. B* **2004**, *578*, 119. [CrossRef]
116. Prelovsek, S.; Leskovec, L. Evidence for X(3872) from DD^* scattering on the lattice. *Phys. Rev. Lett.* **2013**, *111*, 192001. [CrossRef]
117. Padmanath, M.; Lang, C.B.; Prelovsek, S. X(3872) and Y(4140) using diquark-antidiquark operators with lattice QCD. *Phys. Rev. D* **2015**, *92*, 034501. [CrossRef]
118. Matheus, R.D.; Narison, S.; Nielsen, M.; Richard, J.M. Can the X(3872) be a 1++ four quark state? *Phys. Rev. D* **2007**, *75*, 014005. [CrossRef]
119. Narison, S.; Navarra, F.S.; Nielsen, M. On the nature of the X(3872) from QCD. *Phys. Rev. D* **2011**, *83*, 016004. [CrossRef]
120. Gamermann, D.; Oset, E. Isospin breaking effects in the X(3872) resonance. *Phys. Rev. D* **2009**, *80*, 014003. [CrossRef]
121. Gamermann, D.; Nieves, J.; Oset, E.; Ruiz Arriola, E. Couplings in coupled channels versus wave functions: Application to the X(3872) resonance. *Phys. Rev. D* **2010**, *81*, 014029.

122. Danilkin, I.V.; Simonov, Y.A. Dynamical origin and the pole structure of X(3872). *Phys. Rev. Lett.* **2010**, *105*, 102002. [CrossRef] [PubMed]
123. Maiani, L.; Piccinini, F.; Polosa, A.D.; Riquer, V. Diquark-antidiquarks with hidden or open charm and the nature of X(3872). *Phys. Rev. D* **2005**, *71*, 014028. [CrossRef]
124. Tan, Y.; Ping, J. X(3872) in an unquenched quark model. *Phys. Rev. D* **2019**, *100*, 034022. [CrossRef]
125. Hanhart, C.; Kalashnikova, Y.S.; Kudryavtsev, A.E.; Nefediev, A.V. Reconciling the X(3872) with the near-threshold enhancement in the $D^0\bar{D}^{*0}$ final state. *Phys. Rev. D* **2007**, *76*, 034007. [CrossRef]
126. Chen, Y.; Zhu, S.L. The Vector and Axial-Vector Charmonium-like States. *Phys. Rev. D* **2011**, *83*, 034010. [CrossRef]
127. Wallbott, P.C.; Eichmann, G.; Fischer, C.S. $X(3872)$ as a four quark state in a Dyson-Schwinger/Bethe-Salpeter approach. *Phys. Rev. D* **2019**, *100*, 014033. [CrossRef]
128. Bigi, I.; Maiani, L.; Piccinini, F.; Polosa, A.; Riquer, V. Four-quark mesons in non-leptonic B decays: Could they resolve some old puzzles? *Phys. Rev. D* **2005**, *72*, 114016. [CrossRef]
129. Abe, K.; Abe, K.; Adachi, I.; Aihara, H.; Aoki, K.; Arinstein, K.; Asano, Y.; Aso, T.; Aulchenko, V.; Aushev, T. Evidence for $X(3872) \to \gamma J/\psi$ and the sub-threshold decay $X(3872) \to \omega J/\psi$. *arXiv* **2005**, arXiv:hep-ex/0505037.
130. Uhlemann, C.; Kauer, N. Narrow-width approximation accuracy. *Nucl. Phys. B* **2009**, *814*, 195–211. [CrossRef]
131. Voloshin, M. Heavy quark spin selection rule and the properties of the X(3872). *Phys. Lett. B* **2004**, *604*, 69–73. [CrossRef]
132. Voloshin, M. X(3872) diagnostics with decays to $D\bar{D}\gamma$. *Int. J. Mod. Phys. A* **2006**, *21*, 1239–1250. [CrossRef]
133. Dubynskiy, S.; Voloshin, M.B. Pionic transitions from $X(3872)$ to $\chi(cJ)$. *Phys. Rev. D* **2008**, *77*, 014013. [CrossRef]
134. Maiani, L.; Polosa, A.; Riquer, V. Indications of a Four-Quark Structure for the X(3872) and X(3876) Particles from Recent Belle and BABAR Data. *Phys. Rev. Lett.* **2007**, *99*, 182003. [CrossRef] [PubMed]
135. Ivanov, M.A.; Körner, J.G.; Santorelli, P. The J/ψ dissociation cross-sections in a relativistic quark model. *Phys. Rev. D* **2004**, *70*, 014005. [CrossRef]
136. Barnes, T. Charmonium cross-sections and the QGP. *Eur. Phys. J. A* **2003**, *18*, 531.
137. Aubert, B.; Barate, R.; Bona, M.; Boutigny, D.; Couderc, F.; Karyotakis, Y.; Lees, J.P.; Poireau, V.; Tisserand, V.; Zghiche, A.; et al. Search for $B^+ \to X(3872)K^+$, $X_{3872} \to J/\psi\gamma$. *Phys. Rev. D* **2006**, *74*, 071101. [CrossRef]
138. Aubert, B.; Bona, M.; Karyotakis, Y.; Lees, J.P.; Poireau, V.; Prencipe, E.; Prudent, X.; Tisserand, V.; Garra Tico, J.; Grauges, E.; et al. Evidence for $X(3872) \to \psi_{2S}\gamma$ in $B^{\pm} \to X_{3872}K^{\pm}$ decays, and a study of $B \to c\bar{c}\gamma K$. *Phys. Rev. Lett.* **2009**, *102*, 132001. [CrossRef]
139. Bhardwaj, V.; Trabelsi, K.; Singh, J.B.; Choi, S.-K.; Olsen, S.L.; Adachi, I.; Adamczyk, K.; Asner, D.M.; Aulchenko, V.; Aushev, T.; et al. Observation of $X(3872) \to J/\psi\gamma$ and search for $X(3872) \to \psi'\gamma$ in B decays. *Phys. Rev. Lett.* **2011**, *107*, 091803. [CrossRef]
140. Aaij, R.; Adeva, B.; Adinolfi, M.; Affolder, A.; Ajaltouni, Z.; Albrecht, J.; Alessio, F.; Alexander, M.; Ali, S.; Alkhazov, G.; et al. Evidence for the decay $X(3872) \to \psi(2S)\gamma$. *Nucl. Phys. B* **2014**, *886*, 665–680. [CrossRef]
141. Colangelo, P.; De Fazio, F.; Nicotri, S. $X(3872) \to D\bar{D}\gamma$ decays and the structure of X_{3872}. *Phys. Lett. B* **2007**, *650*, 166–171. [CrossRef]
142. Dong, Y.; Faessler, A.; Gutsche, T.; Kovalenko, S.; Lyubovitskij, V.E. X(3872) as a hadronic molecule and its decays to charmonium states and pions. *Phys. Rev. D* **2009**, *79*, 094013.
143. Nielsen, M.; Zanetti, C. Radiative decay of the X(3872) as a mixed molecule-charmonium state in QCD Sum Rules. *Phys. Rev. D* **2010**, *82*, 116002. [CrossRef]
144. Wang, T.; Wang, G. Radiative E1 decays of X(3872). *Phys. Lett. B* **2011**, *697*, 233–237 [CrossRef]
145. Ke, H.; Li, X. What do the radiative decays of X(3872) tell us. *Phys. Rev. D* **2011**, *84*, 114026. [CrossRef]
146. Takizawa, M.; Takeuchi, S.; Shimizu, K. Radiative X(3872) Decays in Charmonium-Molecule Hybrid Model. *Few Body Syst.* **2014**, *55*, 779. [CrossRef]
147. Badalian, A.; Simonov, Y.; Bakker, B. $c\bar{c}$ interaction above threshold and the radiative decay $X(3872) \to J/\psi\gamma$. *Phys. Rev. D* **2015**, *91*, 056001.
148. Takeuchi, S.; Takizawa, M.; Shimizu, K. Radiative Decays of the $X(3872)$ in the Charmonium-Molecule Hybrid Picture. *JPS Conf. Proc.* **2017**, *17*, 112001.
149. Cincioglu, E.; Ozpineci, A. Radiative decay of the $X(3872)$ as a mixed molecule-charmonium state in effective field theory. *Phys. Lett. B* **2019**, *797*, 134856. [CrossRef]

150. Ivanov, M.A.; Locher, M.; Lyubovitskij, V.E. Electromagnetic form-factors of nucleons in a relativistic three quark model. *Few Body Syst.* **1996**, *21*, 131. [CrossRef]
151. Körner, J.G.; Mauser, M.C. One loop corrections to polarization observables. *Lect. Notes Phys.* **2004**, *647*, 212.
152. Dias, J.M.; Navarra, F.S.; Nielsen, M.; Zanetti, C.M. $Z_c^+(3900)$ decay width in QCD sum rules. *Phys. Rev. D* **2013**, *88*, 016004.
153. Wang, Z.G. The magnetic moment of the $Z_c(3900)$ as an axialvector tetraquark state with QCD sum rules. *Eur. Phys. J. C* **2018**, *78*, 297. [CrossRef]
154. Wang, Z.G.; Zhang, J.X. The decay width of the $Z_c(3900)$ as an axialvector tetraquark state in solid quark-hadron duality. *Eur. Phys. J. C* **2018**, *78*, 14. [CrossRef]
155. Deng, C.; Ping, J.; Wang, F. Interpreting $Z_c(3900)$ and $Z_c(4025)/Z_c(4020)$ as charged tetraquark states. *Phys. Rev. D* **2014**, *90*, 054009.
156. Ke, H.W.; Wei, Y.; Li, X.Q. Is $Z_c(3900)$ a molecular state. *Eur. Phys. J. C* **2013**, *73*, 2561. [CrossRef]
157. Wilbring, E.; Hammer, H.W.; Meißner, U.-G. Electromagnetic Structure of the $Z_c(3900)$. *Phys. Lett. B* **2013**, *726*, 326. [CrossRef]
158. Zhang, J.R. Improved QCD sum rule study of $Z_c(3900)$ as a $\bar{D}D^*$ molecular state. *Phys. Rev. D* **2013**, *87*, 116004. [CrossRef]
159. Cui, C.Y.; Liu, Y.L.; Chen, W.B.; Huang, M.Q. Could $Z_c(3900)$ be a $I^G J^P = 1^+ 1^+$ $D^*\bar{D}$ molecular state? *J. Phys. G* **2014**, *41*, 075003. [CrossRef]
160. Patel, S.; Shah, M.; Vinodkumar, P.C. Mass spectra of four quark states in the hidden charm sector. *Eur. Phys. J. A* **2014**, *50*, 131. [CrossRef]
161. Chen, D.Y.; Dong, Y.B. Radiative decays of the neutral $Z_c(3900)$. *Phys. Rev. D* **2016**, *93*, 014003. [CrossRef]
162. Ortega, P.G.; Segovia, J.; Entem, D.R.; Fernandez, F. The Z_c structures in a coupled-channels model. *Eur. Phys. J. C* **2019**, *79*, 78. [CrossRef]
163. Xiao, L.Y.; Wang, G.J.; Zhu, S.L. Hidden-charm strong decays of the Z_c states. *Phys. Rev. D* **2019**, *101*, 054001.
164. Prelovsek, S.; Leskovec, L. Search for $Z_c^+(3900)$ in the 1^{+-} Channel on the Lattice. *Phys. Lett. B* **2013**, *727*, 172. [CrossRef]
165. Chen, Y.; Gong, M.; Lei, Y.; Li, N.; Liang, J.; Liu, C.; Liu, H.; Liu, J.; Liu, L.; Liu, Y.; et al. Low-energy scattering of the $(D\bar{D}^*)^{\pm}$ system and the resonance-like structure $Z_c(3900)$. *Phys. Rev. D* **2014**, *89*, 094506. [CrossRef]
166. Prelovsek, S.; Lang, C.B.; Leskovec, L.; Mohler, D. Study of the Z_c^+ channel using lattice QCD. *Phys. Rev. D* **2015**, *91*, 014504. [CrossRef]
167. Chen, T.; Chen, Y.; Gong, M.; Liu, C.; Liu, L.; Liu, Y.; Liu, Z.; Ma, J.; Werner, M.; Zhang, J. A coupled-channel lattice study on the resonance-like structure $Z_c(3900)$. *Chin. Phys. C* **2019**, *43*, 103103.
168. Ikeda, Y.; Aoki, S.; Doi, T.; Gongyo, S.; Hatsuda, T.; Inoue, T.; Iritani, T.; Ishii, N.; Murano, K.; Sasaki, K. Fate of the Tetraquark Candidate $Z_c(3900)$ from Lattice QCD. *Phys. Rev. Lett.* **2016**, *117*, 242001. [CrossRef]
169. Ikeda, Y. The tetraquark candidate $Z_c(3900)$ from dynamical lattice QCD simulations. *J. Phys. G* **2018**, *45*, 024002. [CrossRef]
170. Liu, C.; Liu, L.; Zhang, K.L. Towards the understanding of $Z_c(3900)$ from lattice QCD. *Phys. Rev. D* **2020**, *101*, 054502. [CrossRef]
171. Braaten, E. How the $Z_c(3900)$ Reveals the Spectra of Quarkonium Hybrid and Tetraquark Mesons. *Phys. Rev. Lett.* **2013**, *111*, 162003. [CrossRef]
172. Voloshin, M.B. $Z_c(3900)$—What is inside? *Phys. Rev. D* **2013**, *87*, 091501. [CrossRef]
173. Zhao, L.; Deng, W.Z.; Zhu, S.L. Hidden-Charm Tetraquarks and Charged Z_c States. *Phys. Rev. D* **2014**, *90*, 094031. [CrossRef]
174. Li, G. Hidden-charmonium decays of $Z_c(3900)$ and $Z_c(4025)$ in intermediate meson loops model. *Eur. Phys. J. C* **2013**, *73*, 2621. [CrossRef]
175. Chen, D.Y.; Liu, X.; Matsuki, T. Reproducing the $Z_c(3900)$ structure through the initial-single-pion-emission mechanism. *Phys. Rev. D* **2013**, *88*, 036008. [CrossRef]
176. Maiani, L.; Riquer, V.; Faccini, R.; Piccinini, F.; Pilloni, A.; Polosa, A.D. A $J^{PG} = 1^{++}$ Charged Resonance in the $Y(4260) \to \pi^+\pi^- J/\psi$ Decay? *Phys. Rev. D* **2013**, *87*, 111102. [CrossRef]
177. Aceti, F.; Bayar, M.; Oset, E.; Martinez Torres, A.; Khemchandani, K.P.; Dias, J.M.; Navarra, F.S.; Nielsen, M. Prediction of an $I = 1$ $D\bar{D}^*$ state and relationship to the claimed $Z_c(3900)$, $Z_c(3885)$. *Phys. Rev. D* **2014**, *90*, 016003. [CrossRef]

178. Agaev, S.S.; Azizi, K.; Sundu, H. Strong $Z_c^+(3900) \to J/\psi\pi^+; \eta_c\rho^+$ decays in QCD. *Phys. Rev. D* **2016**, *93*, 074002. [CrossRef]
179. Albaladejo, M.; Guo, F.K.; Hidalgo-Duque, C.; Nieves, J. $Z_c(3900)$: What has been really seen? *Phys. Lett. B* **2016**, *755*, 337. [CrossRef]
180. Agaev, S.S.; Azizi, K.; Sundu, H. Treating $Z_c(3900)$ and $Z(4430)$ as the ground-state and first radially excited tetraquarks. *Phys. Rev. D* **2017**, *96*, 034026. [CrossRef]
181. Ozdem, U.; Azizi, K. Magnetic and quadrupole moments of the $Z_c(3900)$. *Phys. Rev. D* **2017**, *96*, 074030. [CrossRef]
182. Pilloni, A.; Fernandez-Ramirez, C.; Jackurac, A.; Mathieu, V.; Mikhasenko, M.; Nys, J.; Szczepaniak, A.P. Amplitude analysis and the nature of the $Z_c(3900)$. *Phys. Lett. B* **2017**, *772*, 200. [CrossRef]
183. He, J.; Chen, D.Y. $Z_c(3900)/Z_c(3885)$ as a virtual state from $\pi J/\psi - \bar{D}^*D$ interaction. *Eur. Phys. J. C* **2018**, *78*, 94. [CrossRef]
184. Chen, H.X.; Chen, W. Settling the Zc(4600) in the charged charmoniumlike family. *Phys. Rev. D* **2019**, *99*, 074022. [CrossRef]
185. Chiu, T.-W.; Hsieh, T.-H. Y(4260) on the lattice. *Phys. Rev. D* **2006**, *73*, 094510. [CrossRef]
186. Albuquerque, R.M.; Nielsen, M. QCD sum rules study of the J(PC) = 1− charmonium Y mesons. *Nucl. Phys. A* **2009**, *815*, 53. [CrossRef]
187. Ding, G.J. Are Y(4260) and Z+(2) are D(1) D or D(0) D* Hadronic Molecules? *Phys. Rev. D* **2009**, *79*, 014001. [CrossRef]
188. Wang, Q.; Cleven, M.; Guo, F.K.; Hanhart, C.; Meißner, U.G.; Wu, X.G.; Zhao, Q. Y(4260): Hadronic molecule versus hadro-charmonium interpretation. *Phys. Rev. D* **2014**, *89*, 034001. [CrossRef]
189. Xue, S.R.; Jing, H.J.; Guo, F.K.; Zhao, Q. Disentangling the role of the Y(4260) in $e^+e^- \to D^*\bar{D}^*$ and $D_s^*\bar{D}_s^*$ via line shape studies. *Phys. Lett. B* **2018**, *779*, 402. [CrossRef]
190. Chen, Y.H.; Dai, L.Y.; Guo, F.K.; Kubis, B. Nature of the Y(4260): A light-quark perspective. *Phys. Rev. D* **2019**, *99*, 074016. [CrossRef]
191. Qiao, C.F. One explanation for the exotic state Y(4260). *Phys. Lett. B* **2006**, *639*, 263. [CrossRef]
192. Yuan, C.Z.; Wang, P.; Mo, X.H. The Y(4260) as an omega chi(c1) molecular state. *Phys. Lett. B* **2006**, *634*, 399. [CrossRef]
193. Liu, X.; Li, G. Exploring the threshold behavior and implications on the nature of Y(4260) and $Z_c(3900)$. *Phys. Rev. D* **2013**, *88*, 014013. [CrossRef]
194. Cleven, M.; Wang, Q.; Guo, F.K.; Hanhart, C.; Meißner, U.G.; Zhao, Q. Y(4260) as the first S-wave open charm vector molecular state? *Phys. Rev. D* **2014**, *90*, 074039.
195. Liu, X.; Zeng, X.; Li, X. Possible molecular structure of the newly observed Y(4260). *Phys. Rev. D* **2005**, *72*, 054023.
196. Li, X.; Voloshin, M. Suppression of the S-wave production of $(3/2)^+ + (1/2)^-$ heavy meson pairs in e^+e^- annihilation. *Phys. Rev. D* **2013**, *88*, 034012. [CrossRef]
197. Zhu, S.L. The Possible interpretations of Y(4260). *Phys. Lett. B* **2005**, *625*, 212. [CrossRef]
198. Llanes-Estrada, F.J. Y(4260) and possible charmonium assignment. *Phys. Rev. D* **2005**, *72*, 031503.
199. Kou, E.; Pene, O. Suppressed decay into open charm for the Y(4260) being an hybrid. *Phys. Lett. B* **2005**, *631*, 164. [CrossRef]
200. Close, F.E.; Page, P.R. Gluonic charmonium resonances at BaBar and BELLE? *Phys. Lett. B* **2005**, *628*, 215–222. [CrossRef]
201. Maiani, L.; Riquer, V.; Piccinini, F.; Polosa, A.D. Four quark interpretation of Y(4260). *Phys. Rev. D* **2005**, *72*, 031502.
202. Dong, Y.; Faessler, A.; Gutsche, T.; Lyubovitskij, V.E. Selected strong decay modes of Y(4260). *Phys. Rev. D* **2014**, *89*, 034018. [CrossRef]
203. Wang, Z.G. Tetraquark state candidates: Y(4260), Y(4360), Y(4660) and $Z_c(4020/4025)$. *Eur. Phys. J. C* **2016**, *76*, 387. [CrossRef]
204. Tan, Y.; Ping, J. Y(4626) in a chiral constituent quark model. *Phys. Rev. D* **2020**, *101*, 054010.
205. Zhang, J.R.; Huang, M.Q. The P-wave $[cs][\bar{c}\bar{s}]$ tetraquark state: Y(4260) or Y(4660)? *Phys. Rev. D* **2011**, *83*, 036005. [CrossRef]
206. Martinez Torres, A.; Khemchandani, K.P.; Gamermann, D.; Oset, E. The Y(4260) as a $J/\psi K\bar{K}$ system. *Phys. Rev. D* **2009**, *80*, 094012. [CrossRef]

207. Dubnička, S.; Dubničková, A.; Issadykov, A.; Ivanov, M.A.; Liptaj, A. Y(4260) as four quark state. *arXiv* **2020**, arXiv:2003.04142.
208. Sun, Z.F.; He, J.; Liu, X.; Luo, Z.G.; Zhu, S.L. $Z_b(10610)^\pm$ and $Z_b(10650)^\pm$ as the $B^*\bar{B}$ and $B^*\bar{B}^*$ molecular states. *Phys. Rev. D* **2011**, *84*, 054002. [CrossRef]
209. Bondar, A.E.; Garmash, A.; Milstein, A.I.; Mizuk, R.; Voloshin, M.B. Heavy quark spin structure in Z_b resonances. *Phys. Rev. D* **2011**, *84*, 054010. [CrossRef]
210. Dong, Y.; Faessler, A.; Gutsche, T.; Lyubovitskij, V.E. Decays of Z_b^+ and $Z_b'^+$) as Hadronic Molecules. *J. Phys. G* **2013**, *40*, 015002.
211. Li, X.; Voloshin, M.B. $Z_b(10610)$ and $Z_b(10650)$ decays to bottomonium plus pion. *Phys. Rev. D* **2012**, *86*, 077502. [CrossRef]
212. Cleven, M.; Wang, Q.; Guo, F.K.; Hanhart, C.; Meißner, U.G.; Zhao, Q. Confirming the molecular nature of the $Z_b(10610)$ and the $Z_b(10650)$. *Phys. Rev. D* **2013**, *87*, 074006. [CrossRef]
213. Voloshin, M.B. Enhanced mixing of partial waves near threshold for heavy meson pairs and properties of $Z_b(10610)$ and $Z_b(10650)$ resonances. *Phys. Rev. D* **2013**, *87*, 074011. [CrossRef]
214. Wang, Z.G. Reanalysis of the $Y(3940)$, $Y(4140)$, $Z_c(4020)$, $Z_c(4025)$ and $Z_b(10650)$ as molecular states with QCD sum rules. *Eur. Phys. J. C* **2014**, *74*, 2963. [CrossRef]
215. Baru, V.; Epelbaum, E.; Filin, A.A.; Hanhart, C.; Nefediev, A.V. Spin partners of the $Z_b(10610)$ and $Z_b(10650)$ revisited. *JHEP* **2017**, *1706*, 158. [CrossRef]
216. Baru, V.; Epelbaum, E.; Filin, A.A.; Hanhart, C.; Nefediev, A.V.; Wang, Q. Spin partners W_{bJ} from the line shapes of the $Z_b(10610)$ and $Z_b(10650)$. *Phys. Rev. D* **2019**, *99*, 094013.
217. Ali, A.; Hambrock, C.; Wang, W. Tetraquark Interpretation of the Charged Bottomonium-like states $Z_b^{+-}(10610)$ and $Z_b^{+-}(10650)$ and Implications. *Phys. Rev. D* **2012**, *85*, 054011.
218. Wang, Z.G.; Huang, T. The $Z_b(10610)$ and $Z_b(10650)$ as axial-vector tetraquark states in the QCD sum rules. *Nucl. Phys. A* **2014**, *930*, 63.
219. Agaev, S.S.; Azizi, K.; Sundu, H. Spectroscopic parameters and decays of the resonance $Z_b(10610)$. *Eur. Phys. J. C* **2017**, *77*, 836. [CrossRef]
220. Ozdem, U.; Azizi, K. Magnetic dipole moment of $Z_b(10610)$ in light-cone QCD. *Phys. Rev. D* **2018**, *97*, 014010. [CrossRef]
221. Li, M.T.; Wang, W.L.; Dong, Y.B.; Zhang, Z.Y. $Z_b(10650)$ and $Z_b(10610)$ States in A Chiral Quark Model. *J. Phys. G* **2013**, *40*, 015003. [CrossRef]
222. Chen, D.Y.; Liu, X.; Matsuki, T. Interpretation of $Z_b(10610)$ and $Z_b(10650)$ in the ISPE mechanism and the Charmonium Counterpart. *Chin. Phys. C* **2014**, *38*, 053102. [CrossRef]
223. Swanson, E.S. Z_b and Z_c Exotic States as Coupled Channel Cusps. *Phys. Rev. D* **2015**, *91*, 034009. [CrossRef]
224. Kang, X.W.; Guo, Z.H.; Oller, J.A. General considerations on the nature of $Z_b(10610)$ and $Z_b(10650)$ from their pole positions. *Phys. Rev. D* **2016**, *94*, 014012. [CrossRef]
225. Wang, Q.; Baru, V.; Filin, A.A. Hanhart, C.; Nefediev, A.V.; Wynen, J.-L. Line shapes of the $Z_b(10610)$ and $Z_b(10650)$ in the elastic and inelastic channels revisited. *Phys. Rev. D* **2018**, *98*, 074023. [CrossRef]
226. Voloshin, M.B. Radiative transitions from Upsilon(5S) to molecular bottomonium. *Phys. Rev. D* **2011**, *84*, 031502.
227. Zhang, J.R.; Zhong, M.; Huang, M.Q. Could $Z_b(10610)$ be a $B^*\bar{B}$ molecular state? *Phys. Lett. B* **2011**, *704*, 312.
228. Cui, C.; Liu, Y.; Huang, M. Investigating different structures of the $Z_b(10610)$ and $Z_b(10650)$. *Phys. Rev. D* **2012**, *85*, 074014. [CrossRef]
229. Yang, Y.; Ping, J.; Deng, C.; Zong, H.S. Dynamical study of the $Z_b(10610)$ and $Z_b(10650)$ as molecular states. *J. Phys. G* **2012**, *39*, 105001. [CrossRef]
230. Danilkin, I.V.; Orlovsky, V.D.; Simonov, Y.A. Hadron interaction with heavy quarkonia. *Phys. Rev. D* **2012**, *85*, 034012. [CrossRef]
231. Chen, D.Y.; Liu, X.; Zhu, S.L. Charged bottomonium-like states $Z_b(10610)$ and $Z_b(10650)$ and the $Y(5S) \to Y(2S)\pi^+\pi^-$ decay. *Phys. Rev. D* **2011**, *84*, 074016. [CrossRef]
232. Chen, D.; Liu, X. $Z_b(10610)$ and $Z_b(10650)$ structures produced by the initial single pion emission in the $Y(5S)$ decays. *Phys. Rev. D* **2011**, *84*, 094003. [CrossRef]
233. Liu, X.; Chen, D. Charged Bottomonium-Like Structures $Z_b(10610)$ and $Z_b(10650)$. *Few Body Syst.* **2013**, *54*, 165–170. [CrossRef]

234. Cleven, M.; Guo, F.K.; Hanhart, C.; Meißner, U.G. Bound state nature of the exotic Z_b states. *Eur. Phys. J. A* **2011**, *47*, 120. [CrossRef]
235. Guo, T.; Cao, L.; Zhou, M.Z.; Chen, H. The possible candidates of tetraquark: $Z_b(10610)$ and $Z_b(10650)$. *arXiv* **2011**, arXiv:1106.2284.
236. Navarra, F.S.; Nielsen, M.; Richard, J.M. Exotic Charmonium and Bottomonium-like Resonances. *J. Phys. Conf. Ser.* **2012**, *348*, 012007. [CrossRef]
237. Li, G.; Shao, F.; Zhao, C.W.; Zhao, Q. $Z_b/Z_b' \to \Upsilon\pi$ and $h_b\pi$ decays in intermediate meson loops model. *Phys. Rev. D* **2013**, *87*, 034020. [CrossRef]
238. Ohkoda, S.; Yasui, S.; Hosaka, A. Decays of $Z_b \to \Upsilon\pi$ via triangle diagrams in heavy meson molecules. *Phys. Rev. D* **2014**, *89*, 074029. [CrossRef]
239. Dias, J.M.; Aceti, F.; Oset, E. Study of $B\bar{B}^*$ and $B^*\bar{B}^*$ interactions in $I = 1$ and relationship to the $Z_b(10610)$, $Z_b(10650)$ states. *Phys. Rev. D* **2015**, *91*, 076001. [CrossRef]
240. Huo, W.S.; Chen, G.Y. The nature of Z_b states from a combined analysis of $Y(5S) \to h_b(mP)\pi^+\pi^-$ and $Y(5S) \to B^{(*)}\bar{B}^{(*)}\pi$. *Eur. Phys. J. C* **2016**, *76*, 172. [CrossRef]
241. Gutsche, T.; Lyubovitskij, V.E.; Schmidt, I. Tetraquarks in holographic QCD. *Phys. Rev. D* **2017**, *96*, 034030. [CrossRef]

© 2020 by the authors. Licensee MDPI, Basel, Switzerland. This article is an open access article distributed under the terms and conditions of the Creative Commons Attribution (CC BY) license (http://creativecommons.org/licenses/by/4.0/).

Review

Charge Asymmetry in Top Quark Pair Production

Roman Lysák

FZU—Institute of Physics of the Czech Academy of Sciences, Na Slovance 1999/2, 18221 Prague, Czech Republic; lysak@fzu.cz

Received: 16 June 2020; Accepted: 21 July 2020; Published: 2 August 2020

Abstract: The top quark is the heaviest elementary particle known. It has been proposed many times that new physics beyond the current theory of elementary particles may reveal itself in top quark interactions. The charge asymmetry in the pair production of a fermion and its antiparticle has been known for many decades. Early measurements of such asymmetry in top quark pair production showed a disagreement with the prediction by more than 3 standard deviations. Many years of an effort on both experimental and theoretical side have allowed to understand the top quark pair charge asymmetry better and to bring back the agreement between the measurements and the theory. In this article, these efforts are reviewed together with the discussion about a potential future of such measurements.

Keywords: top quark; pair production; charge asymmetry; forward–backward asymmetry

1. Introduction

The Standard Model (SM) of particles is a quantum field theory which describes strong and electroweak interactions [1–3]. During the past about 40 years, it has been successfully tested in a large number of experiments which performed numerous measurements. However, the SM has its shortcomings. For example, it has too many free parameters, there is an absence of the explanation for the observed amount of dark matter [4], and the prediction for the matter–antimatter asymmetry is way too low compared to the observation [5]. There have been many theoretical attempts to overcome SM shortcomings. On the other hand, the experimentalists have been trying to find a discrepancy between predictions and measurements. This would serve as a hint for a more complex theory going beyond the Standard Model (BSM) framework.

The top quark is one of the fundamental fermions, spin-half particles, in the SM. It has a large mass ($m_t = 173 \pm 0.4$ GeV [6]), much larger than a mass of any other quark or lepton (the next heaviest quark, b quark is about 40 times lighter). This means the top quark may play a special role in BSM theories or the BSM physics may reveal first in the interactions involving the top quark [7,8]. Another consequence of its large mass is that it has a very short lifetime so it has no time to hadronize. Top quark properties are thus transferred to its decay products. From an experimental point of view, it is important that top quark properties can be studied without a complication from the hadronization, unlike with any other quark.

The top quark has been observed in the experiments at only two accelerators: in proton–antiproton ($p\bar{p}$) collisions at the Tevatron in Fermilab, USA and in proton–proton (pp) collisions at the Large Hadron Collider (LHC) at the European Organization for Nuclear Research (CERN), Switzerland. The top quark was observed for the first time in 1995 at the Tevatron in a data taking period called 'Run I' at a center-of-mass energy of interactions of $\sqrt{s} = 1.8$ TeV by CDF and D0 experiments [9,10]. The Run I took place during 1992–1996 and the amount of data collected per experiment corresponded to about 100 pb^{-1} of the integrated luminosity. Only a few tens of top–antitop ($t\bar{t}$) pair candidate events were collected at both experiments. The second data period (Run II) at the Tevatron happened during 2001–2011 at a bit larger energy of $\sqrt{s} = 1.96$ TeV. Overall, about a hundred times more data (10 fb^{-1})

were collected by each experiment. This amount of data allowed detailed measurements of top quark properties although a lot of the measurements have been statistically limited. The LHC started its operation in 2008, but after the incident a few days later, the first collisions at $\sqrt{s}=7$ TeV happened only in 2010. The center-of-mass energy of pp collisions (and the luminosity) has gradually risen from $\sqrt{s}=7$ TeV in 2010 (5 fb^{-1}) to $\sqrt{s}=8$ TeV in 2011–2012 (20 fb^{-1}) and $\sqrt{s}=13$ TeV in 2015–2018 (150 fb^{-1}) with the shutdown happening in 2013–2014. The data taking period from 2010–2012 is called "Run 1" while the second data taking period between 2015–2018 is called "Run 2". At present, there is another accelerator shutdown which is planned for years 2019–2021. Given the much higher energy of interactions and much larger luminosity at the LHC compared to the Tevatron, many more top quarks have been produced which allowed for much more detailed measurements of top quark properties.

One of the top quark properties which has been studied is a charge asymmetry in the top quark pair production. This means there is a difference in the angular distribution for top and antitop quarks with respect to a given direction. It is a small effect in the SM [11–32] which could be greatly enhanced by various BSM models [33–39]. The initial measurements at the Tevatron observed larger asymmetries than predicted by the SM at that time [40–44]. A few deviations larger than two standard deviations (SD) were observed by both experiments, with the largest deviation of more than 3 SD observed by the CDF experiment at a large invariant mass of the top quark pair [42].

The unexpectedly large measured charge asymmetries started a huge interest in both theoretical and experimental communities in studying this effect in a much more detail. Theoretical physicists calculated the asymmetry more precisely within the SM [17–22,24–26,28–32] and also tried to explain it with many new BSM models, see Refs. [45,46] and references therein. A few years ago the full next-to-next-leading order (NNLO) prediction in quantum chromodynamics (QCD) for the top quark pair production [47,48] and later for the $t\bar{t}$ charge asymmetry became available [30–32]. The experiments studied the underlying effect at both the Tevatron and the LHC, using different channels, studying various observables, and measuring the asymmetry in more detail differentially. The experiments at the Tevatron and the LHC are complementary. They can not measure the exact same asymmetry, rather two different observables based on the same underlying cause. There are advantages and disadvantages to perform the measurements at both colliders. The advantage at the Tevatron is that the predicted asymmetry ($\approx$10%) is about an order of magnitude larger compared to the LHC ($\approx$1%). On the other hand, the disadvantage at the Tevatron compared to the LHC is a limited data statistics. The non-zero forward–backward asymmetry has been already observed ($\geq$5 SD) at the Tevatron a few years ago [49], while one of the LHC experiments, A Toroidal LHC Apparatus (ATLAS), has been able to see the evidence ($\geq$3 SD) for a non-zero charge asymmetry for the first time only the last year [50].

Given that large theoretical and experimental progress in the $t\bar{t}$ charge asymmetry during the past more than 10 years, the review of these studies is in order which this article tries to address. In the next section, the basic description of the top quark charge asymmetry and its various definitions are provided. Section 3 gives a brief overview of theoretical predictions for the charge asymmetry expected in the SM at various orders in the perturbative theory and also for various BSM models. In Section 4, the review of both Tevatron and LHC measurements is presented. In Section 5 follows a discussion of current results and the outlook for next measurements at the LHC and future colliders with the conclusion being in Section 6.

2. Charge Asymmetry in $t\bar{t}$ Production

In this section, the top quark production within the SM is described. Afterwards, the charge asymmetry is discussed within the quantum electrodynamics (QED), the electroweak (EW) theory, the quantum chromodynamics, and specifically in the top quark pair production. Finally, a few different definitions of the asymmetry will be mentioned which have been used in the measurements and theoretical predictions.

2.1. Top Quark Production in the SM at Hadron Colliders

At hadron colliders, the dominant production of the top quark is via the top–antitop ($t\bar{t}$) pair production through strong interactions described by QCD. At the lowest, leading order (LO) of the perturbative QCD, the top-quark pair (when talking about the top quark pair, it is always assumed the top–antitop pair, unless stated otherwise) production is possible through the quark–antiquark annihilation ($q\bar{q} \to t\bar{t}$) or the gluon fusion ($gg \to t\bar{t}$). Feynman diagrams of the LO $t\bar{t}$ production are shown in Figure 1. At higher orders, e.g., next-to-leading (NLO) or next-to-next-to-leading (NNLO), the $t\bar{t}$ production is possible also through the quark–gluon interaction ($gq \to t\bar{t}$ or $g\bar{q} \to t\bar{t}$).

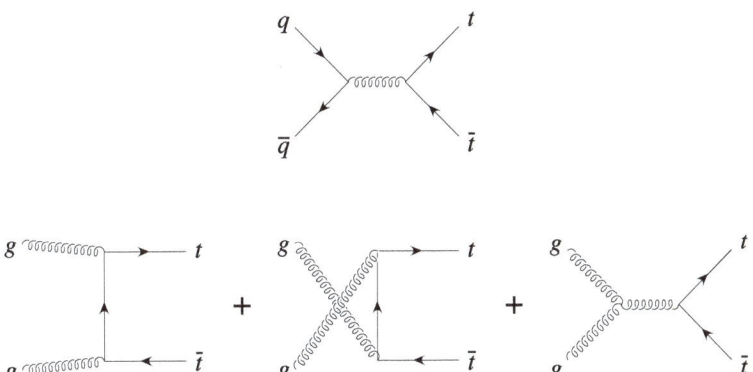

Figure 1. The Feynman diagrams of leading order processes contributing to the top quark pair production at hadron colliders.

The top (antitop) quark alone can not be produced in QCD due to the flavour conservation. However, the flavour is not conserved in weak interactions. Therefore, a single top quark can be produced, see Figure 2. Since the strong coupling constant α_s is the largest of all couplings, the $t\bar{t}$ pair production has a cross-section larger than the single-top production even though there are two top quarks produced.

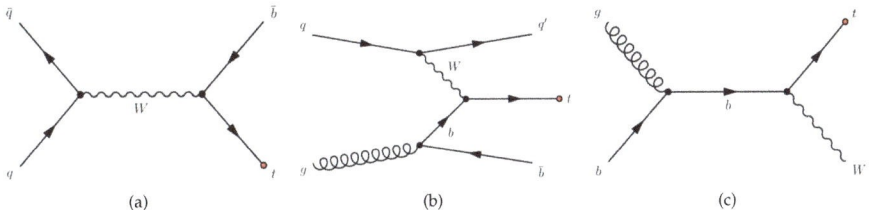

Figure 2. The Feynman diagrams for the single-top quark production: (**a**) s-channel, (**b**) t-channel, and (**c**) Wt-channel.

The $t\bar{t}$ pair production mechanism is quite different at the Tevatron and at the LHC. In $p\bar{p}$ collisions at the Tevatron in Run II at $\sqrt{s} = 1.96$ TeV, the dominant $t\bar{t}$ production is through the $q\bar{q}$ annihilation (85% $q\bar{q}$ and 15% gg at LO in QCD [6]). At the LHC in pp interactions, it is almost opposite. The gluon fusion production channel is dominant, being about 80–90% at LO when going from $\sqrt{s} = 7$ TeV to $\sqrt{s} = 14$ TeV [6].

There are a few $t\bar{t}$ inclusive cross-section predictions available at NNLO or higher in QCD. The initial NNLO calculation became available in Ref. [47]. This included later the higher-order soft-gluon corrections through the resummation at the next-to-next-leading-logarithm (NNLL) accuracy, see [48] and references therein. Based on the above NNLO calculation and adding next-to-next-to-next-to-leading-order (N³LO) soft-gluon corrections by applying a different method, the approximate N³LO (aN³LO) prediction became available [51]. Recently, there was performed another independent calculation of the $t\bar{t}$ production at NNLO QCD using the MATRIX framework [52–54].

The cross-section predictions of the $t\bar{t}$ pair production are shown in Table 1 for both the Tevatron and the LHC, for different center-of-mass energies, and for different calculations. The values are very similar and consistent between different calculations although different parton distribution function (PDF) sets were used for different predictions. The comparison of measured $t\bar{t}$ inclusive cross-sections at the Tevatron and at the LHC with the NNLO + NNLL theoretical predictions is shown in Figure 3. The agreement between predictions and measurements is excellent.

Table 1. The predicted next-to-next-to-leading order (NNLO) $t\bar{t}$ production cross-sections in pb for various energies at the Tevatron and at the Large Hadron Collider (LHC) and for different available calculations. The uncertainties include the factorization and renormalization scale and parton distribution function (PDF)+α_s uncertainties. The assumed top quark mass is always $m_t = 173.3$ GeV except for NNLO + next-to-next-leading-logarithm (NNLL) prediction at the LHC where it is $m_t = 173.2$ GeV.

Collider	$\sqrt{s}$ [TeV]	NNLO + NNLL [48]	aN³LO [51]	NNLO [54]
Tevatron	1.96	$7.16^{+0.20}_{-0.23}$	7.37 ± 0.39	
LHC	7	174^{+10}_{-11}	174^{+11}_{-12}	
LHC	8	248^{+13}_{-14}	248^{+14}_{-15}	
LHC	13	816^{+39}_{-45}	810^{+38}_{-36}	794^{+28}_{-45}

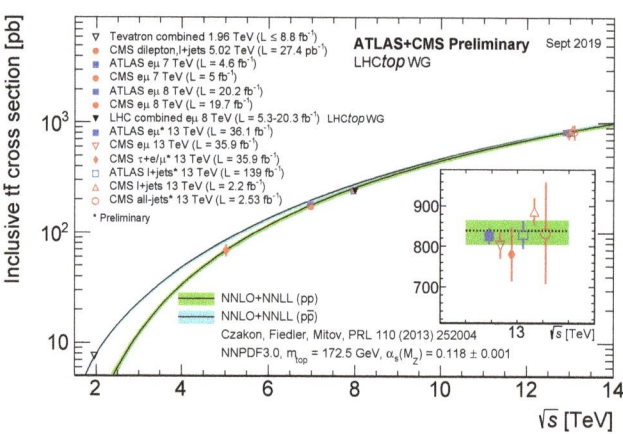

Figure 3. Measured inclusive cross-sections at the Tevatron and at the LHC compared to NNLO + NNLL predictions [55].

The top quark decays almost always into $t \to W + b$. The $t\bar{t}$ decay channels are thus characterized by decays of W boson which could be leptonic $W \to \ell\nu$ or hadronic $W \to q\bar{q}'$. The $t\bar{t}$ decay chain is shown in Figure 4. There are three decay channels according to the number of charged leptons (the inclusion of the charge-conjugate mode is implied): the dilepton ($t\bar{t} \to \ell^+\nu b\ell^-\bar{\nu}\bar{b}$) (11%), lepton+jets ($\ell$+jets, $t\bar{t} \to \ell^+\nu b q\bar{q}'\bar{b}$) (44%), and all-hadronic channel ($t\bar{t} \to q\bar{q}'b q\bar{q}'\bar{b}$) (44%). The quark is color particle which hadronize to create a spray of colorless final state particles (mostly hadrons) flying in about the same direction, a jet.

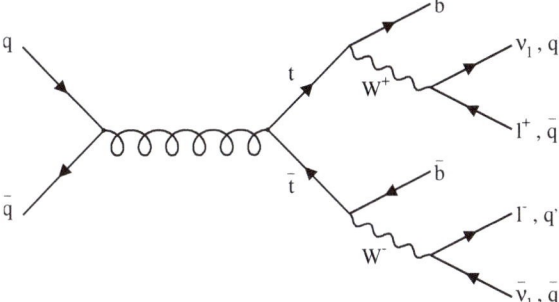

Figure 4. The top–antitop quark pair decay chain.

2.2. Charge Asymmetry in QED and EW Theory

The angular asymmetry in the differential cross-section of the pair production is the difference in production rates for a fermion and an antifermion flying along a given direction. In QED, it was noticed and calculated a long time ago for e.g., the $e^+e^- \to \mu^+\mu^-$ production [56]. At LO in QED, the $\mu^+\mu^-$ pair production is symmetric under the transformation $\mu^+ \leftrightarrow \mu^-$, i.e., under the charge conjugation (C), with respect to the incoming e^+ and e^- beams. The asymmetry is present at NLO due to the interference of processes that differ under C-conjugation, i.e., between the lowest order and two-photon box graphs and between C-odd and C-even breamshtrahlung diagrams, see Figure 5. The overall effect is that the positive muons μ^+ fly a little bit more often in the same direction as incoming positive electrons e^+ while negative μ^- fly preferentially in the direction of negative e^-. It should be stressed that no parity-violating interactions are involved. The QED asymmetry prediction has been confirmed in the experiment [57], see Figure 6.

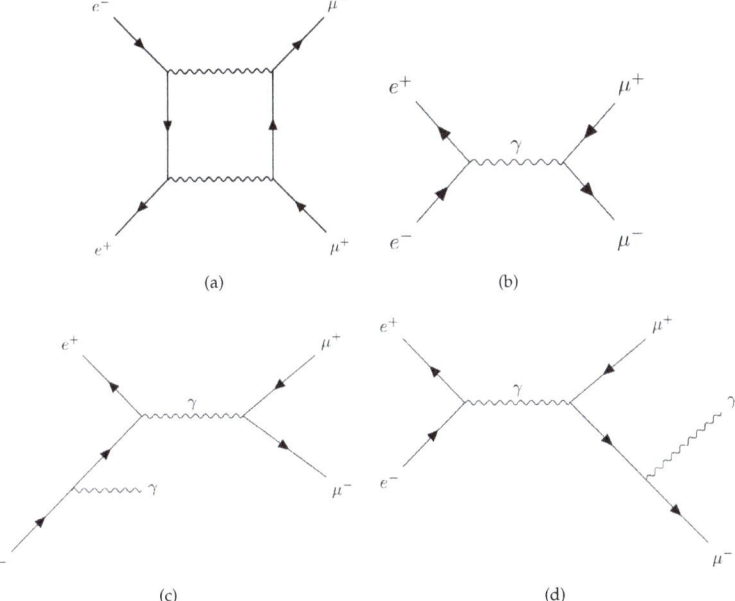

Figure 5. Diagrams of processes contributing to the quantum electrodynamic (QED) charge asymmetry for the $e^+e^- \to \mu^+\mu^-$ production: the box diagram in (**a**) interfering with the leading order (LO) diagram in (**b**), and breamshtrahlung diagrams with C-odd in (**c**) and C-even state in (**d**) [56].

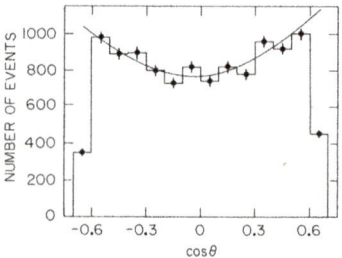

Figure 6. The measured angular distribution for the $e^+e^- \to \mu^+\mu^-$ production together with the QED prediction [57]. The angle θ is the angle between the incoming e^+ direction and the outgoing μ^+ direction.

In the electroweak theory, the angular asymmetry is already predicted at LO due to the Z-boson axial-vector coupling to fermions [58]. It was precisely measured at LEP experiments for $e^+e^- \to q\bar{q}$, $q = c, b$ reactions [59].

2.3. Charge Asymmetry in QCD

Similarly to QED, there is no charge asymmetry at LO in the QCD production of $q\bar{q} \to Q\bar{Q}$ while it is expected at NLO [60]. The asymmetry is thus of the order of α_s relative to the dominant production process. The corresponding QCD diagrams similar to QED diagrams contribute and again the asymmetry is induced through the interference between the amplitudes which are relatively odd under the $t \leftrightarrow \bar{t}$, i.e., the interference of box and LO Born diagrams and the interference of final-state and initial-state radiation diagrams, see Figure 7. The interference of virtual (box) diagrams and LO (Born) diagrams (Figure 7) contributes to a positive asymmetry while the interference of the diagrams with real-corrections has a negative asymmetry with the former to be dominant. The overall net effect is thus a positive asymmetry.

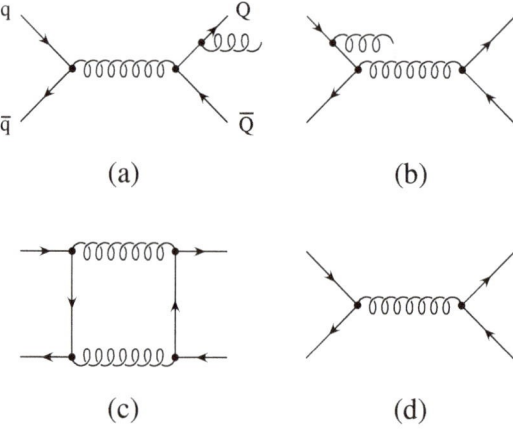

Figure 7. The diagrams contributing to the QCD charge asymmetry in the production of heavy quarks at hadron colliders: interference of final-state (a) with initial-state (b) gluon breamshtrahlung plus interference of the box (c) with the Born diagram (d) [13].

The asymmetry is present in the production for all quark pairs, not only top quark pairs. For light quarks (u,d) when the initial and outgoing quark is the same, the t-channel $q\bar{q} \to q\bar{q}$ must be considered too [60]. The charge asymmetry in the $pp \to b\bar{b}$ production was measured at the LHC by the LHCb experiment [61], while the forward–backward asymmetry in the $p\bar{p} \to b\bar{b}$ production was measured at the Tevatron by the CDF experiment [62,63] and by the D0 experiment in the production of $B^{\pm}$ mesons [64]. At the Tevatron, the $b\bar{b}$ production is dominated by the gg fusion unlike the top quark pair production due to the much lower b-quark mass. Therefore, the asymmetry is expected to be much smaller for $b\bar{b}$ compared to $t\bar{t}$. The A_{FB} measurement was performed at the CDF experiment at both low and high $m_{b\bar{b}}$, see Figure 8. Both CDF measurements and also LHCb measurement are consistent with the SM predictions, although typically with quite large uncertainties. The D0 measurement shows the discrepancy of about 3 SD between the measurement and the NLO QCD estimate from MC@NLO with a large theoretical uncertainty for the prediction which suggests that more precise prediction is needed to interpret this result.

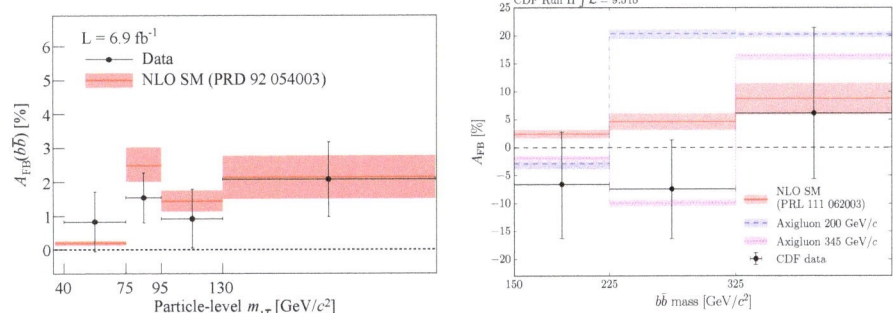

Figure 8. The $b\bar{b}$ forward–backward asymmetry measured by the CDF experiment at low (**left**) and high (**right**) invariant mass of $b\bar{b}$ pair [62,63].

2.4. Top Quark Pair Charge Asymmetry

From the above description in Section 2.3, it follows that the $q\bar{q} \to t\bar{t}$ production is charge symmetric at LO and becomes asymmetric at NLO, i.e., the production of the top and antitop quark along a given direction is different. For the $gg \to t\bar{t}$ production, the initial state is symmetric and thus no asymmetry is predicted. For the $qg(\bar{q}g) \to t\bar{t}$ production, the asymmetry is also expected due to interference terms.

Moreover, there is an asymmetry already at LO in the EW production of the top quark pair $q\bar{q} \to \gamma^*/Z \to t\bar{t}$. In addition, QCD–EW interference terms contribute to the asymmetry. However, since the EW production of the $t\bar{t}$ pair is small compared to the QCD production, its contribution to the charge asymmetry is subdominant although important as it will be seen later.

The fraction of $gg \to t\bar{t}$ is increasing with increasing of the energy $\sqrt{s}$. Therefore, the overall asymmetry in pp or $p\bar{p}$ collisions is decreasing as energy of collisions is increasing.

The $t\bar{t}$ charge asymmetry is quite different at the Tevatron and at the LHC. As mentioned above, $q\bar{q}$ is the dominant production process at the Tevatron, and it is $p\bar{p}$ collider, so the axis of initial quark largely coincides with the axis of initial proton. The asymmetry in $q\bar{q} \to t\bar{t}$ is thus largely preserved in $p\bar{p} \to t\bar{t}$ collisions. At the LHC, the dominant production process is $gg \to t\bar{t}$ which has no asymmetry. Therefore, the charge asymmetry is largely suppressed. Moreover, since pp is a charge symmetric initial state, there is no overall charge asymmetry in $pp \to t\bar{t}$. However, the interacting initial quark is a valence quark, so it has in average a larger longitudinal momentum compared to an antiquark from the sea of quarks in the proton. Since top quarks fly a bit more often in the quarks direction, as a consequence, the top quarks will fly more in forward/backward direction compared to more central antitops.

2.5. Asymmetry Definitions

In general, it is important to state how the asymmetry is defined, in which frame and which observable is used for the definition.

As was already mentioned, the charge asymmetry present in a process $q\bar{q} \to t\bar{t}$ means that in the center-of-mass frame, the number of top quarks flying in a certain direction is different compared to the number of antitop quarks. Therefore, the asymmetry can be defined as

$$A_C^{t\bar{t}}(\cos\theta) = \frac{N_t(\cos\theta) - N_{\bar{t}}(\cos\theta)}{N_t(\cos\theta) + N_{\bar{t}}(\cos\theta)}, \tag{1}$$

where $N_t(\cos\theta)$ and $N_{\bar{t}}(\cos\theta)$ are the number of top quarks and antitop quarks flying along $\cos\theta$ direction, where θ is computed with respect to the direction of the initial quark (and not the antiquark) in both cases. Since the asymmetry is defined as the ratio of two numbers, it is often quoted as a percentage. Such convention is adopted also in this article.

Assuming CP-invariance in strong interactions, $N_{\bar{t}}(\cos\theta) = N_t(-\cos\theta)$, and the charge asymmetry definition in Equation (1) can be rewritten using only the top quark without a need for the antitop quark:

$$A_C^{t\bar{t}}(\cos\theta) = \frac{N_t(\cos\theta) - N_t(-\cos\theta)}{N_t(\cos\theta) + N_t(-\cos\theta)}. \tag{2}$$

This can be then used to define the inclusive charge asymmetry by summing events from opposite hemispheres and it becomes forward–backward asymmetry:

$$A_{FB}^{t\bar{t}} = \frac{N(\cos\theta > 0) - N(\cos\theta < 0)}{N(\cos\theta > 0) + N(\cos\theta < 0)}, \tag{3}$$

where $N(\cos\theta > 0)$ is the number of events with the top quark fulfilling condition $\cos\theta > 0$.

The experiments at the Tevatron and at the LHC use a right-handed coordinate system with its origin at the nominal interaction point in the center of the detector and the z-axis along the beam pipe. At the Tevatron experiments, the z-axis points along the proton direction, while the y-axis points vertically upward and the x-axis points radially outwards. At LHC experiments, the x-axis points radially inward toward the center of the LHC, while the y-axis points vertically upward. Cylindrical coordinates $(r; \phi)$ are used in the transverse plane, ϕ being the azimuthal angle around the beam pipe. The polar angle is calculated with respect to z-axis. The pseudorapidity is defined in terms of the polar angle θ as $\eta = -\ln(\tan\theta/2)$, while the rapidity y is defined as $y = 1/2 \ln[(E + p_z)/(E - p_z)]$.

Instead of $\cos\theta$, other observables can be also used to define the asymmetry, for example the rapidity (y_t) or the pseudorapidity (η_t) of the top quark, or $\Delta y_{t\bar{t}} = y_t - y_{\bar{t}}$. Clearly, the inclusive forward–backward asymmetry stays the same in $q\bar{q}$ center-of-mass frame independently of which variable is used.

Experimentally, the top and antitop quark are reconstructed using their decay products registered within a detector. There are numerous methods which have been developed for such task in the past [65–74]. For charge asymmetry measurements, it is needed to determine which of them is the top quark and the antitop quark. In the dilepton channel, it is relatively simple. The quark which decay has assigned the positive lepton is labeled as the top quark while the quark which has assigned negative lepton is labeled as the antitop quark. In the ℓ+jets channel, there is just one final state lepton. After the $t\bar{t}$ reconstruction, the hadronic (t_h) and leptonic (t_ℓ) top quark is labeled depending on which one has the hadronic and leptonic decay assigned, respectively. If the final state lepton has a positive electromagnetic charge, the leptonic top quark is labeled as the top quark and the hadronic top quark as the antitop quark. If the final state lepton has a negative charge, the leptonic top quark is labeled as the antitop quark.

Asymmetries can be measured also for decay products of top quarks. They will carry the information about a direction of flight of the top quarks due to the boost given by top quarks, but the direction will not be 100% correlated. Therefore, the asymmetry in decay products will be a little bit diluted. The potential decay products are W bosons, b quarks, light quarks, or charged leptons and neutrinos from W boson decay. The good candidates must have a well reconstructed direction of flight and an electromagnetic charge must be well measured for them. The light quarks are almost hopeless since they are reconstructed as jets which direction is not precisely measured and more importantly the charge determination is very hard. Better candidates would be b-jets since there are methods to measure the b-jet charge [75–77] but these are not very precise. The W boson is also not particularly good object because it can be reconstructed from either jets (hard to get a charge) or from the charged lepton and the neutrino (hard to precisely measure a direction). Therefore, the best candidates for measuring the asymmetry are charged leptons, since the reconstruction of both their charge and direction is excellent. This is also reason why the all-hadronic channel is not used for charge-asymmetry measurements.

Similarly as for the top quark, more variables can be used to define the leptonic asymmetry. Typically, only electrons and muons are used in measurements since the τ lepton reconstruction is more complicated. For these leptons, the pseudorapidity is used rather then the rapidity since it is easier to be measured. For practical purposes, the values are the same, given their large energies compared to their mass. Therefore, the η of leptons or $\Delta\eta = \eta_{\ell^+} - \eta_{\ell^-}$ are used in the definition of the asymmetry.

2.5.1. Asymmetry Definitions for Tevatron

At the Tevatron $p\bar{p}$ collider, where $q\bar{q} \to t\bar{t}$ is the dominant production process, the valence quark from the proton interacts with the valence antiquark from the antiproton most of the time. Therefore, the charge and forward–backward asymmetry in $q\bar{q} \to t\bar{t}$ is mostly preserved in the $p\bar{p} \to t\bar{t}$ production although it is a little bit smaller due to a symmetric contribution of the gg fusion.

The Tevatron measurements are performed in the laboratory frame ($p\bar{p}$ center-of-mass frame) or in the $t\bar{t}$ rest frame. There are advantages and disadvantages for both frames. The advantage of the $t\bar{t}$ rest frame is that the asymmetry is larger than the asymmetry in the laboratory frame. At the Tevatron in Run II, the laboratory frame asymmetry is diluted by 30% [42]. That is because in a given $q\bar{q} \to t\bar{t}$ interaction, the interacting quark and antiquark has in general a different longitudinal momentum in the laboratory frame which will give a boost to the $t\bar{t}$ system. It can then happen that even that the top quark flies in the forward direction in the $t\bar{t}$ rest frame, it will fly in the backward direction in the laboratory frame. For that reason, there is an advantage to use the $\Delta y_{t\bar{t}}$ variable. It is a Lorentz invariant, so it is independent of the $t\bar{t}$ longitudinal motion and it is simply related to the top quark rapidity in the $t\bar{t}$ rest frame: $y_t^{t\bar{t}} = 1/2\Delta y$. The advantage of the laboratory frame is that it can be defined to rely only on the measured hadronically decaying top quark rapidity y_{t_h} which has a much better resolution compared to the leptonically decaying top quark which includes only indirectly and partially (no z component) measured neutrino. Consequently, the statistical precision of A_{FB} is better in the $p\bar{p}$ frame.

The forward–backward asymmetry in the laboratory frame is usually defined using the rapidity of the top quark in the laboratory ($p\bar{p}$) frame, y_t:

$$A_{FB}^{p\bar{p}} = \frac{N(y_t > 0) - N(y_t < 0)}{N(y_t > 0) + N(y_t < 0)} = \frac{N(-q_\ell y(t_h) > 0) - N(-q_\ell y(t_h) < 0)}{N(-q_\ell y(t_h) > 0) + N(-q_\ell y(t_h) < 0)}, \quad (4)$$

where the rapidity of the hadronic top $y(t_h)$ is used. This has the advantage that it has a better resolution than Δy, but the disadvantage is that it measures the diluted laboratory frame asymmetry. Similarly, another variable can be used in the forward–backward asymmetry definition: $\cos\theta = q_\ell \cdot \cos\alpha_p$ is used, where θ is the polar angle between the top quark and the proton beam while α_p is the polar angle between the hadronic top quark and the proton beam.

The $t\bar{t}$ rest frame asymmetry is defined using $\Delta y = y_t - y_{\bar{t}}$

$$A_{FB}^{t\bar{t}} = \frac{N(\Delta y > 0) - N(\Delta y < 0)}{N(\Delta y > 0) + N(\Delta y < 0)} = \frac{N(q_\ell \cdot (y_{t_\ell} - y_{t_h}) > 0) - N(q_\ell \cdot (y_{t_\ell} - y_{t_h}) < 0)}{N(q_\ell \cdot (y_{t_\ell} - y_{t_h}) > 0) + N(q_\ell \cdot (y_{t_\ell} - y_{t_h}) < 0)}, \quad (5)$$

where q_ℓ is the electric charge of the lepton, and y_{t_ℓ}, y_{t_h} is the rapidity of the hadronically, leptonically decaying top, or antitop quark. The disadvantage is that Δy has a worse resolution compared to t_h since it combines the uncertainties of both quark reconstructions, including neutrino-related complications of the t_ℓ quark system.

For the leptonic defined asymmetry in the laboratory frame, the asymmetry is usually defined as:

$$A_{FB}^{\ell} = \frac{N(q_\ell \eta_\ell > 0) - N(q_\ell \eta_\ell < 0)}{N(q_\ell \eta_\ell > 0) + N(q_\ell \eta_\ell < 0)}. \quad (6)$$

Similarly to the top quark asymmetry in the $t\bar{t}$ rest frame defined using Δy, the asymmetry in the dilepton channel can be defined this way:

$$A_{FB}^{\ell\ell} = \frac{N(\Delta \eta_{\ell\ell} > 0) - N(\Delta \eta_{\ell\ell} < 0)}{N(\Delta \eta_{\ell\ell} > 0) + N(\Delta \eta_{\ell\ell} < 0)} = \frac{N(\eta_{\ell+} - \eta_{\ell-} > 0) - N(\eta_{\ell+} - \eta_{\ell-} < 0)}{N(\eta_{\ell+} - \eta_{\ell-} > 0) + N(\eta_{\ell+} - \eta_{\ell-} < 0)}. \quad (7)$$

The A_{FB}^{ℓ} is related to A_{FB}^{pp} but the effect is smaller thanks to the dilution due to leptons not following the top quark direction precisely, which in turn is smaller than $A_{FB}^{t\bar{t}}$. The $A_{FB}^{\ell\ell}$ is related to $A_{FB}^{t\bar{t}}$ but a bit smaller due to the same reason. Therefore, there is an advantage in using $A_{FB}^{\ell\ell}$ since its value is not diluted by laboratory frame. However, the disadvantage is that $A_{FB}^{\ell\ell}$ can be measured only in the dilepton channel, which has the smallest statistics.

2.5.2. Asymmetry Definitions for LHC

As it was already pointed out above, top quarks are more often produced in the forward/backward direction while antitops are produced more often in the central region. This can be explored to define the edge-central charge asymmetry, which will be called just simply the charge asymmetry in the following. Most often, the variable $\Delta|y| = |y_t| - |y_{\bar{t}}|$ is used to define the charge asymmetry:

$$A_C = \frac{N(\Delta|y| > 0) - N(\Delta|y| < 0)}{N(\Delta|y| > 0) + N(\Delta|y| < 0)}. \quad (8)$$

For leptons, $\Delta|\eta| = |\eta_{\ell+}| - |\eta_{\ell-}|$ is used instead to define the dileptonic asymmetry:

$$A_C^{\ell\ell} = \frac{N(\Delta|\eta| > 0) - N(\Delta|\eta| < 0)}{N(\Delta|\eta| > 0) + N(\Delta|\eta| < 0)}. \quad (9)$$

3. Theory Overview

The overview of theoretical predictions of the charge asymmetry in the top quark pair production at hadron colliders is presented. In the first part, the evolution of SM predictions is described while in the second part are mentioned various BSM models which could affect the $t\bar{t}$ charge asymmetry.

3.1. SM Predictions

First, it should be noted that two charge asymmetry definitions are used in theoretical calculations. In the first calculation (unexpanded), the most precise calculation available for the numerator and the denominator in the asymmetry definition is used. In such case, the denominator is effectively calculated at the higher order than the numerator since the asymmetry is non-zero starting only at NLO. For example, using the NLO calculation, the total cross-section (denominator) is at NLO while the numerator is effectively at LO. Therefore, it has been argued that it is better to define the charge asymmetry in a way that both the numerator and the denominator are at the same order, e.g., using the expansion in α_s (expanded definition). For example, use the LO cross-section in the denominator for the asymmetry calculation at NLO. In such way, uncalculated higher-order corrections should be at about the same level in both the numerator and the denominator. Using these two definitions provide quite different predictions for the asymmetry at NLO which is due to the large change between the LO and NLO cross-section. However, at NNLO, these definitions already provide very similar values, therefore it does not matter much which definition is used.

The initial attempt at predicting the forward–backward asymmetry in the pair production of top quarks was made more than 30 years ago even before the top quark was discovered [11]. The top quark mass used in the calculation was 45 GeV. In this calculation, only contributions from the initial and final state gluon radiation interference were considered, i.e., only the asymmetry in the $t\bar{t}$+jet process at LO was calculated. This was not the full NLO QCD correction to the inclusive $t\bar{t}$ asymmetry. This calculation required the introduction of cuts on the energy and the rapidity of gluons to avoid singularities. The predicted asymmetry was negative: up to about -2% for a given kinematic criteria on heavy quarks and gluons.

After the discovery of the top quark, but well before first measurements became available, the charge asymmetry was studied in a more detail using the full NLO QCD prediction [12,13]. This means including the interference terms of Born and virtual box corrections. It turned out that these contributions to the asymmetry were larger than the initial and final state radiation interference and they were in the opposite direction. The overall prediction for the forward–backward asymmetry thus changed the sign compared to the first partial NLO prediction and was positive. At the Tevatron and $\sqrt{s} = 1.8$ TeV, the asymmetry was predicted to be up to 15% in $q\bar{q} \to t\bar{t}$ process in certain kinematic regions while the integrated forward–backward asymmetry was about 7–8% in the $t\bar{t}$ rest frame, see Figure 9, and about 4–5% in the $p\bar{p}$ laboratory frame.

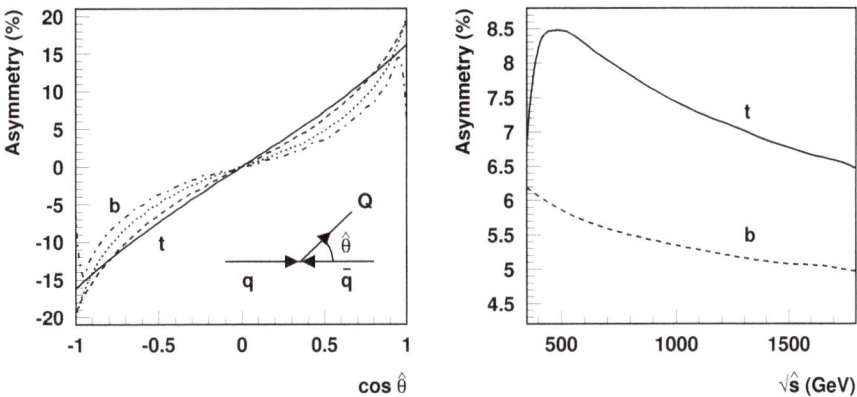

Figure 9. The differential charge asymmetry (**left**) in the $t\bar{t}$ pair production $q\bar{q} \to t\bar{t}$ for the fixed partonic center-of-mass energy $\sqrt{\hat{s}} = 400$ GeV. The integrated charge asymmetry (**right**) for $q\bar{q} \to t\bar{t}$ as a function of $\sqrt{\hat{s}}$ [12].

As it was mentioned already, the asymmetry appears first at NLO in the inclusive $t\bar{t}$ production. This means that using the NLO QCD prediction, the asymmetry is known only at leading order accuracy since numerator has only the leading order contribution. In the $t\bar{t}$+jet production, the asymmetry is already present at LO QCD. Therefore, the NLO calculation of such process provides a true NLO prediction of the asymmetry. It was shown that NLO QCD contributions to the $t\bar{t}$+jet production provide very large corrections to the $t\bar{t}$+jet asymmetry which is then drastically reduced from about -8% at LO to about -2% at NLO for jet $p_T > 20$ GeV for the Tevatron Run II [14,15].

It was not clear at that time whether a similar shift would not happen for the inclusive $t\bar{t}$ prediction. Therefore, the NNLO prediction for the inclusive $t\bar{t}$ process was highly desirable.

It was believed for quite some time that EW corrections are small similarly as EW corrections to the inclusive cross-section are small due to $\alpha_s \gg \alpha$. It turned out this was not the case. It was found that EW corrections of the order $\mathcal{O}(\alpha^2)$ and $\mathcal{O}(\alpha\alpha_s^2)$ have a surprisingly large effect. In general, about 20% of the enhancement with respect to the NLO QCD prediction is observed when including EW corrections [20,21,24,25]. Similarly, leptonic and dileptonic asymmetries were computed for the Tevatron and for the LHC too [23,24].

Furthermore, the higher-order corrections from the soft gluon QCD resummation have been studied at various accuracies. Initially, the soft gluon corrections at next-to-leading-logarithm (NLL) level were computed [16]. It was found that the asymmetry is stable with respect to these corrections: the inclusive asymmetry changed from 6.7% to 6.6% at the Tevatron Run II. Later on, the next-to-next-to-leading logarithm (NNLL) corrections were computed in [18,19]. Here, the change in the inclusive asymmetry is from $(7.4^{+0.7}_{-0.6})$% at NLO to $(7.3^{+1.1}_{-0.7})$% in Ref. [19], so again negligible change, while a modest change from 4.0% to 5.2% was found in Ref. [18].

Moreover, further understanding of soft-gluon emissions came from parton shower studies. It was shown that a coherent QCD radiation in the $t\bar{t}$ production leads to a forward–backward asymmetry that grows more negative with the increasing transverse momentum of the pair [78].

Finally, full NNLO QCD corrections were calculated [30]. These provided large, 27%, increase relative to the NLO QCD prediction for the inclusive asymmetry. The evolution of various calculations is shown in Figure 10. The EW contributions considered here are $\mathcal{O}(\alpha_s^2\alpha)$ and $\mathcal{O}(\alpha^2)$. The detailed studies of NNLO QCD predictions for various kinematic distributions have been presented in Ref. [31].

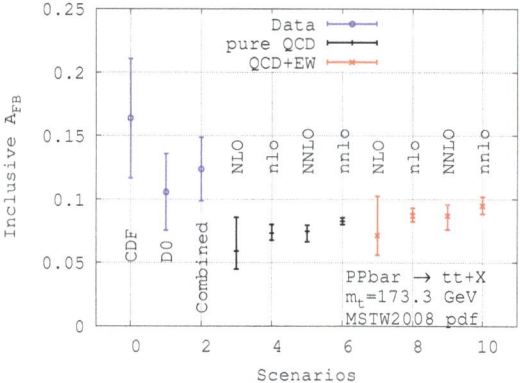

Figure 10. Various levels predictions of the inclusive forward–backward asymmetry at the Tevatron compared to the CDF and D0 measurements. Capital letters (NLO, NNLO) correspond to the unexpanded definition, while small letters (nlo, nnlo) to the expanded definition [30].

Later on, the 'complete NLO' corrections were added to the NNLO QCD prediction [32]. The complete NLO contributions include NLO QCD corrections at $\mathcal{O}(\alpha_s^3)$, the NLO EW at $\mathcal{O}(\alpha_s^2\alpha)$ as well as contributions at $\mathcal{O}(\alpha_s\alpha^2)$ and $\mathcal{O}(\alpha^3)$ together with LO corrections at $\mathcal{O}(\alpha_s\alpha)$ and $\mathcal{O}(\alpha^2)$. The comparison of various predictions of the charge asymmetry A_C with the experiments for the LHC at $\sqrt{s} = 8$ TeV is shown in Figure 11.

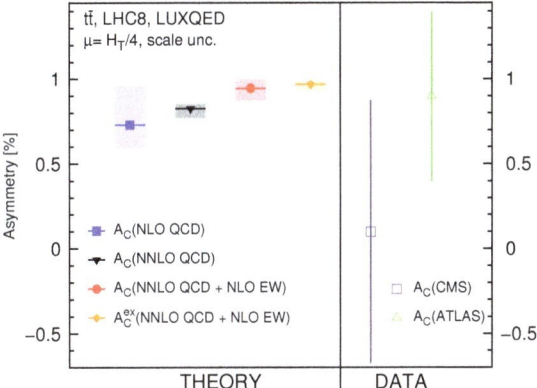

Figure 11. Various predictions of the inclusive charge asymmetry for the LHC at $\sqrt{s} = 8$ TeV compared to the A Toroidal LHC ApparatuS (ATLAS) and the Compact Muon Solenoid (CMS) experiment measurements. A_C corresponds to the unexpanded definition while A_C^{ex} corresponds to the expanded definition [32].

Similarly as for the inclusive cross-section, the calculation of N³LO soft-gluon contributions together with the inclusion of above NNLO QCD calculation allowed to obtain the aN³LO prediction of A_{FB} [29]. The increase in the A_{FB} at the Tevatron due to N³LO soft-gluon contributions is about 5% compared to the pure NNLO QCD calculation.

As it was already mentioned above, the independent NNLO QCD prediction for the $t\bar{t}$ process has become available recently with the MATRIX program which provides fully differential $t\bar{t}$ predictions. The Δy distribution was obtained and the forward–backward asymmetry for the Tevatron calculated for the following set of parameters: $m_t = 172.5$ GeV, $\mu_R = \mu_F = m_t$, NNPDF3.1 NNLO with $\alpha_s(M_Z) = 0.118$ parton distribution function [79]. The result is: $A_C^{t\bar{t}} = (7.4^{+0.3}_{-0.8})\%$. It should be noted that the uncertainty estimated here is only approximate since the MATRIX program provides only minimal and maximal deviations in the cross-section for scale variations for a given bin and the maximal potential difference to estimate the scale uncertainty was used. Nevertheless, the result is in excellent agreement with the above mentioned pure NNLO QCD prediction $(7.49^{+0.49}_{-0.86})\%$ from Ref. [30]. Similarly, the charge asymmetry at the LHC $\sqrt{s} = 7$ TeV was calculated $A_C = (0.95 \pm 0.08)\%$.

The usual theoretical predictions set a renormalization and factorization scale to some value typical for the process, e.g., for the $t\bar{t}$ process it is typically a top quark mass. The uncertainty is then evaluated by changing the renormalization and factorization scale by a factor of two which is essentially just a consensus within the theoretical community, but has no deep foundation within the theory. The alternative calculation of the charge asymmetry is based on the Principle of Maximum Conformality (PMC) scale-setting approach where the renormalization scale is automatically determined and the corresponding uncertainty is essentially eliminated [80,81]. The PMC predictions were computed at NLO QCD with partial NNLO terms and also including NLO EW corrections (aNNLO + NLO EW). They were computed for the Tevatron and for the LHC [26–28]. The large difference between the PMC prediction and the conventional scale-setting NNLO prediction is seen for the Tevatron $A_{FB}(m_{t\bar{t}} > 450$ GeV$)$ where the PMC predicts $A_{FB}(m_{t\bar{t}} > 450$ GeV$) = 29.9\%$ which is much larger compared to the NNLO prediction $A_{FB}(m_{t\bar{t}} > 450$ GeV$) \approx 11\%$ [31]. It should be noted that the PMC method has a residual scale dependence due to the unknown perturbative terms which could be relatively large in the $t\bar{t}$ pair production [82] while the updates of the PMC method try to overcome this limitation [83,84].

When the direct theoretical prediction is not available, e.g., in a specific fiducial phase space, the charge asymmetry predictions are calculated using Monte Carlo (MC) programs or generators of particle collisions. Such programs have typically only NLO QCD corrections implemented, e.g., MC@NLO [85], POWHEG [86], or MCFM [87].

Summary of SM Predictions

The SM inclusive predictions at various orders, in various frames, and using different definitions are shown in Tables 2 and 3. At the Tevatron, the predictions have increased significantly by almost factor of two when going from NLO in the laboratory frame to NNLO QCD + NLO EW in the $t\bar{t}$ frame. At the LHC, the charge asymmetry decreases with the increase of the energy of interactions due to the increase of the symmetric gg production process fraction.

Table 2. The summary of Standard Model (SM) predictions for $t\bar{t}$ and leptonic forward–backward asymmetries at the Tevatron at various levels of the perturbation theory. Some predictions are in the laboratory frame (lab) while some are in the $t\bar{t}$ rest frame ($t\bar{t}$). Some of the predictions are using the unexpanded definition while the others use the expanded (ex) definition.

Prediction	$A_{FB}^{t\bar{t}}$ [%]	A_{FB}^{ℓ} [%]	$A_{FB}^{\ell\ell}$ [%]
NLO QCD [12,13]	4–5 (lab)		
NLO QCD [30]	$5.89^{+2.70}_{-1.40}$ ($t\bar{t}$)		
NLO QCD [30]	$7.34^{+0.68}_{-0.58}$ ($t\bar{t}$, ex)		
NLO QCD [23]	$4.9^{+0.5}_{-0.4}$ (lab, ex)		
NLO QCD [23]	$7.6^{+0.8}_{-0.5}$ ($t\bar{t}$, ex)		
NLOW [23]	$5.1^{+0.5}_{-0.3}$ (lab, ex)		
NLOW [23]	$8.0^{+0.7}_{-0.5}$ ($t\bar{t}$, ex)		
NLO QCD + EW [20,21,24,25]	5–6 (lab)		
NLO QCD [24]		3.1 ± 0.3 (lab, ex)	4.0 ± 0.4 (ex)
NLO QCD + EW [24]	$5.77^{+0.40}_{-0.31}$ (lab, ex)	3.8 ± 0.3 (lab, ex)	4.8 ± 0.4 (ex)
NLO QCD + EW [24]	$8.75^{+0.58}_{-0.48}$ ($t\bar{t}$, ex)		
NLO QCD + NNLL [30]	$7.24^{+1.04}_{-0.67}$ ($t\bar{t}$, ex)		
NNLO [30]	$7.49^{+0.49}_{-0.86}$ ($t\bar{t}$)		
NNLO(MATRIX)	$7.4^{+0.3}_{-0.8}$ ($t\bar{t}$)		
NNLO [30]	$8.28^{+0.27}_{-0.26}$ ($t\bar{t}$, ex)		
aN³LO QCD [29]	8.7 ± 0.2 ($t\bar{t}$, ex)		
NNLO QCD + EW [30]	9.5 ± 0.7 ($t\bar{t}$, ex)		
aN³LO QCD + EW [29]	10.0 ± 0.6 ($t\bar{t}$, ex)		
PMC [28]	12.5 ($t\bar{t}$, ex)		

Table 3. The summary of SM predictions for charge asymmetry at various levels of perturbation theory at the LHC for different center-of-mass energies. All of these predictions are in the laboratory frame. Some of the predictions are using the unexpanded definition while the others use the expanded (ex) definition.

Prediction	$\sqrt{s}$ [TeV]	$A_C^{t\bar{t}}$ [%]	$A_C^{\ell\ell}$ [%]
NLO [24]	7	1.07 ± 0.04 (ex)	0.61 ± 0.03
NLO+EW [24]	7	1.23 ± 0.05 (ex)	0.70 ± 0.03
NLO+EW [21]	7	1.15 ± 0.06 (ex)	
NLO+EW ($\Delta|\eta|$) [21]	7	1.36 ± 0.08 (ex)	
NNLO (MATRIX)	7	0.95 ± 0.08	
PMC [27]	7	$1.15^{+0.01}_{-0.03}$ (ex)	
NLO [24]	8	0.96 ± 0.04 (ex)	0.55 ± 0.03
NLO+EW [24]	8	1.11 ± 0.04 (ex)	0.64 ± 0.03
NLO [32]	8	$0.73^{+0.23}_{-0.13}$	
NLO [32]	8	$0.96^{+0.11}_{-0.09}$ (ex)	
NLO+EW [32]	8	$0.86^{+0.25}_{-0.14}$	
NLO+EW [32]	8	$1.13^{+0.10}_{-0.08}$ (ex)	
NNLO [32]	8	$0.83^{+0.03}_{-0.06}$	
NNLO [32]	8	$0.85^{+0.02}_{-0.04}$ (ex)	
NNLO+EW [32]	8	$0.95^{+0.05}_{-0.07}$	
NNLO+EW [32]	8	$0.97^{+0.02}_{-0.03}$ (ex)	
PMC [27]	8	$1.03^{+0.01}_{-0.00}$ (ex)	
NLO+EW [25]	13	$0.75^{+0.04}_{-0.05}$ (ex)	0.55 ± 0.03 (ex)
NNLO+EW [88]	13	$0.64^{+0.06}_{-0.05}$	
NLO [24]	14	0.58 ± 0.03 (ex)	0.36 ± 0.02 (ex)
NLO+EW [24,25]	14	$0.66^{+0.05}_{-0.04}$ (ex)	0.43 ± 0.02 (ex)
PMC [27]	14	$0.62^{+0.00}_{-0.02}$ (ex)	

It was realized already very early in the initial predictions [11–13,60] that the charge asymmetry depends on the initial/final state gluon radiation due to the interference of these diagrams contributing to the asymmetry. It is therefore expected the asymmetry depends on p_T of the final-state jet which is related to $p_{T,t\bar{t}}$ (the size of p_T of the jet will be the same as p_T of $t\bar{t}$, the direction will be opposite in the transverse plane). The contribution from such interference to the asymmetry is negative, so the larger p_T the more negative asymmetry is expected. The variable $|y_{t\bar{t}}|$ is sensitive to the ratio of the contributions from the $q\bar{q}$ and gg initial states. The charge-symmetric gg initial state produces more central $t\bar{t}$ events while $q\bar{q}$ contributes more in the forward direction. Therefore, it is expected the asymmetry will rise with the increasing value of Δy. The charge asymmetry is expected to also rise for the $m_{t\bar{t}}$ variable since the $q\bar{q}$ initial state is enhanced for larger values of this variable. Finally, the charge asymmetry is expected to rise steeply for high boost of the $t\bar{t}$ system along the longitudinal axis [33]. It is due to the much higher average momentum fractions for quarks than for antiquarks in pp collisions. Requiring the high boost of $t\bar{t}$ system thus increases the $q\bar{q}$ fraction and consequently also the charge asymmetry. As a consequence, the predictions of differential asymmetries as a function of the above mentioned variables were calculated (and also measured), see Figure 12.

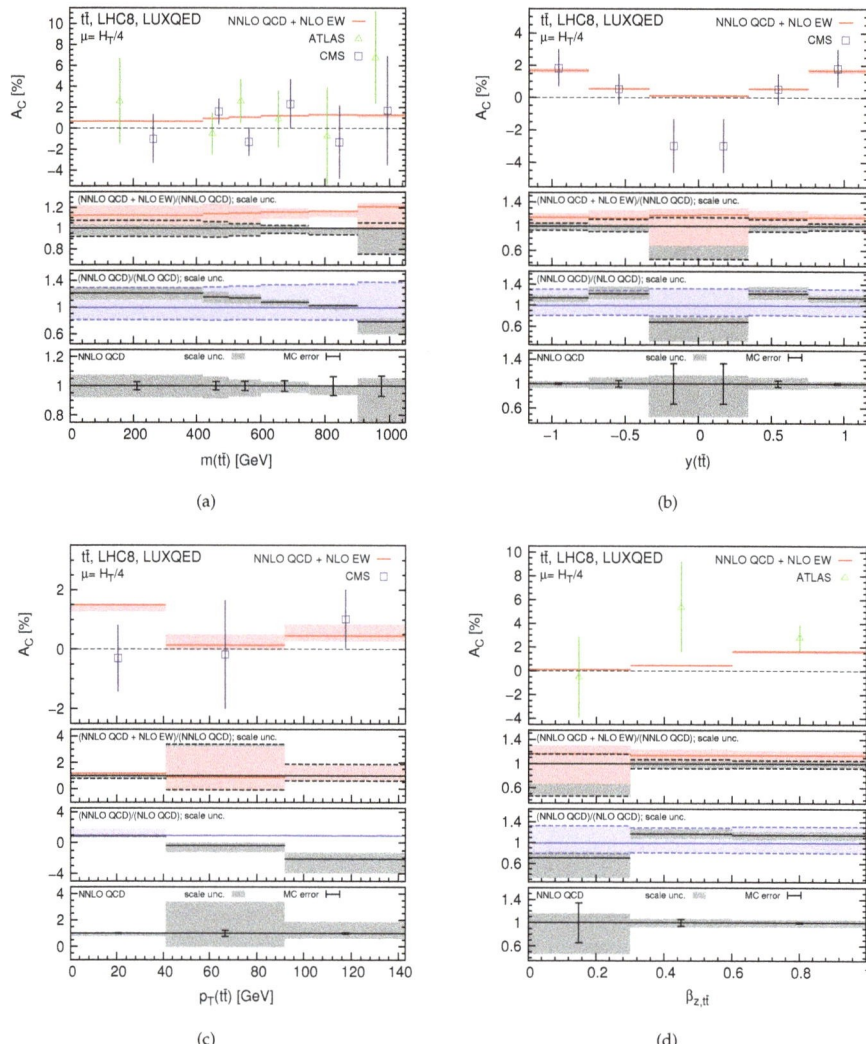

Figure 12. The NNLO predictions for the differential charge asymmetry as a function of $m_{t\bar{t}}$ in (**a**), $y_{t\bar{t}}$ in (**b**), $p_{T,t\bar{t}}$ in (**c**), and $\beta_{z,t\bar{t}}$ in (**d**) at the LHC at $\sqrt{s} = 8$ TeV [32].

3.2. BSM Models

After CDF claimed an evidence for 3 SD deviation from the SM prediction at that time in one of their A_{FB} measurements (for large $m_{t\bar{t}} > 450$ GeV), there have been lots of new physics BSM models proposed to explain such measurement, see Refs. [33–39,45,46].

The BSM models also give us a strong reason to measure all different combinations of asymmetries, e.g., A_C, A_{FB}, A_{FB}^{ℓ}. While in the SM, the A_{FB} and A_C have the same underlying cause and there is a specific relation between A_{FB} and A_C, this relation can be largely changed in BSM models. It was shown that using for example the axigluon model, it is possible to obtain a negative A_C at the LHC for positive A_{FB} asymmetry at the Tevatron [36,37]. The correlation between A_{FB} and A_{FB}^{ℓ} is given in the SM. It is due to the fact that there is about zero top quark polarization in $t\bar{t}$ events, i.e., there is an equal number of positive and negative helicity top quarks produced. Models with different top quark polarization could change the A_{FB} and A_{FB}^ℓ relations in both directions. This was studied in Refs. [46,89] for axigluon and W' models, where it was shown that for the same A_{FB} there was different A_{FB}^ℓ predicted.

There are a few models which can change the charge asymmetry [34,35]:

- axigluons (a color octet vector $\mathcal{G}_\mu$): massive gluons with axial currents ('axigluons'). Similarly to EW theory with the axial current which has a massless photon and a massive Z boson and there is an asymmetry due to the $\gamma - Z$ interference already at LO, the interference between gluon and axigluon in the s-channel mediating $q\bar{q} \to t\bar{t}$ process produces a charge asymmetry;
- Z' (a neutral vector boson $\mathcal{B}_\mu$): a flavour violating Z' exchanged in the t-channel in $u\bar{u} \to t\bar{t}$;
- W' (a charged boson $\mathcal{B}_\mu^1$): a boson with right-handed couplings exchanged in the t-channel in $d\bar{d} \to t\bar{t}$;
- ω^4 (color-triplet scalar): a color triplet with right-handed flavour-violating tu couplings exchanged in the u-channel in $u\bar{u} \to t\bar{t}$;
- Ω^4 (color-sextet scalar): similarly as above, a color sextet with right-handed flavour-violating $t - u$ couplings exchanged in the u-channel. There may be diagonal uu, tt couplings, in contrast with the ω^4 triplet above;
- ϕ (scalar isodoublet): a color-singlet Higgs-like isodoublet, which contains neutral and charged scalars, coupling the top quark to the first generation and exchanged in the t-channel.

The diagrams showing potential contributions from BSM models are shown in Figure 13. The potential values of the charge asymmetry at the LHC and the forward–backward asymmetry at the Tevatron for the above models with various parameters are shown in Figure 14.

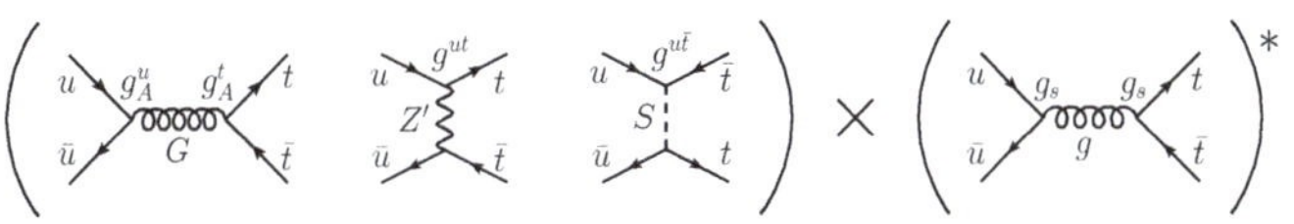

$$\left(\begin{array}{c} u \overset{g_A^u}{\underset{G}{\times}} \overset{g_A^t}{t} \\ \bar{u} \end{array} \begin{array}{c} u \overset{g^{ut}}{\underset{Z'}{\times}} t \\ \bar{u} \end{array} \begin{array}{c} u \overset{g^{ut}}{\underset{S}{\times}} \bar{t} \\ \bar{u} \end{array} \right) \times \left(\begin{array}{c} u \overset{g_s}{\underset{g}{\times}} \overset{g_s}{t} \\ \bar{u} \end{array} \right)^*$$

Figure 13. The interference of various beyond the Standard Model (BSM) particles which contribute to the charge asymmetry with the gluons [90].

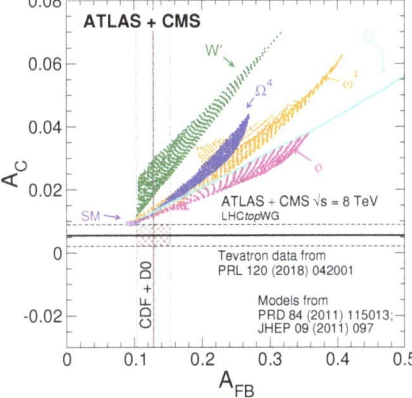

Figure 14. The measured inclusive charge asymmetry A_C at the LHC at $\sqrt{s} = 8$ TeV (horizontal line) plotted against the forward–backward asymmetry A_{FB} (vertical lines) at the Tevatron. The data are compared with the SM prediction at NNLO QCD + NLO electroweak (EW) and predictions incorporating various potential BSM contributions: a W' boson, a heavy axigluon (G_μ), a scalar isodoublet (f), a color triplet scalar (ω^4), and a color sextet scalar (Ω^4) [91].

4. Experimental Measurements

The measurements performed at the Tevatron and at the LHC are reviewed in this section. At the Tevatron, being the $p\bar{p}$ collider, it was possible to measure forward–backward asymmetry. At the LHC, being pp collider, the edge–central charge asymmetry A_C has been measured.

All A_{FB} measurements at the Tevatron were performed in Run II data taking period. Two general-purpose experiments were collecting the data: CDF [92] and D0 [93]. The charge asymmetry measurements at the LHC have been performed in both the Run 1 and Run 2 data taking periods by both general-purpose experiments, ATLAS [94] and the Compact Muon Solenoid (CMS) [95].

At the LHC, there are more experiments running. The LHCb experiment, designed to study b-quark interactions, is one of them and it also observed the top quark [96]. However, up to now, only cross-section measurements of the top quark have been performed at the LHCb and no $t\bar{t}$ charge asymmetry related studies.

As it was mentioned above, it is important to measure all possible combinations of the asymmetries, i.e., measure both A_{FB} at the Tevatron and A_C at the LHC and also both $t\bar{t}$ and leptonic asymmetries, since the relation between them is model dependent. Experiments at the Tevatron and at the LHC therefore performed full set of these measurements where possible, in both ℓ+jets and dilepton channels.

Most of the measurements follow this typical procedure. First, selection criteria are applied to select the sample which is enhanced in $t\bar{t}$ events. The backgrounds are estimated using MC or data driven methods. For the $t\bar{t}$ asymmetry, the kinematic reconstruction of top and antitop quark 4-momenta is performed using the measured top quark decay products. For lepton-based asymmetries, this is not needed. The kinematic variable which is used to define the asymmetry is calculated. At the Tevatron, this is mostly Δy (for the top quark asymmetry), η_ℓ (for the lepton-based asymmetry), and $\Delta \eta_{\ell\ell}$ (for the dileptonic asymmetry). At the LHC, this is mostly $\Delta|y|$ (for the top quark asymmetry) and $\Delta|\eta|$ (for the dileptonic asymmetry). The kinematic distribution of the observable of interest is plotted for data, see e.g., Figure 15a. This corresponds to 'reco level' before the background subtraction. Subsequently, the expected distribution for the background process is subtracted from the distribution in data, corresponding to 'reco level' after background subtraction, see e.g., Figure 15b. The resulting distribution is assumed to correspond to the $t\bar{t}$ distribution after the reconstruction and event selection.

The 'reco level' has a disadvantage that it includes detector resolution and acceptance effects, so such results can be only compared to MC generator predictions which pass detector simulation and reconstruction and can not be directly compared to results from other experiments nor to the direct theoretical predictions. In order to be able to compare to the latest, most precise predictions, such distribution needs to be corrected for detector resolution effects and for the event selection acceptance and inefficiency. This correction is typically performed by the procedure of unfolding. Removing the detector effects brings the distributions to the 'parton' or 'particle' level, see e.g., Figure 15c. When extrapolating to the full phase space and to the level of top quarks, it is called a parton level. This allows to compare experimental results directly to theoretical calculations or to other experiments. Most of the results from the experiments are at parton level and it will be not mentioned unless it is otherwise. When unfolding to the level of final state stable particles and typically requiring some fiducial cuts on final state particles such as p_T or $|\eta|$ which are similar to event selection criteria, it is called a particle level. The advantage of the 'particle fiducial level' is that there is a much smaller degree of the extrapolation to an unmeasured phase-space compared to the parton level and consequently modeling uncertainties are typically smaller. Moreover, it is less ambiguous than the parton level, since the 'top quark' definition can differ between different MC generators which are necessary when performing the unfolding. However, not many predictions are available at such level.

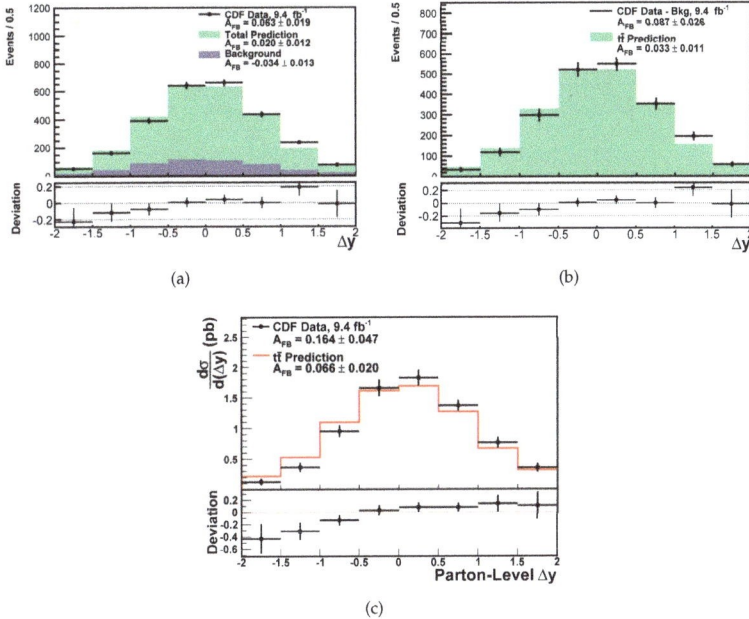

Figure 15. The Δy distribution at the reco level in (a), at the reco level after the background subtraction in (b), and at the parton level in (c), compared to the prediction [97].

The asymmetry can be calculated at each level by counting the events at positive and negative side of the x-axis with the statistical uncertainty properly calculated taking into account possible correlations between the bins which arise due to e.g., systematic uncertainties of the unfolding procedure. Since the asymmetry is expected to depend on a few different observables within the SM and even differently in BSM models, the experiments also performed lots of differential measurements of the asymmetry mostly as a function of $m_{t\bar{t}}, p_{T,t\bar{t}}, \Delta y_{t\bar{t}}$, and $\beta_{z,t\bar{t}}$.

There are also some measurements which use different methods to obtain the asymmetry and these will be mentioned later when describing a specific measurement.

4.1. Forward–Backward Asymmetry Measurements at the Tevatron

The CDF and D0 experiments started to perform measurements of A_{FB} from the beginning of Run II. The initial measurements at the Tevatron were performed in the ℓ+jets channel with about 10–20% of eventual Run II statistics [40,41]. The next set of measurements were performed with about half of full Run II statistics by both CDF [42] and D0 [43,44]. These already included also the measurement of leptonic asymmetries.

Both CDF and D0 collaborations performed the full set of measurements in both ℓ+jets and dilepton channels with full Run II statistics [72,97–103]. Moreover, there was performed the combination of the CDF and D0 measurements [49].

All these measurements will be described in the following.

4.1.1. Initial Measurements

The first measurement related to the charge asymmetry in the top quark pair production was performed by the D0 experiment using 0.9 fb^{-1} of the integrated luminosity [40]. It was performed only at reco level in the $t\bar{t}$ rest frame using Δy. The inclusive asymmetry was not measured, only the A_{FB} as a function of the number of reconstructed jets. The measured values are listed in Table 4 together with predictions from MC@NLO generator where events passed the full simulation and reconstruction as data.

Table 4. The MC@NLO predictions and measured forward–backward asymmetries at reco level as a function of the number of jets in the D0 measurement using 0.9 fb^{-1} [40].

Number of Jets	$A_{FB}^{t\bar{t}}$(MC@NLO) [%]	$A_{FB}^{t\bar{t}}$(data) [%]
≥ 4	0.8 ± 1.0	12 ± 8(stat.) ± 1(syst.)
4	2.3 ± 1.0	19 ± 9(stat.) ± 2(syst.)
≥ 5	-4.9 ± 1.1	-16^{+15}_{-17}(stat.) ± 3(syst.)

In the CDF measurement using 1.9 fb^{-1} [41], the acceptance and reconstruction effects were already corrected for and parton-level asymmetries were measured. The asymmetry was measured using two observables, $-q_\ell \cdot \cos\theta_p$ and Δy, which measured the asymmetry in $p\bar{p}$ and $t\bar{t}$ rest frame, respectively. Measured distributions for these variables are shown in Figure 16. Inclusive asymmetries were measured to be $A^{p\bar{p}} = (17 \pm 7(\text{stat.}) \pm 4(\text{syst.}))\% = (17 \pm 8)\%$ in the $p\bar{p}$ frame and $A_{FB}^{t\bar{t}} = (24 \pm 13(\text{s}$ ... consistent (within 2 S ... about 30% higher pre ... ber of jets at reco lev ...

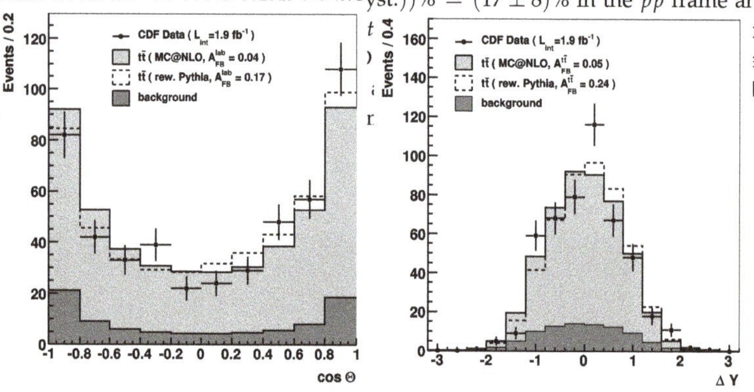

Figure 16. The production angle $\cos\theta$ (left) and Δy (right) distribution at the reco level for the A_{FB} measurements in the $p\bar{p}$ and $t\bar{t}$ frame, respectively. The solid line is the prediction for $t\bar{t}$ with MC@NLO generator and $\sigma_{t\bar{t}} = 8.2$ pb, plus the expected non-$t\bar{t}$ backgrounds. The dashed curve shows the prediction when $t\bar{t}$ is reweighted according to the form $1 + A_{FB}\cos\alpha$ using measured values of A_{FB} [41].

To summarize these initial Tevatron measurements: only the $t\bar{t}$ forward–backward asymmetry was measured in the ℓ+jets channel only. Both experiments showed larger than at that time and even presently predicted asymmetries although they were really limited by a small data sample and so still consistent with the predictions. The CDF results pointed to the expected frame dependence. The measurement as a function of the number of jets in both CDF and D0 pointed to the expected trend of decreasing asymmetry with the increase in the number of jets.

4.1.2. Measurements with Half of Run II Statistics

Both CDF and D0 performed the measurement of the $t\bar{t}$ asymmetry with about a half of the full Run II statistics. CDF performed only the $t\bar{t}$ asymmetry measurement in the ℓ+jets channel while D0 performed the measurement of $A_{FB}^{t\bar{t}}$, A_{FB}^{ℓ}, and $A_{FB}^{\ell\ell}$ in the ℓ+jets and dilepton channel.

The CDF measurement [42] was performed with 5.3 fb^{-1} in both laboratory and $t\bar{t}$ rest frames using y_t and Δy, respectively. The distributions at the reco level are shown in Figure 17 where data are compared to MC@NLO generator predictions. The inclusive measurements at different levels are summarized in Table 5. The measured $A_{FB}^{p\bar{p}}$ asymmetry exceeds the MC@NLO prediction by more than two standard deviations at all correction levels. The $A_{FB}^{t\bar{t}}$ asymmetries are similar in a magnitude to the

$A_{FB}^{p\bar{p}}$ but they are less significant because of the larger relative uncertainties. The $A_{FB}^{t\bar{t}}$ is also measured as a function of Δy and $m_{t\bar{t}}$ in two bins, see Figure 18. At high values of Δy and $m_{t\bar{t}}$, the asymmetries are higher than predictions available at that time. While the difference was less than 2 SD for high Δy, it was about 3.4 SD for high $m_{t\bar{t}}$ bin $((47.5 \pm 10.1(\text{stat.}) \pm 4.9(\text{syst.}))\% = (47.5 \pm 11.4)\%$ in data, while $(8.8 \pm 1.3)\%$ for NLO QCD in MCFM). It should be noted that in this measurement, CDF also tested the CP-invariance assumption by calculating charge separated asymmetries. The asymmetries in both laboratory and $t\bar{t}$ rest frame are equal and opposite within uncertainties, as expected.

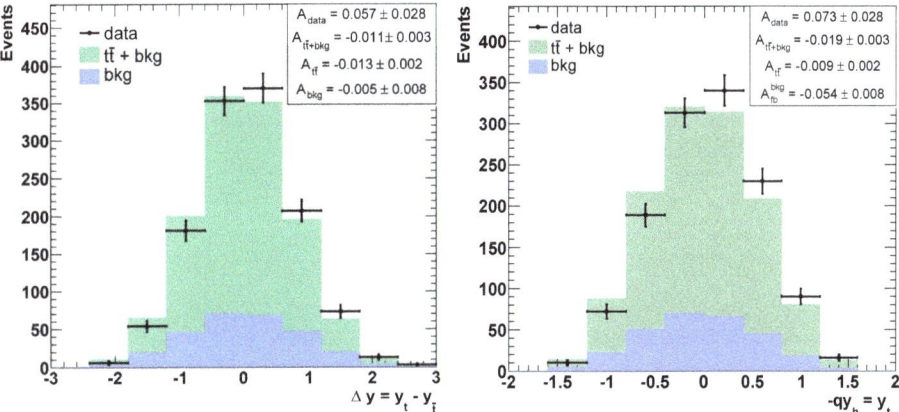

Figure 17. The Δy (**left**) and y_t (**right**) distribution at the reco level corresponding to the CDF measurement performed in the ℓ+jets channel using 5.3 fb^{-1} [42].

Table 5. The summary of inclusive asymmetries in $t\bar{t}$ and $p\bar{p}$ rest frames at the reco level with and without including the background, and at the parton level corresponding to the CDF measurement using 5.3 fb^{-1}. Uncertainties include statistical, systematic, and theoretical uncertainties [42].

Sample	Level	$A_{FB}^{t\bar{t}}$ [%]	$A_{FB}^{p\bar{p}}$ [%]
data	reco (with background)	5.7 ± 2.8	7.3 ± 2.8
MC@NLO	reco (with background)	1.7 ± 0.4	0.1 ± 0.3
data	reco (without background)	7.5 ± 3.7	11.0 ± 3.9
MC@NLO	reco (without background)	2.4 ± 0.5	1.8 ± 0.5
data	parton	15.8 ± 7.4	15.0 ± 5.5
MCFM	parton	5.8 ± 0.9	3.8 ± 0.6

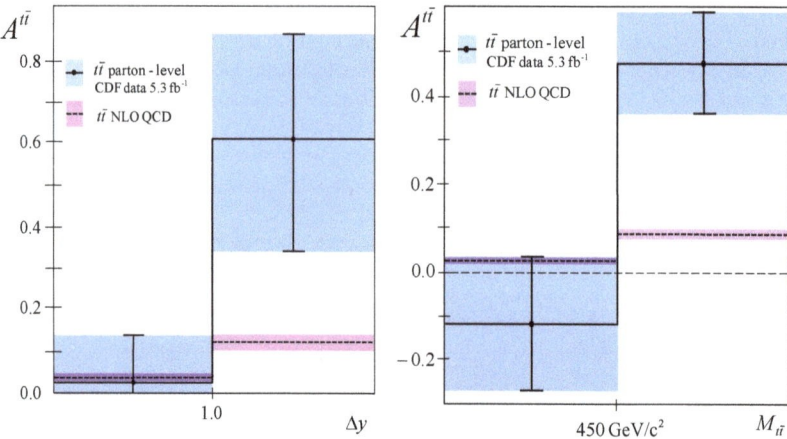

Figure 18. Parton level asymmetries as a function of Δy (**left**) and $m_{t\bar{t}}$ (**right**) compared to the SM prediction of MCFM. The negative going uncertainty for $\Delta y < 1.0$ is suppressed [42].

D0 performed a similar measurement in the ℓ+jets channel using 5.4 fb^{-1} [43]. Both $t\bar{t}$ and leptonic asymmetries are measured using Δy and $q \cdot y_\ell$ distributions, respectively, see Figure 19. The measured inclusive asymmetry is $A_{FB}^{t\bar{t}} = (19.6 \pm 6.5)\%$ and $A_{FB}^\ell = (15.2 \pm 4.0)\%$ which disagree with the NLO QCD prediction from MC@NLO ($A_{FB} = (5.0 \pm 0.1)\%$ and $A_{FB}^\ell = (2.1 \pm 0.1)\%$) by about 2.4 SD and 3.2 SD, respectively. The differential $t\bar{t}$ asymmetry measured as a function of $m_{t\bar{t}}$ and Δy only at the reco level is summarized in Table 6.

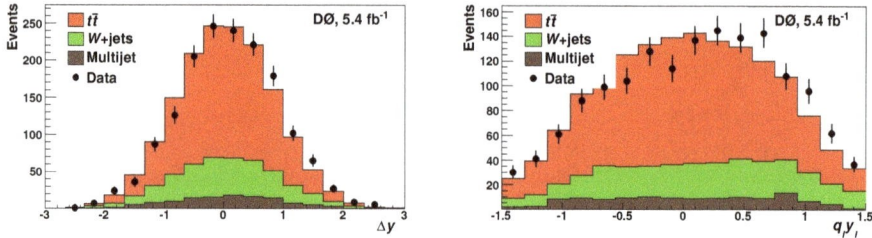

Figure 19. The reconstructed Δy (**left**) and the charge-signed lepton rapidity (**right**) corresponding to the D0 measurement in the ℓ+jets channel [43].

Table 6. The reco level $A_{FB}^{t\bar{t}}$ by subsample in the D0 ℓ+jets measurement using 5.4 fb^{-1} [43].

Subsample	$A_{FB}^{t\bar{t}}$ (data) [%]	$A_{FB}^{t\bar{t}}$ (MC@NLO) [%]
$m_{t\bar{t}} < 450$ GeV	7.8 ± 4.8	1.3 ± 0.6
$m_{t\bar{t}} > 450$ GeV	11.5 ± 6.0	4.3 ± 1.3
$|\Delta y| < 1.0$	6.1 ± 4.1	1.4 ± 0.6
$|\Delta y| > 1.0$	21.3 ± 9.7	6.3 ± 1.6

In the dilepton channel, the leptonic and dileptonic asymmetry is measured by D0 using 5.4 fb^{-1} [44]. The leptonic asymmetry is measured to be $A_{FB}^\ell = (5.8 \pm 5.1(\text{stat.}) \pm 1.3(\text{syst.}))\%$ while the dileptonic asymmetry is $A_{FB}^{\ell\ell} = (5.3 \pm 7.9(\text{stat.}) \pm 2.9(\text{syst.}))\%$. The combination with the result in the ℓ+jets channel yields $A_{FB}^\ell = (11.8 \pm 3.2)\%$.

To summarize mid-term measurements: these measurements were performed in both ℓ+jet and dilepton channels and both $t\bar{t}$ and leptonic based asymmetries were measured. Moreover, the $t\bar{t}$

asymmetry is measured as a function of $m_{t\bar{t}}$ and Δy. The inclusive $t\bar{t}$ asymmetry is observed by both CDF and D0 to be larger than the predictions available at that time by 1.4 – 2.2 SD depending on the experiment, the frame, and the prediction. The $t\bar{t}$ asymmetry at high $y_{t\bar{t}}$ and $m_{t\bar{t}}$ region was measured by CDF to be larger than predicted by up to 3.4 SD at high $m_{t\bar{t}}$ values while D0 did not see any significant disagreement. The leptonic asymmetry was measured only by D0 and was again higher than the prediction available at that time, mostly for the ℓ+jets channel where the deviation is in between 2.6 and 3.3 SD depending on the prediction ($(2.1\pm0.1)\%$ in MC@NLO and $(4.7\pm0.1)\%$ in MC@NLO + NLO EW corrections, see [43,44]). Moreover, $A_{FB}^{\ell\ell}$ was also measured to be consistent with the prediction, but with significant uncertainties. The dominant uncertainty in these measurements was still the statistical uncertainty. Although the amount of data analyzed increased by factor 3–6, the systematic uncertainties were largely improved by about factor of two. It was clear that the progress on both the experimental side, to improve the statistical precision, and on the theory side, to make more reliable uncertainties, was needed.

4.1.3. Measurements with Full Statistics

The measurements with the full Tevatron Run II statistics were performed by CDF and D0 in both ℓ+jets and dilepton channels and for the whole set of $A_{FB}^{t\bar{t}}$, A_{FB}^{ℓ}, and $A_{FB}^{\ell\ell}$ asymmetries.

CDF measured $t\bar{t}$ asymmetry in the ℓ+jets channel using 9.4 fb^{-1} [97] and in the dilepton channel using 9.1 fb^{-1} [72]. The Δy distribution in the ℓ+jets channel is plotted in Figure 15 where also the inclusive asymmetry measurements are summarized at various levels. The measured inclusive asymmetry in the dilepton channel is $A_{FB}^{t\bar{t}} = (12\pm11(\text{stat.})\pm7(\text{syst.}))\% = (12\pm13)\%$. The asymmetry is also measured as a function of $|\Delta y|$, and $m_{t\bar{t}}$ in the ℓ+jets channel, see Figure 20. The dependencies on these kinematic variables are linear and the slope is higher by 2.8 SD and 2.4 SD than expected from the NLO QCD prediction by POWHEG. In the dilepton channel, only the differential asymmetry as a function of Δy is measured in two bins ($A_{FB}^{t\bar{t}}(|\Delta y|<0.5) = (12\pm3.9)\%$ and $A_{FB}^{t\bar{t}}(|\Delta y|>0.5) = (13\pm17)\%$.

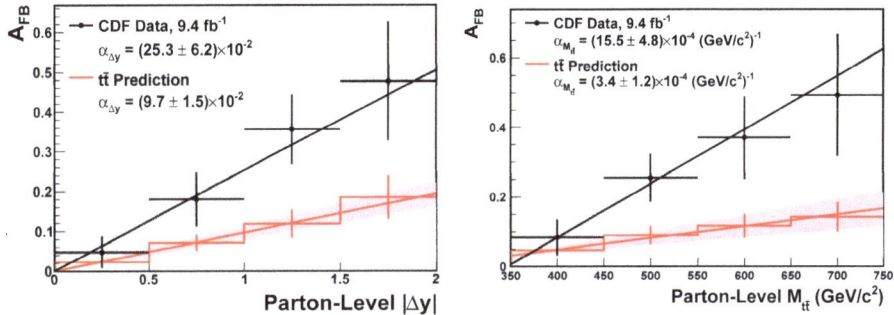

Figure 20. The forward–backward asymmetry as a function of $|\Delta y|$ (**left**) and $m_{t\bar{t}}$ (**right**) with a best-fit line superimposed. The shaded region represents the theoretical uncertainty on the slope of the prediction [97].

D0 measured the $t\bar{t}$ asymmetry in the ℓ+jets channel using full Run II statistics of 9.7 fb^{-1} [102]. The inclusive asymmetry is measured $A_{FB}^{t\bar{t}} = (10.6\pm3.0)\%$. The dependence of the asymmetry on the $|\Delta y|$ and $m_{t\bar{t}}$ was also measured, see Figure 21. The linear fit is performed to these dependencies and the slope is measured to be 15.4 ± 4.3 and $3.9\pm4.4\times10^{-4}$, respectively.

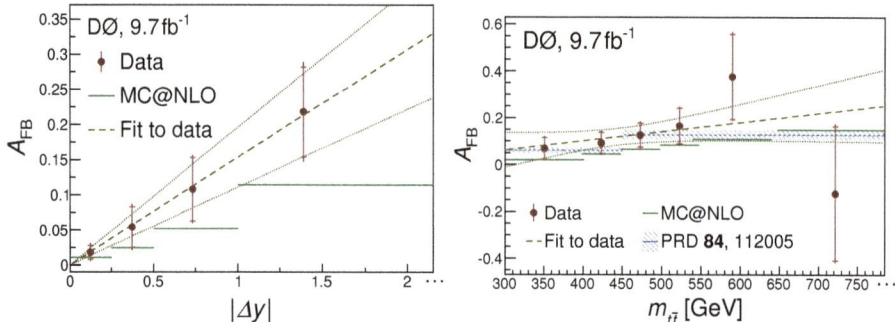

Figure 21. The $A_{\rm FB}$ dependence on $|\Delta y|$ (left) and on $m_{t\bar{t}}$ (right). The dashed line shows the fit to the data with the dotted lines indicating the fit uncertainty. The x coordinate of each datum point is the observed average of $|\Delta y|$ in the corresponding bin [102].

In the dilepton channel, D0 measured simultaneously $A_{\rm FB}^{t\bar{t}}$ and the top quark polarization using 9.7 fb^{-1} [103]. If the top quark polarization is fixed to its expected SM value, the measured value of asymmetry is $A_{\rm FB}^{t\bar{t}} = (17.5 \pm 5.6 ({\rm stat.}) \pm 3.1 ({\rm syst.}))\%$.

In order to study the source of unexpectedly large forward–backward asymmetry in more detail, CDF measured the cross-section as a function of the top quark production angle $d\sigma/d\cos\theta_t$ [98]. The shape of such differential distribution is characterized by Legender polynomials and the Legender moments $a_1 - a_8$ are measured. For the $q\bar{q} \to t\bar{t}$ process at LO, it is expected that there are non-zero a_0 and a_2 moments. $gg \to t\bar{t}$ is expected to add only small contributions to all even-degree Legender moments. The measured Legender moments $a_1 - a_4$ are shown in Figure 22, the remaining ones are consistent with zero within large uncertainties. A good agreement within the uncertainties with the NLO SM prediction is observed for the moments $a_2 - a_8$, but a_1 showed an excess with respect to the prediction: $a_1 = 0.40 \pm 0.12$ vs. NLO SM: $0.15^{+0.07}_{-0.03}$. It means the excess was observed in the differential cross-section in the term linear in $\cos\theta$.

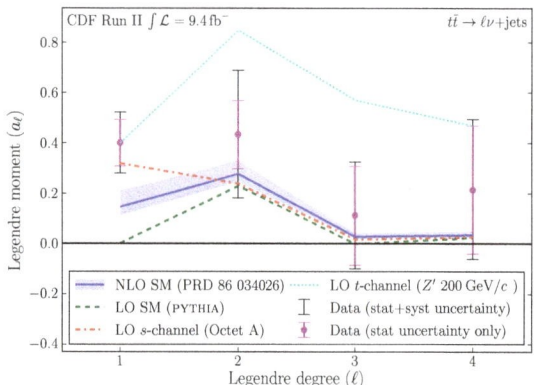

Figure 22. Measured Legendre moments $a_1 - a_4$, with various theory predictions overlaid [98].

The leptonic measurements have been performed in both ℓ+jets and dilepton channels by both CDF and D0 experiments. In the ℓ+jets channel, the leptonic asymmetry is measured by CDF using 9.4 fb^{-1} [99] while D0 uses 9.7 fb^{-1} [104]. Both CDF and D0 measure the single-leptonic asymmetry using the charge-weighted rapidity qy_ℓ. The CDF measurement separates the rapidity distribution into symmetric part, which is largely independent on the model, and the antisymmetric part which encapsulates the possible variation between the modes. The inclusive asymmetry is measured to be $A^\ell_{FB} = (10.5 \pm 2.4(\text{stat.})^{+2.2}_{-1.7}(\text{syst.}))\% = (10.5^{+3.2}_{-2.9})\%$ which is 2.3 SD away from the NLO QCD + NLO EW prediction $((3.8 \pm 0.3)\%)$. D0 measured the asymmetry in restricted region of $|y_\ell| < 1.5$: $A^\ell_{FB} = (4.2^{+2.9}_{-3.0})\%$. The asymmetry is measured also as a function of p_T of lepton and $|y_t|$.

In the dilepton channel, both CDF using 9.1 fb^{-1} [100] and D0 using 9.7 fb^{-1} [101] measure both single-lepton and dilepton asymmetries using $q\eta_\ell$ and $\Delta\eta_\ell$, respectively. In the CDF measurement, similarly to previous measurement in the ℓ+jets channel, the pseudorapidity distributions are splitted into symmetric and antisymmetric parts. The results are $A^\ell_{FB} = (7.2 \pm 5.2(\text{stat.}) \pm 3.0(\text{syst.}))\% = (7.2 \pm 6.0)\%$ and $A^{\ell\ell}_{FB} = (7.6 \pm 7.2(\text{stat.}) \pm 3.9(\text{syst.}))\% = (7.6 \pm 8.2)\%$. D0 measured the inclusive asymmetries using the distributions shown in Figure 23. The measured values are $A^\ell_{FB} = (4.4 \pm 3.7(\text{stat.}) \pm 1.1(\text{syst.}))\%$ and $A^{\ell\ell}_{FB} = (12.3 \pm 5.4(\text{stat.}) \pm 1.5(\text{syst.}))\%$. The dependence of asymmetries on $|q\eta_\ell|$ and $|\Delta\eta_\ell|$ is also measured but only in the fiducial parton level phase-space (leptons must have $|\eta| < 2$ and $|\Delta\eta| < 2.4$).

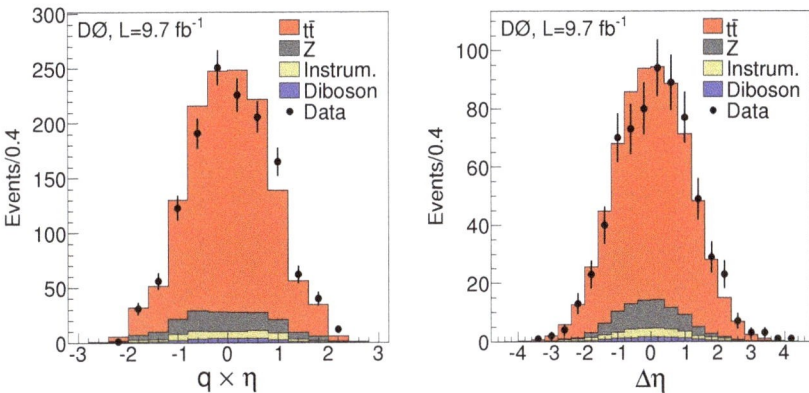

Figure 23. The reco level distribution of $q \cdot \eta$ (left) and $\Delta\eta = \eta_{\ell^+} - \eta_{\ell^-}$ (right). The error bars indicate the statistical uncertainty on the data [101].

4.1.4. Full Dataset Combinations

Both CDF and D0 performed individual combinations of their $t\bar{t}$ and single-lepton asymmetry measurements from both ℓ+jets and dilepton channels using the full Run II data statistics of the actual experiment [72,100,103,104]. Moreover, they performed together the Tevatron combinations of all their results from both channels [49].

CDF combined inclusive $t\bar{t}$ asymmetry is $A^{t\bar{t}}_{FB} = (16.0 \pm 4.5(\text{stat.} + \text{syst.}))\%$ while the D0 combination is $A^{t\bar{t}}_{FB} = (11.8 \pm 2.5(\text{stat.}) \pm 1.3(\text{syst.}))\%$. The combined single-leptonic asymmetry at CDF is $A^\ell_{FB} = (9.0^{+2.8}_{-2.6}(\text{stat.} + \text{syst.}))\%$ while the D0 combination is $(4.7 \pm 2.3(\text{stat.}) \pm 1.5(\text{syst.}))\%$. For the differential $A^{t\bar{t}}_{FB}$ as a function of Δy, rather then combining the data, the combined fit of the slope to both CDF ℓ+jets and dilepton data was performed. The result is $\alpha = 0.227 \pm 0.057$ which is 2 SD larger than the NNLO QCD prediction of $0.114^{+0.006}_{-0.012}$ [31].

The Tevatron combination of the CDF and D0 inclusive $t\bar{t}$ forward–backward asymmetry is $A^{t\bar{t}}_{FB} = (12.8 \pm 2.1(\text{stat.}) \pm 1.14(\text{syst.}))\% = (12.8 \pm 2.5)\%$. The precision of the combination is such that the A_{FB} is measured with a significance of 5 SD from zero asymmetry. The combined inclusive single-lepton asymmetry is $A^\ell_{FB} = (7.3 \pm 1.6(\text{stat.}) \pm 1.12(\text{syst.}))\%$, while the combined dileptonic asymmetry is $A^{\ell\ell}_{FB} = (10.8 \pm 4.3(\text{stat.}) \pm 1.6(\text{syst.}))\%$. All inclusive combined measurements together with the individual measurements used as the inputs to the combination and the theoretical predictions are summarized in Figure 24.

Differential measurements of $A^{t\bar{t}}_{FB}$ as a function of $m_{t\bar{t}}$ were measured only in the ℓ+jets channel and combined together, see Figure 25. For the combination, the data are fitted by a linear function.

The obtained slope of $9.71 \pm 3.28 \times 10^{-4}$ GeV^{-1} is compatible with NNLO QCD + NLO EW prediction of $5.11^{+0.42}_{-0.64} \times 10^{-4}$ GeV^{-1} at the level of 1.3 SD. The differential $t\bar{t}$ asymmetry as a function of Δy is available from CDF for both ℓ+jets and dilepton channels, and from D0 for the ℓ+jets channel, see Figure 25. Since the choice of the binning differs for these measurements, the simultaneous fit to a linear function with zero offset was performed for all available measurements employing the correlations. The slope parameter is measured to be 0.187 ± 0.038 which is compatible with NNLO QCD + NLO EW prediction of $0.129^{+0.006}_{-0.012}$ at the level of 1.5 SD. The individual CDF and D0 measurements of A^{ℓ}_{FB} as a function of $|q_\ell \eta_\ell|$ together with the individual measurements of $A^{\ell\ell}_{FB}$ as a function of $|\Delta \eta|$ are shown in Figure 26 without any quantitative comparison to the prediction. Looking at the plots, there can not be seen any striking disagreement with NLO QCD + NLO EW prediction.

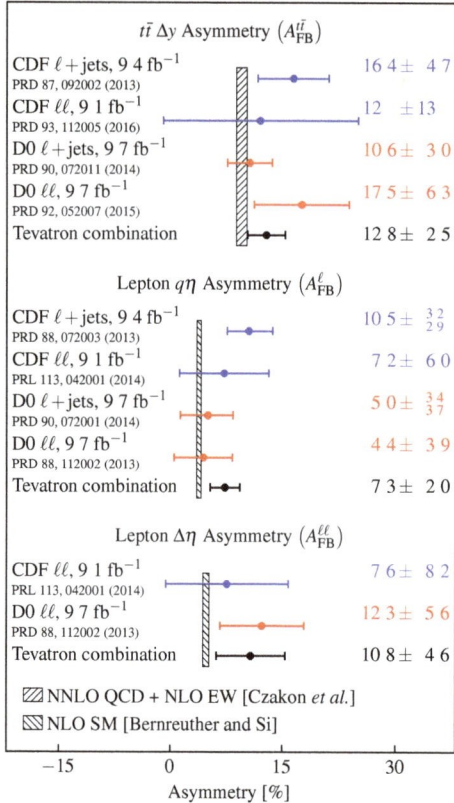

Figure 24. Summary of inclusive forward–backward asymmetries used in the Tevatron combination together with their combination [49].

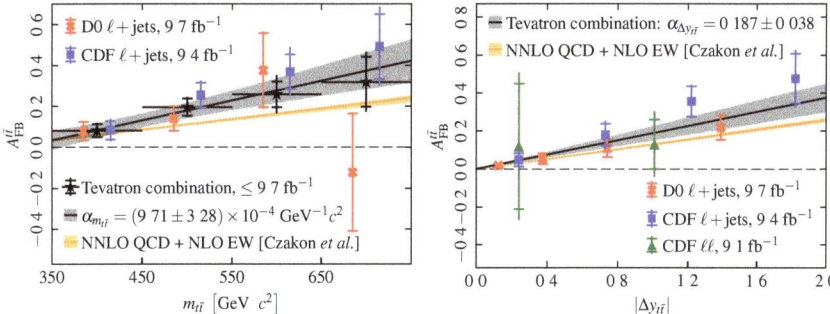

Figure 25. The dependence of A_{FB} as a function of $m_{t\bar{t}}$ (**left**). The individual measurements and the Tevatron combination are shown together with the NNLO QCD + NLO EW prediction. The dependence of A_{FB} as a function of Δy (**right**). Here, the individual measurements are shown together with the simultaneous fit and the NNLO QCD + NLO EW prediction [49].

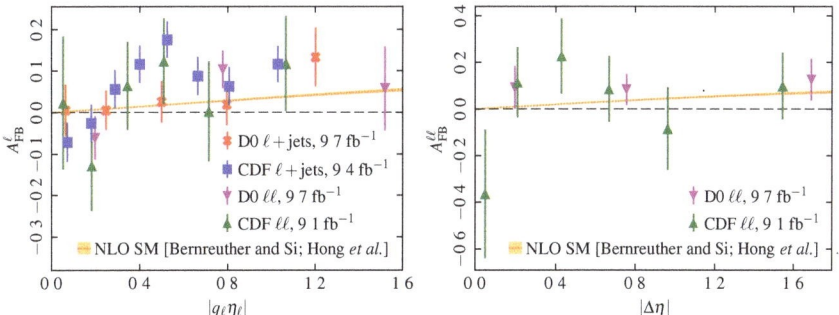

Figure 26. The individual CDF and D0 measurements of A_{FB}^{ℓ} as a function of $|q_\ell \eta_\ell|$ (**left**) and of $A_{FB}^{\ell\ell}$ as a function of $|\Delta \eta|$ (**right**) together with the NLO QCD prediction [49].

4.1.5. Summary and Discussion of Tevatron Measurements

All inclusive Tevatron measurements of forward–backward asymmetry at the parton level in the $t\bar{t}$ rest frame are summarized in Table 7. These should be compared to the latest NNLO QCD + EW prediction of $A_{FB} = (9.5 \pm 0.7)\%$ [30] and the aN^{3}LO QCD + EW prediction of $A_{FB} = (10.0 \pm 0.6)\%$ [29], while for the leptonic and dileptonic asymmetries only NLO QCD + EW predictions exist: $A_{FB}^{\ell} = (3.8 \pm 0.3)\%$ and $A_{FB}^{\ell\ell} = (4.8 \pm 0.4)\%$ [24].

The inclusive $A_{FB}^{t\bar{t}}$ are all consistent between them for different channels and experiments, and are consistent with both NNLO QCD + EW and aN^{3}LO QCD + EW predictions. The maximum deviation is 1.6 SD for D0 ℓ+jets measurement using 5.4 fb^{-1} while the final Tevatron combination is 1.3 SD higher when compared to NNLO QCD + EW prediction. Similar conclusions hold also for the inclusive dileptonic asymmetry $A_{FB}^{\ell\ell}$ with the maximum deviation at 1.3 SD in the D0 dileptonic measurement using 9.7 fb^{-1}. However, it should be noted that all measurements are consistently higher than both NNLO QCD + EW and aN^{3}LO QCD + EW predictions. For the inclusive leptonic asymmetry A_{FB}^{ℓ} again all measurements are higher than the prediction. The biggest deviation is for the D0 measurement in the ℓ+jets channel using 5.4 fb^{-1} at the level of 2.8 SD, while this has been lowered significantly with the full Run II statistics. The final Tevatron combination is only about 1.7 SD higher than prediction. It should be also noted that the prediction here is only at NLO QCD + EW level, so with the potential NNLO it is expected the deviation will be even lower. The dominant uncertainty in inclusive measurements is the statistical uncertainty.

Table 7. Summary of Tevatron measurements of inclusive forward–backward asymmetries. For a given measurement, if there is just one uncertainty, it is combined statistical and systematic uncertainty. If there are two uncertainties, the first one is always statistical and the second one is systematic uncertainty.

Experiment, Channel	$\mathcal{L}[\text{fb}^{-1}]$	$A_{FB}^{t\bar{t}}[\%]$	$A_{FB}^{\ell}[\%]$	$A_{FB}^{\ell\ell}[\%]$
CDF, ℓ+jets	1.9	$24 \pm 13 \pm 4$		
CDF, ℓ+jets	5.3	15.8 ± 7.4		
D0, ℓ+jets	5.4	19.6 ± 6.5	15.2 ± 4.0	
D0, dilepton	5.4		$5.8 \pm 5.1 \pm 1.3$	$5.3 \pm 7.9 \pm 2.9$
D0, combination	5.4		11.8 ± 3.2	
CDF, ℓ+jets	9.4	$16.4 \pm 3.9 \pm 2.6$	$10.5^{+3.2}_{-2.9}$	
CDF, dil	9.1	$12 \pm 11 \pm 7$	$7.2 \pm 5.2 \pm 3.0$	$7.6 \pm 7.2 \pm 3.9$
D0, ℓ+jets	9.7	10.6 ± 3.0	$5.0^{+3.4}_{-3.7}$	
D0, dil	9.7	$17.5 \pm 5.6 \pm 3.1$	$4.4 \pm 3.7 \pm 1.1$	$12.3 \pm 5.4 \pm 1.5$
CDF, combination	9.7	16.0 ± 4.5	$9.0^{+2.8}_{-2.6}$	
D0, combination	9.7	$11.8 \pm 2.5 \pm 1.3$		
Tevatron, combination	9.7	$12.8 \pm 2.1 \pm 1.4$	$7.3 \pm 1.6 \pm 1.2$	$10.8 \pm 4.3 \pm 1.6$

The forward–backward asymmetry differential measurements at the Tevatron were performed as a function of $m_{t\bar{t}}$ and rapidity related observable $|\Delta y_{t\bar{t}}|$, $|y_\ell|$, and $\Delta \eta_{\ell\ell}$. The $m_{t\bar{t}}$ and $|\Delta y_{t\bar{t}}|$ dependencies are a bit stronger than expected but the agreement is within 2 SD.

4.2. LHC Measurements

Both ATLAS and CMS experiments started to perform the measurements of the charge asymmetry from the beginning of Run 1 at the energy $\sqrt{s} = 7$ TeV. Initially, only the measurements in the ℓ+jets channel and with a small luminosity of about 1 fb^{-1} were performed by ATLAS [65] and CMS [69].

Next measurements at $\sqrt{s} = 7$ TeV were performed with the full statistics of 2011 year ($\approx$5 fb^{-1}) in both ℓ+jets and dilepton channels by both ATLAS [105,106] and CMS collaborations [107,108]. Afterwards, the measurements with about four times larger statistics (full 2012 year, $\approx$20 fb^{-1}) at $\sqrt{s} = 8$ TeV were again performed in both dilepton and ℓ+jets channels and by both ATLAS [109–111] and CMS [112–114] collaborations.

In Run 2 at $\sqrt{s} = 13$ TeV, CMS performed the measurement in both ℓ+jets [115] and dilepton [116] channels using partial dataset (2015+2016 years, $\approx$36 fb^{-1}). On the other hand, ATLAS performed the measurement in the ℓ+jets channel with the full Run 2 statistics (years 2015–2018, $\approx$140 fb^{-1}) [50]. However, it is for now only the preliminary measurement.

Moreover, ATLAS and CMS combined their measurements from Run 1 at $\sqrt{s} = 7$ TeV and $\sqrt{s} = 8$ TeV [91]. In the following, all these measurements will be briefly described.

4.2.1. Measurements at $\sqrt{s} = 7$ TeV

The initial LHC measurement of A_C was performed by CMS using 1.09 fb^{-1} of luminosity [69]. The $A_C^{t\bar{t}}$ was measured using distribution $\Delta|\eta| = |\eta_t| - |\eta_{\bar{t}}|$ and $\Delta y^2 = (y_t - y_{\bar{t}})(y_t + y_{\bar{t}})$ shown in Figure 27. Using 12,757 data events with the expected background of 2520 ± 246 events, the inclusive asymmetry is measured to be $A_C^{t\bar{t},\eta} = (-1.7 \pm 3.2(\text{stat.})^{+2.5}_{-3.6}(\text{syst.}))\%$, and $A_C^{t\bar{t},y} = (-1.3 \pm 2.8(\text{stat.})^{+2.9}_{-3.1}(\text{syst.}))\%$, consistent with the QCD NLO + EW predictions of $(1.36 \pm 0.08)\%$ and $(1.15 \pm 0.06)\%$ [21], respectively.

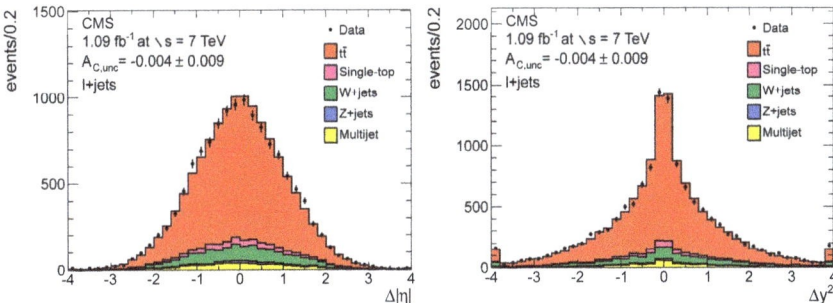

Figure 27. Reconstructed $\Delta|\eta|$ (**left**) and Δy^2 (**right**) distributions for the ℓ+jets channel. The outermost bins include the overflows [69].

Similarly, the initial ATLAS measurement of the charge asymmetry was performed using 1.04 fb^{-1} of luminosity [65]. The measured inclusive asymmetry is $A_C^{t\bar{t}} = (-1.9 \pm 2.8(\text{stat.}) \pm 2.4(\text{syst.}))\%$ which is consistent with the NLO QCD + NLO EW prediction $A_C = (1.23 \pm 0.05)\%$ [24]. The differential measurement of the asymmetry as a function of $m_{t\bar{t}}$ was measured only in two bins and had large uncertainties.

The measurements at $\sqrt{s} = 7$ TeV using full statistics were performed by both ATLAS and CMS in both ℓ+jets and dilepton channels. In the ℓ+jets channel, both ATLAS using 4.7 fb^{-1} [105] and CMS using 5.0 fb^{-1} [107] measure the asymmetry inclusively and and also differentially as a function of $m_{t\bar{t}}$, $y_{t\bar{t}}$, $p_{T,t\bar{t}}$. The inclusive results are $A_C^{t\bar{t}} = (0.6 \pm 1.0(\text{stat.} + \text{syst.}))\%$ for ATLAS and $A_C^{t\bar{t}} = (0.4 \pm 1.0(\text{stat.}) \pm 1.1(\text{syst.}))\%$ for CMS. The differential asymmetries are shown in Figures 28 and 29.

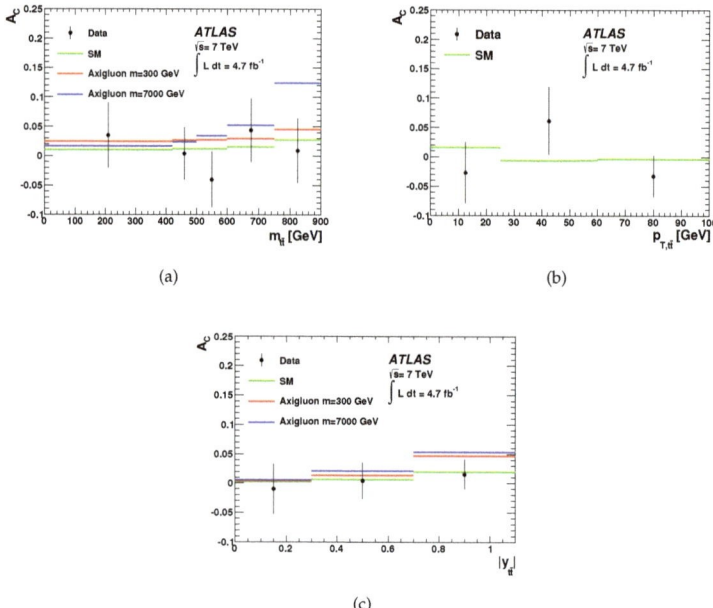

Figure 28. Distributions of A_C as a function of $m_{t\bar{t}}$ in (**a**), $p_{T,t\bar{t}}$ in (**b**), and $y_{t\bar{t}}$ in (**c**). The measured A_C values are compared with the NLO QCD + EW predictions (SM) [24] and the predictions for a color-octet axigluon [105].

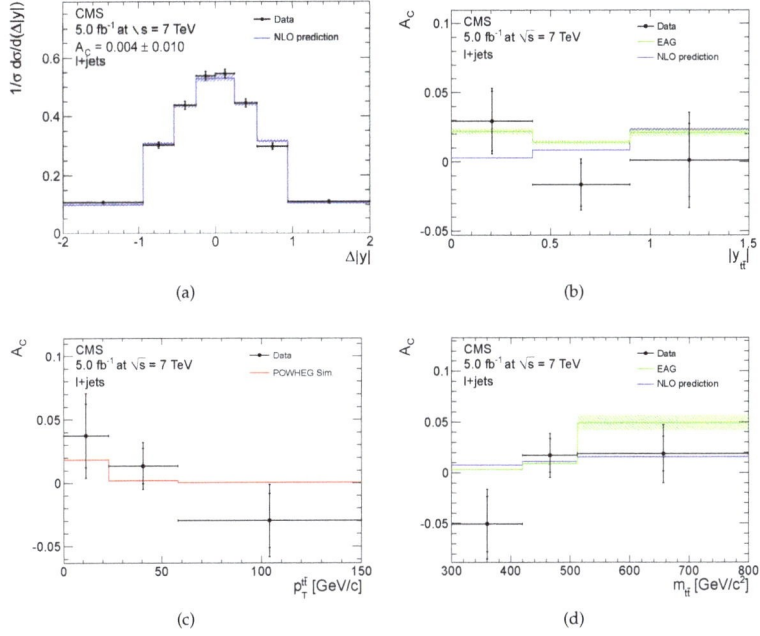

Figure 29. The unfolded $\Delta|y|$ distribution in (**a**), the charge asymmetry as a function of $y_{t\bar{t}}$ in (**b**), $p_{T,t\bar{t}}$ in (**c**), and $m_{t\bar{t}}$ in (**c**). The measured values are compared to NLO QCD + EW calculations of Ref. [21], and to the predictions of a model featuring an effective axial-vector coupling of the gluon (EAG) [117]. The error bars on the differential asymmetry values indicate the statistical and total uncertainties [107].

In the dilepton channel, both ATLAS using 4.6 fb^{-1} [106] and CMS using 5.0 fb^{-1} [108] measure the $t\bar{t}$ and dileptonic asymmetry inclusively: $A_C^{t\bar{t}} = (2.1 \pm 2.5(\text{stat.}) \pm 1.7(\text{syst.}))\%$, $A_C^{\ell\ell} = (2.4 \pm 1.5(\text{stat.}) \pm 0.9(\text{syst.}))\%$ in ATLAS while $A_C^{t\bar{t}} = (-1.0 \pm 1.7(\text{stat.}) \pm 0.8(\text{syst.}))\%$, $A_C^{\ell\ell} = 0.9 \pm 1.0(\text{stat.}) \pm 0.6(\text{syst.}))\%$ in CMS. The comparison of the inclusive ATLAS $A_C^{t\bar{t}}$ and $A_C^{\ell\ell}$ measurements to the theory prediction is shown in Figure 30. CMS also measured the dileptonic asymmetry as a function of $t\bar{t}$ mass, rapidity, and transverse momentum, see Figure 31.

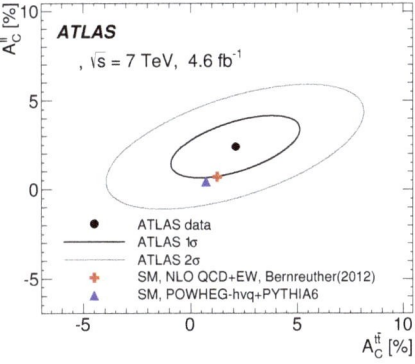

Figure 30. The comparison of correlated inclusive $A_C^{\ell\ell}$ and $A_C^{t\bar{t}}$ measurements to the NLO QCD+EW prediction [24] and the prediction of the POWHEG+ PYTHIA generator [106].

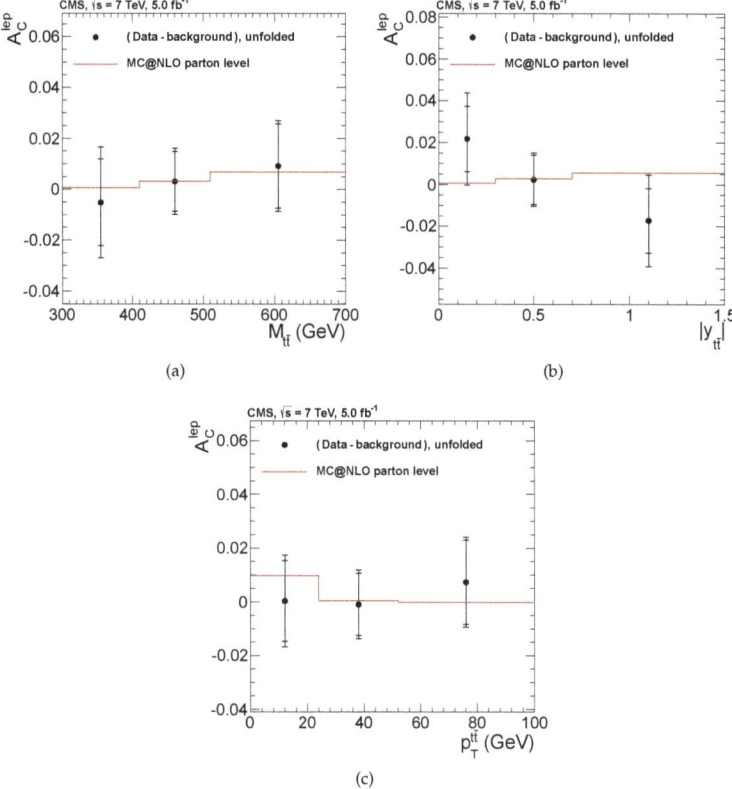

Figure 31. The dependence of $A_C^{\ell\ell}$ on $m_{t\bar{t}}$ in (**a**), $y_{t\bar{t}}$ in (**b**), and $p_{T,t\bar{t}}$ in (**c**). The inner and outer error bars represent the statistical and total uncertainty, respectively [108].

To summarize the measurements at $\sqrt{s} = 7$ TeV: both ATLAS and CMS measurements are in agreement with predictions. The most precise measurement of the $t\bar{t}$ charge asymmetry has a total uncertainty of 1.0%. The initial measurements with partial statistics had large ($\approx$3%) and about the same statistical and systematic uncertainties. This is a bit different to the Tevatron measurements where statistical uncertainty was dominant in most of the measurements. This is mostly due to the large $t\bar{t}$ sample already available for initial measurements and the fact these were measurements at the very early of the LHC running, so the detectors were not well understood yet. The full statistics measurements improved both statistical and systematic precision considerably. They are now mostly limited by the statistical uncertainty, especially dilepton measurements. The differential measurements are also in agreement with the predictions, they are mostly statistically limited and not really even able to disfavor BSM models.

4.2.2. Measurements at $\sqrt{s} = 8$ TeV

The measurements at $\sqrt{s} = 8$ TeV with the full statistics were performed by both ATLAS and CMS in both ℓ+jets and dilepton channels.

ATLAS performed two measurements in the ℓ+jets channel using 20.3 fb^{-1} [109,110]. In the first analysis, the asymmetry is measured inclusively ($A_C = (0.9 \pm 0.5(\text{stat.} + \text{syst.}))\%$) and also differentially as a function of $m_{t\bar{t}}$, $y_{t\bar{t}}$, $p_{T,t\bar{t}}$, and $\beta_{z,t\bar{t}}$, see Figure 32 using standard unfolding procedure [109]. The inclusive measurement is compatible with the NNLO QCD + NLO EW prediction

$(0.97^{+0.02}_{-0.03})$%. The second measurement focused on a large $t\bar{t}$ invariant mass region ($m_{t\bar{t}} > 0.75$ TeV, another requirement is $|\Delta|y|| < 2$) using reconstruction techniques specifically designed for the decay topology of highly boosted top quarks. In such cases, hadronicaly decaying top quarks are reconstructed as single large-radius jets with a specific jet substructure. In such phase space, the asymmetry is measured to be $A_C^{t\bar{t}} = (4.2 \pm 3.2(\text{stat.} + \text{syst.}))$%. A differential measurement as a function of $m_{t\bar{t}}$ is also performed, see Figure 33.

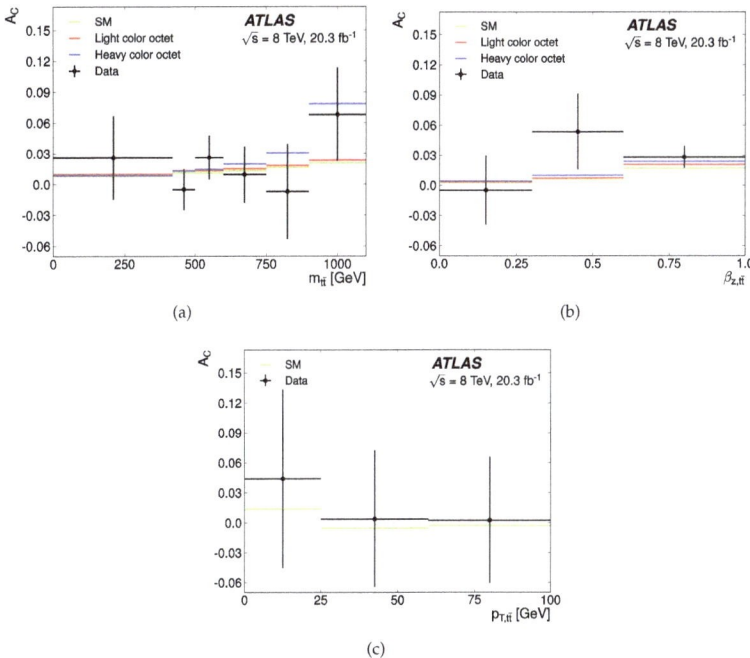

Figure 32. Measured A_C values as a function of $m_{t\bar{t}}$ in (a), $\beta_{z,t\bar{t}}$ in (b), and $p_{T,t\bar{t}}$ in (c), compared with NLO QCD + NLO EW predictions [24] and with the right-handed color octets with masses below the $t\bar{t}$ threshold [109].

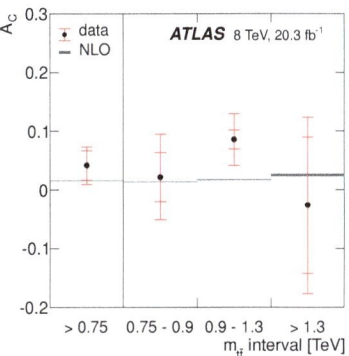

Figure 33. A summary of the charge asymmetry measurements for different ranges of $m_{t\bar{t}}$. The error bars on the data indicate the modeling and unfolding systematic uncertainties, shown as the inner bar, and the total uncertainty [110].

CMS also performed two measurements in the ℓ+jets channel using 19.7 fb^{-1} [112] and 19.6 fb^{-1} [113], respectively. In the first measurement [112], the asymmetry is measured inclusively ($A_C^{t\bar{t}} = (0.10 \pm 0.68\text{(stat.)} \pm 0.37\text{(syst.)})\%$) and also differentially as a function of $m_{t\bar{t}}$, $y_{t\bar{t}}$, and $p_{T,t\bar{t}}$. Moreover, CMS performed here the first LHC measurement at the particle level in the fiducial phase space mimicking the selection criteria. The inclusive fiducial ($A_C^{t\bar{t},fid} = (-0.35 \pm 0.72 \pm 0.31)\%$) and differential measurements as a function of $m_{t\bar{t}}$, $y_{t\bar{t}}$, and $p_{T,t\bar{t}}$, see Figure 34, are consistent with NLO QCD + EW prediction (inclusive prediction is $(1.01 \pm 0.10)\%$).

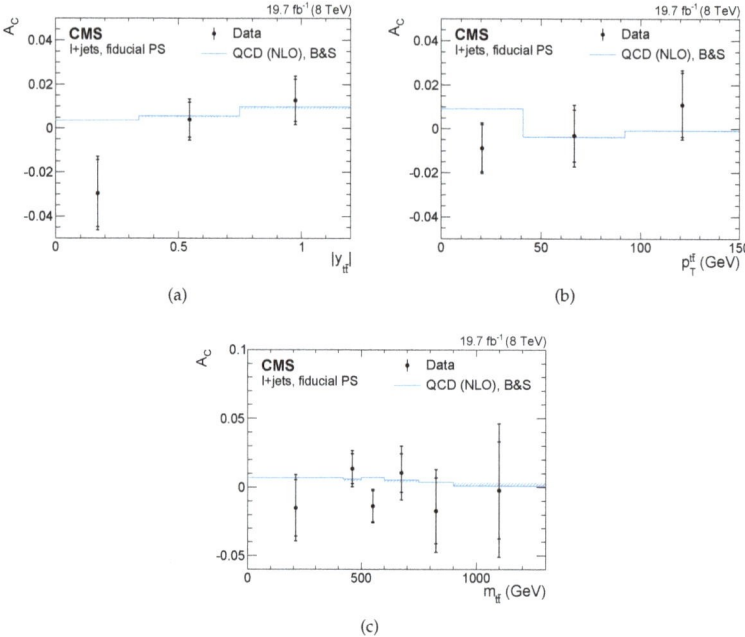

Figure 34. The charge asymmetry as a function of $y_{t\bar{t}}$ in (**a**), $p_{T,t\bar{t}}$ in (**b**), and $m_{t\bar{t}}$ in (**c**) measured at the particle level in the fiducial phase space. The inner bars indicate the statistical uncertainties, while the outer bars represent the statistical and systematic uncertainties added in quadrature [112].

The second CMS measurement in the ℓ+jets channel used a template method [113]. In this method, templates based on the SM were created for symmetric and antisymmetric components of the measured distribution ($Y_{t\bar{t}} = \tanh \Delta |y|$) for various $t\bar{t}$ production processes, see Figure 35. Fitting data to these templates, see Figure 36, the inclusive asymmetry was measured: $A_C^{t\bar{t}} = (0.33 \pm 0.26 \pm 0.33)\%$ which was the most precise measurement of A_C at that time. However, the disadvantage of this measurement was that it was more model dependent on SM predictions compared to usual unfolding measurements.

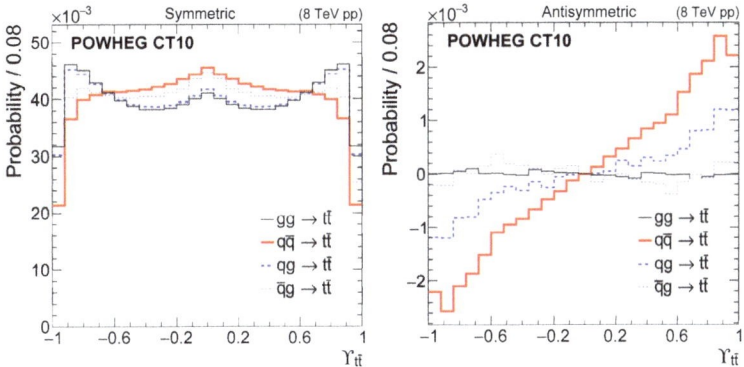

Figure 35. The symmetric (**left**) and antisymmetric (**right**) components of the binned probability distributions in the observable $Y_{t\bar{t}}$, constructed using POWHEG generator for different $t\bar{t}$ initial processes [113].

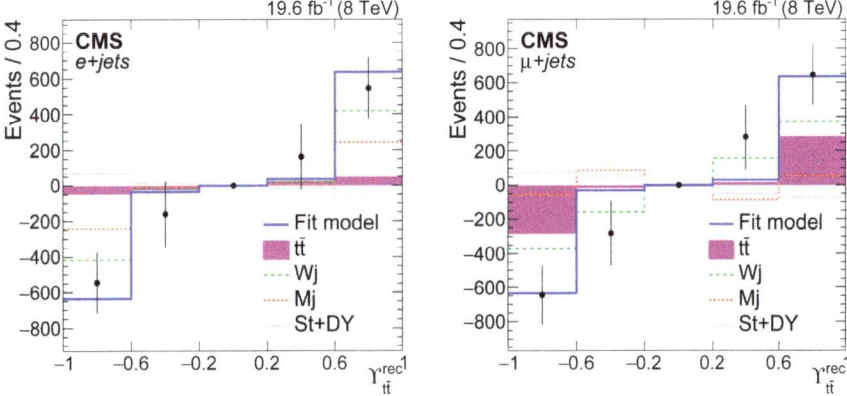

Figure 36. The antisymmetric $t\bar{t}$ contribution is measured in the $Y_{t\bar{t}}^{rec}$ distribution. The antisymmetric component of the $Y_{t\bar{t}}^{rec}$ distribution is shown here. The thick line shows the antisymmetric component of the fit model. The measurements are performed independently in the *e*+jets (**left**) and *μ*+jets (**right**) channels [113].

In the dilepton channel using 20.3 fb^{-1}, ATLAS measured the $t\bar{t}$ and dileptonic asymmetry at parton level in the full phase space and at the particle level in the fiducial phase space [111]. Both, the inclusive measurements at parton level ($A_C^{t\bar{t}} = (2.1 \pm 1.6(\text{stat.} + \text{syst.}))\%$, $A_C^{\ell\ell} = (0.8 \pm 0.6(\text{stat.} + \text{syst.}))\%$) and particle level ($A_C^{t\bar{t}} = (1.7 \pm 1.8(\text{stat.} + \text{syst.}))\%$, $A_C^{\ell\ell} = (0.6 \pm 0.5(\text{stat.} + \text{syst.}))\%$) are consistent with the predictions. The differential measurements in two bins were measured as a function of $m_{t\bar{t}}$, $p_{T,t\bar{t}}$, and $\beta_{z,t\bar{t}}$ for both $A_C^{t\bar{t}}$ and $A_C^{\ell\ell}$ in both full and fiducial phase spaces. The summary of dileptonic asymmetry measurements in the fiducial phase space is in Figure 37. The difference between the results at the parton and particle level is small given that the $t\bar{t}$ modeling systematics is not a dominant uncertainty.

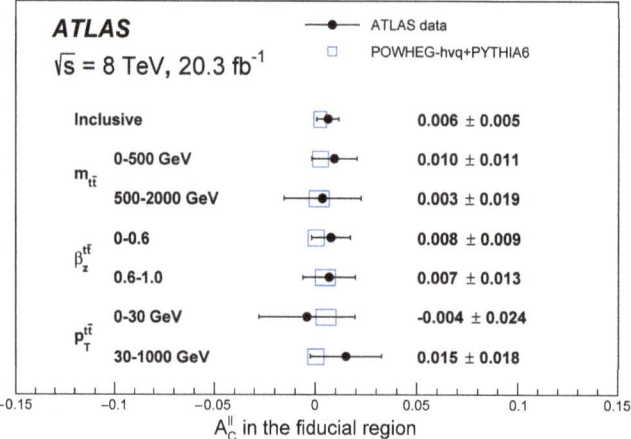

Figure 37. Summary of the measurements for the dileptonic asymmetry in the fiducial volume. The predictions shown in blue are obtained using POWHEG + PYTHIA at NLO [111].

CMS also measured the asymmetry in the dilepton channel using 19.5 fb^{-1} [114]. It measured the $t\bar{t}$ and dileptonic asymmetry inclusively ($A_C^{t\bar{t}} = (1.1 \pm 1.1(\text{stat.}) \pm 0.7(\text{syst.}))\%$, $A_C^{\ell\ell} = (0.3 \pm 0.6(\text{stat.}) \pm 0.3(\text{syst.}))\%$) and also differentially as a function of $m_{t\bar{t}}$, $y_{t\bar{t}}$, and $p_{T,t\bar{t}}$.

ATLAS and CMS combined their measurements performed at $\sqrt{s} = 7$ TeV and $\sqrt{s} = 8$ TeV [91]. Only measurements of the $t\bar{t}$ asymmetry in the ℓ+jets channel are combined. The measurements in the dilepton channel were statistically limited and their inclusion would not improve the overall uncertainty. The combination of inclusive measurements at $\sqrt{s} = 7$ TeV and $\sqrt{s} = 8$ TeV yielded $A_C^{t\bar{t}} = (0.5 \pm 0.7(\text{stat.}) \pm 0.6(\text{syst.}))\%$ and $A_C^{t\bar{t}} = (0.55 \pm 0.23(\text{stat.}) \pm 0.25(\text{syst.}))\%$, respectively. The CMS template measurement at $\sqrt{s} = 8$ TeV [113] was used in the combination for the inclusive measurement while CMS unfolding measurement at $\sqrt{s} = 8$ TeV [112] was used for the combination of differential measurements as a function of $m_{t\bar{t}}$. The summary of the inclusive Tevatron forward–backward and LHC 8 TeV charge asymmetry measurements together with the predictions of various BSM models is shown in Figure 14. The combined ATLAS+CMS charge asymmetry as a function of the invariant mass of the $t\bar{t}$ system in comparison with theoretical predictions for the SM and two versions of a color-octet model is shown in Figure 38.

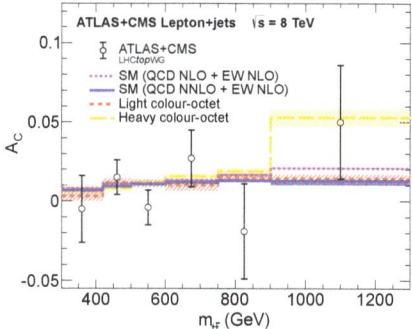

Figure 38. The combined ATLAS+CMS charge asymmetry as a function of $m_{t\bar{t}}$ in comparison with theoretical predictions for the SM [24,32] and two versions of a color-octet model [91].

In summary, the measurements at $\sqrt{s} = 8$ TeV provided a significant progress compared to $\sqrt{s} = 7$ TeV measurements. The measurements still agree with the SM prediction. Both the statistical and systematic uncertainties decreased almost at the same rate. The most precise individual inclusive measurement had an uncertainty of about 0.42% while the combined 8 TeV measurement had a precision of 0.33%. For the first time, the statistical uncertainty was no longer dominating the uncertainty in all measurements. The systematic uncertainties were smaller or similar to the statistical ones in the CMS template measurement, most of the ATLAS dilepton measurements, the ATLAS high $m_{t\bar{t}}$ measurement, and the LHC combination at $\sqrt{s} = 8$ TeV. The first fiducial level measurements at particle level were performed although their advantage was not yet much visible due to the fact that $t\bar{t}$ modeling systematics were still not dominant uncertainties. In addition, a specific measurement at high $m_{t\bar{t}}$ was performed.

4.2.3. Measurements at $\sqrt{s} = 13$ TeV

CMS performed already two measurements at $\sqrt{s} = 13$ TeV using a partial Run 2 dataset of 35.9 fb^{-1}.

In the dilepton channel [116], the normalized distribution of $\Delta|y|_{t\bar{t}}$ is measured at parton and particle level while the distribution of $\Delta|\eta|_{\ell\ell}$ is measured at particle level, see Figure 39. Using these distributions, charge asymmetries are obtained: $A_C^{t\bar{t}}(parton) = (1.0 \pm 0.9(\text{stat.} + \text{syst.}))\%$, $A_C^{t\bar{t}}(particle) = (0.8 \pm 0.9(\text{stat.} + \text{syst.}))\%$, and $A_C^{\ell\ell}(particle) = (-0.5 \pm 0.4(\text{stat.} + \text{syst.}))\%$, which are compared to various SM predictions in Figure 40.

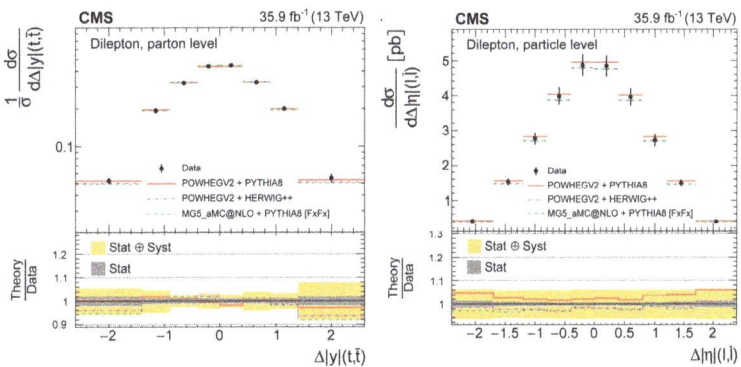

Figure 39. The normalized differential $t\bar{t}$ production cross-section as a function of $\Delta|y|$ at the parton level in the full phase space (**left**) and as a function of $\Delta|\eta|$ in the fiducial phase space at the particle level (**right**) [116].

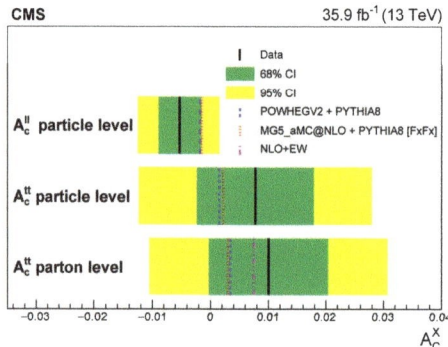

Figure 40. The results of the A_C extraction from integrating normalized parton level and particle level differential cross-section measurements as a function of $\Delta|y|$ and $\Delta|\eta|$ are shown [116].

CMS also measured the forward–backward asymmetry in the ℓ+jets channel at $\sqrt{s} = 13$ TeV using 35.9 fb^{-1} [115]. This is a bit different measurement compared to all the other LHC measurements. The approximate forward–backward asymmetry $A_{FB}^{(1)}$ is determined instead of edge–central charge asymmetry as measured in all the other LHC measurements. The template method is used based on $m_{t\bar{t}}$, $x_F = 2p_L/\sqrt{s}$, and $\cos\theta*$ variables, where p_L is the scaled longitudinal momentum p_L of the $t\bar{t}$ system in the laboratory frame, and $\theta*$ is the production angle of the top quark relative to the direction of the initial-state parton in the $t\bar{t}$ center-of-mass frame. The $q\bar{q} \to t\bar{t}$ differential cross-section in $\cos\theta$ can be expressed as a linear combination of symmetric and antisymmetric functions, where the antisymmetric function can be approximated as a linear function of $\cos\theta$ and parameter $A_{FB}^{(1)}$. Such approximation describes the LO terms and interference terms expected from an s-channel resonance with chiral couplings. In such approximation, $A_{FB} = A_{FB}^{(1)}$. The generator level distributions for the above mentioned variables for the $t\bar{t}$ production initiated by different processes are shown in Figure 41. The application of fitting procedure yields $A_{FB}^{(1)} = (4.8^{+9.5}_{-8.7}(\text{stat.})^{+2.0}_{-2.9}(\text{syst.}))\%$. The result is consistent with the NLO QCD [13,21,118] and NNLO QCD prediction [32], although the statistical uncertainty is quite large.

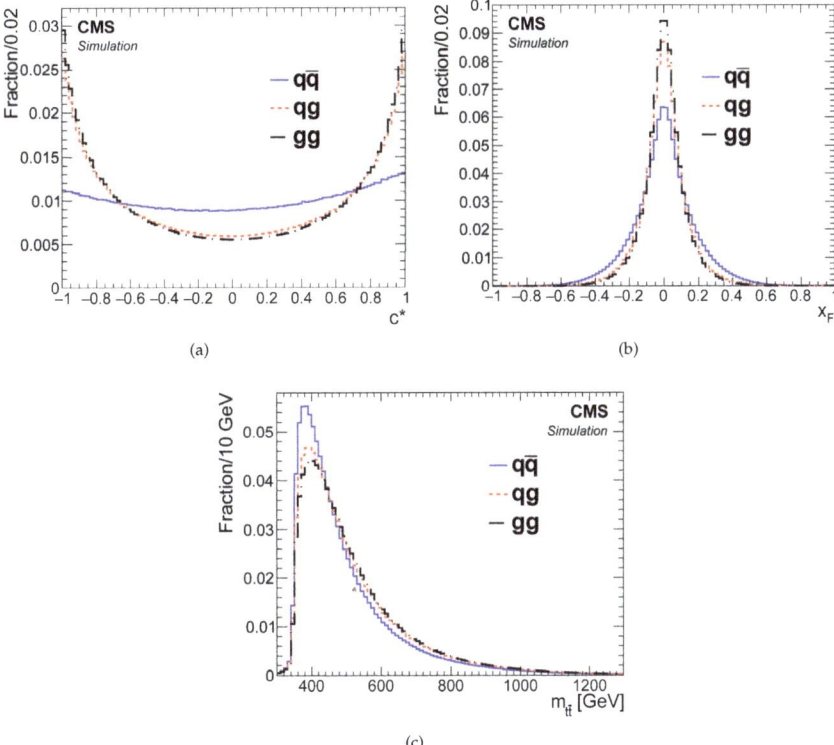

Figure 41. The generator-level cos θ (labeled here as c^*) in (a), x_F in (b), and $m_{t\bar{t}}$ normalized distributions in (c) for the subprocesses $q\bar{q}$, qg, and gg. These distributions correspond to the CMS measurement in the ℓ+jets channel performed at $\sqrt{s} = 13$ TeV using 35.9 fb^{-1} [115].

ATLAS already performed a preliminary A_C measurement in the ℓ+jets channel using the full Run 2 statistics (139 fb^{-1}) [50]. Altogether, more than four millions of $t\bar{t}$ candidates were selected in data events with the expected background of about 15%. The asymmetry is measured to be $A_C = (0.60 \pm 0.15(\text{stat.} + \text{syst.}))\%$, consistent with the NNLO QCD + NLO EW prediction of $(0.64^{+0.05}_{-0.06})\%$. Differential measurements in $m_{t\bar{t}}$ and $\beta_{z,t\bar{t}}$ were also performed, see Figure 42. Moreover, the charge asymmetry measurement was interpreted in the framework of an effective field theory (EFT). In EFT formalism the SM Lagrangian is extended with operators that encode the new physics phenomena. The Warsaw basis includes a complete set of dimension-six operators [119]. The charge asymmetry is affected by the difference $C^- = C^1 - C^2$, where $C^1 = C^1_u = C^1_d$ and $C^2 = C^2_u = C^2_d$ are Wilson coefficients which are obtained from seven four-fermion operators in Warsaw basis by using a flavour-specific linear combination [120]. The constrains on C^- are shown in Figure 43.

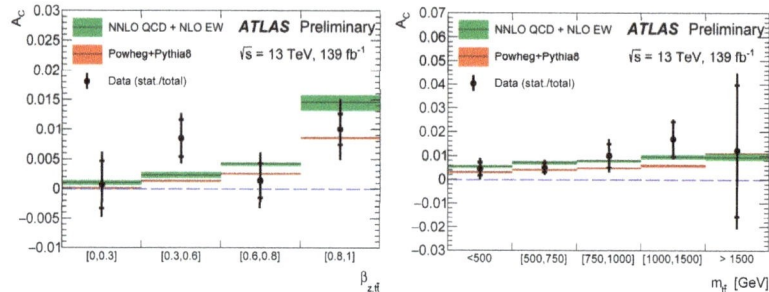

Figure 42. Differential charge asymmetry measurements as a function of $\beta_{z,t\bar{t}}$ (left) and $m_{t\bar{t}}$ (right) [50].

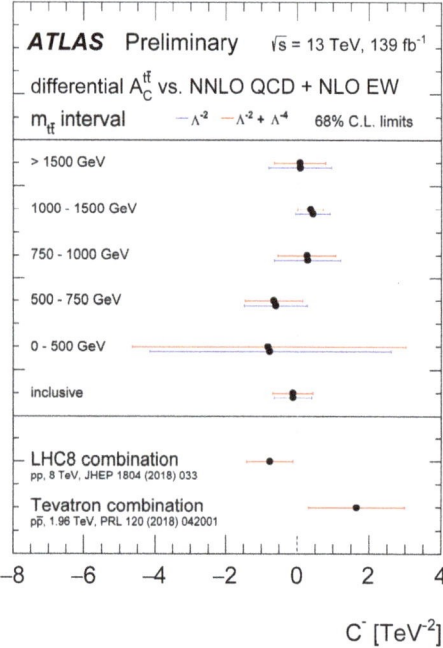

Figure 43. Constraints on linear combination C^-/Λ^2 of Wilson coefficients of dimension 6 operators from inclusive and $m_{t\bar{t}}$ differential charge asymmetry measurements [50].

4.2.4. Summary of LHC Measurements

All inclusive charge asymmetry measurements performed at the LHC are summarized in Table 8. The $A_C^{t\bar{t}}$ asymmetries should be compared with NLO QCD including electroweak corrections prediction [24] $(1.23 \pm 0.05)\%$ at $\sqrt{s} = 7$ TeV, NNLO QCD + NLO EW prediction [32] $(0.97^{+0.02}_{-0.03})\%$ at $\sqrt{s} = 8$ TeV and $(0.64^{+0.06}_{-0.05})\%$ at $\sqrt{s} = 13$ TeV. The $A_C^{\ell\ell}$ asymmetries should be compared with NLO QCD + EW prediction $(0.70 \pm 0.03)\%$ at $\sqrt{s} = 7$ TeV [24], $(0.64 \pm 0.03)\%$ at $\sqrt{s} = 8$ TeV [24], and NLO QCD + EW prediction $(0.55 \pm 0.03)\%$ at $\sqrt{s} = 13$ TeV [25].

All LHC measurements at all energies are well within 2 SD consistent with the SM prediction. The measurements at $\sqrt{s} = 7$ TeV are limited by the statistics with all of them at least to have the absolute uncertainty of 1%. The exception is the combination of ATLAS and D0 in the ℓ+jets channel which has the total uncertainty of about 0.9%. At $\sqrt{s} = 8$ TeV, there are already many measurements which have comparable statistical and total systematic uncertainty. The most precise is the combination of the ATLAS and CMS ℓ+jets channel measurements which has the overall uncertainty of about 0.34%

with the dominant systematic uncertainties due to calibration of jets and signal modeling. Finally at $\sqrt{s} = 13$ TeV, the full statistics measurement are not yet available except for the preliminary ATLAS ℓ+jets measurement. This measurement is already very precise at the absolute level of 0.15%, very well consistent with NNLO QCD + NLO EW prediction, and differs from zero by 4 standard deviations. This is the first evidence for non-zero charge asymmetry at the LHC. The early measurements are not precise enough to be able to observe the expected decrease of the asymmetry with the energy of interactions.

The leptonic asymmetries have for now uncertainties larger than 0.4% (particle level) and are all consistent with SM predictions.

The differential measurements are also consistent with the SM prediction. Most of the time, the statistical uncertainties are dominant, although in the latest ATLAS measurement at $\sqrt{s} = 13$ TeV the total systematic uncertainties are comparable to statistical uncertainties except for high $m_{t\bar{t}}$ bins.

Table 8. Summary of inclusive $t\bar{t}$ and leptonic charge asymmetry measurements performed at the LHC. For a given measurement, if there is just one uncertainty, it is combined statistical and systematical uncertainty. If there are two uncertainties, the first one is statistical and the second one is systematic uncertainty. All measurements used $\Delta|y|$ variable except for the measurement with * which used $\Delta|\eta|$. All measurements were performed at the parton level except for the measurement with ** which was performed at particle level.

Experiment, Channel	$\sqrt{s}$ [TeV]	L [fb^{-1}]	$A_C^{t\bar{t}}$[%]	$A_C^{\ell\ell}$[%]
CMS, ℓ+jets	7	1.1	$-1.7 \pm 3.2^{+2.5}_{-3.6}$ *	
ATLAS, ℓ+jets	7	1.1	$-1.9 \pm 2.8 \pm 2.4$	
CMS, ℓ+jets	7	5.0	$0.4 \pm 1.0 \pm 1.1$	
CMS, dil	7	5.0	$-1.0 \pm 1.7 \pm 0.8$	$0.9 \pm 1.0 \pm 0.6$
ATLAS, ℓ+jets	7	4.7	0.6 ± 1.0	
ATLAS, dil	7	4.6	$2.1 \pm 2.5 \pm 1.7$	$2.4 \pm 1.5 \pm 0.9$
LHC, combination	7	5.0	$0.5 \pm 0.7 \pm 0.6$	
CMS, ℓ+jets	8	19.7	$0.10 \pm 0.68 \pm 0.37$	
CMS, ℓ+jets(template)	8	19.6	$0.33 \pm 0.26 \pm 0.33$	
CMS, dil	8	19.5	$1.1 \pm 1.1 \pm 0.7$	$0.3 \pm 0.6 \pm 0.3$
ATLAS, ℓ+jets	8	20.3	0.9 ± 0.5	
ATLAS, dil	8	20.3	2.1 ± 1.6	0.8 ± 0.6
LHC, combination	8	20.3	$0.55 \pm 0.23 \pm 0.25$	
CMS, dilepton	13	35.9	1.0 ± 0.9	-0.5 ± 0.4 **
ATLAS, ℓ+jets	13	139	0.60 ± 0.15	

5. Discussion and Outlook

It is clear from the description in Sections 3 and 4 that the long path and large effort in improving the theory and experiments has paid off. Although, some may be unhappy that tensions between theoretical calculations and experimental measurements mostly disappeared, the understanding of the $t\bar{t}$ charge asymmetry is much better now.

On the theoretical side, the progress has been enormous from only a partial NLO prediction for A_{FB} at the Tevatron which predicted negative asymmetry, through the full NLO prediction in the laboratory frame of about 5%, to the latest full NNLO QCD + NLO EW prediction for both A_{FB} at the Tevatron and A_C at the LHC and the aN³LO QCD + NLO EW prediction at the Tevatron. At the Tevatron, the predicted asymmetry is about 10% while it is around 1% at the LHC. Moreover, differential asymmetries have been also calculated at NNLO QCD + NLO EW too as a function of many variables such as $m_{t\bar{t}}$, Δy, $p_{T,t\bar{t}}$, $\beta_{z,t\bar{t}}$, and $\cos\theta$. The leptonic asymmetry has been calculated at NLO+EW order.

On the experimental side, there has been performed a full set of measurements for various observables. The very early measurements were performed just at the reco level. Later, this has been improved to perform measurements at the parton level and lately also at the particle level. There are now available not only inclusive measurements of both forward–backward and charge asymmetries, but also detailed differential measurements as a function of a few variables such as $m_{t\bar{t}}$, $p_{T,t\bar{t}}$, Δy, $\beta_{z,t\bar{t}}$. All inclusive $t\bar{t}$ asymmetry measurements of CDF, D0, ATLAS, CMS show a very good agreement with the NNLO QCD + NLO EW prediction with the largest disagreement of about 1.6 SD. The leptonic asymmetry measurements with the full Tevatron dataset and at the LHC also agree with the NLO QCD + EW prediction with the largest disagreement of about 2.3 SD for the CDF leptonic asymmetry measurement. However, it should be mentioned that all inclusive Tevatron measurements are higher than the NNLO QCD + NLO EW prediction, so it is possible that some non-negligible correction is still not calculated. At the LHC, the asymmetries both higher and lower compared to the best prediction have been measured. At the Tevatron, the non-zero forward–backward asymmetry ($\delta A_{FB}/A_{FB} = 20\%$) has been observed now (with a significance of about 5 SD) and the leptonic asymmetry is measured with the relative precision of about 26%. For A_C at the LHC at $\sqrt{s} = 13$ TeV, the evidence (significance of at least 3 SD) of non-zero charge asymmetry has been obtained and the relative precision is about 25%. Given that the dileptonic asymmetry has not been measured yet with the full LHC Run 2 statistics, the fact that dileptonic asymmetry is supposed to be smaller than A_C, and the fact it can be measured only in the dilepton channel, its relative precision is for now only around 80%. Most of the inclusive measurements at both the Tevatron and the LHC have been statistically limited although the statistical and total systematic uncertainties are about the same in the LHC combination at $\sqrt{s} = 7$ TeV and $\sqrt{s} = 8$ TeV and in the latest measurement at $\sqrt{s} = 13$ TeV. The A_{FB} and A_C asymmetries and their leptonic versions have been measured also differentially as a function of a few variables. Most of the measurements have been statistically limited, but this starts to change with the full LHC Run 2 statistics. The Tevatron results are very probably final, since the data taking finished already in 2011.

The LHC running will continue, mostly at the energy of $\sqrt{s} = 14$ TeV and about 20 times more data (3000 fb^{-1}) are expected to be delivered by the end of the LHC lifetime. This will allow to improve the statistical uncertainty by at least a factor of 4–5 and the systematic uncertainties will become dominant. Based on the ATLAS measurement at $\sqrt{s} = 13$ TeV, it can be expected the dominant systematic uncertainties will be the $t\bar{t}$ modeling, the jet energy calibration related uncertainties and the W+jets background modeling. These systematic uncertainties will become dominant also for differential measurements and this will allow to measure them in a more detail using more bins and the larger range. Eventually, the dileptonic asymmetry should be more precisely measured because the leptons are more precisely measured than top quarks and typically have smaller systematic related uncertainties. Moreover, it is expected that another LHC experiment, the LHCb, will be able to observe a non-zero $t\bar{t}$ charge asymmetry at the high-luminosity LHC [121]. Additionally, there is a possibility to measure different types of asymmetries, such as energy asymmetry between the top and antitop quarks [122].

At the potential Future Circular Collider (FCC) in pp collisions at $\sqrt{s} = 100$ TeV, the charge asymmetry is greatly diluted by the dominance of the gg initial state. The SM expected value is $A_C = 0.12\%$ [123] which will make it very hard to measure. However, the asymmetry is enhanced in associated processes $t\bar{t} + Z$, $t\bar{t} + \gamma$ and mainly in $t\bar{t} + W$, where the asymmetry is enhanced by about a factor of ten due to the $t\bar{t} + W$ process being dominated by a $q\bar{q}$ initial state [123,124]. A relative statistical precision of about 3% is expected in the determination of A_C in the $t\bar{t} + W$ process [124].

At the linear e^+e^- collider, the EW based forward–backward asymmetry in $e^+e^- \to t\bar{t}$ is expected [125,126]. The preliminary studies for the potential International Linear Collider at $\sqrt{s} = 500$ GeV show that for the large asymmetry of about 40% (depending on the polarization of the beams), the expected relative precision of about 2% can be achieved [125].

The asymmetry measurements should also help in the model independent search for a new BSM physics within the effective field theory approach by constraining the EFT coefficients related to the top quark production.

6. Conclusions

As the heaviest known elementary particle, the top quark and studies of its properties is a promising portal to the new physics beyond the Standard Model. The charge asymmetry in the $t\bar{t}$ production is the effect which is predicted to be present at higher orders in perturbative quantum chromodynamics and by necessity to be small, but it is highly enhanced in various theories beyond the Standard Model.

After unexpectedly large values of the forward–backward asymmetry in the top quark pair production were observed in initial measurements at the Tevatron, a lot of attention has been paid to it by the experimental and theoretical community. This allowed to perform precise and detailed tests of the SM at high energies. At present, the prediction is known at full next-to-next-to-leading order in perturbative QCD with complete next-to-leading order electroweak corrections. The full statistics Tevatron forward–backward and the LHC charge asymmetry results for the inclusive and differential measurements agree with the predictions very well, mostly within two standard deviations, with the largest deviation of about 2.3 standard deviation. The predicted forward–backward asymmetry at the Tevatron of about 10% is now measured with a relative precision of 20%. At the LHC, although the effect is much smaller ($\approx 1\%$), the relative precision of the latest measurement is already at the level of about 25%.

In the coming years at the LHC and potential future colliders, it can be expected that more measurements will be performed at higher energies and in the processes like $t\bar{t} + W$ boson where the relative precision at the level of a few percent can be potentially achieved. Moreover, there is a possibility to measure a very large $t\bar{t}$ asymmetry in electroweak interactions at the lepton collider in polarized beams with a relative precision of a couple of percent. This will allow to precisely test the present theory at high energies and to potentially observe the presence of BSM effects or to constrain the BSM physics either by excluding particular models or by constraining parameters of effective theories.

Funding: This research was supported by the project LTT17018 of Ministry of Education, Youth and Sports of Czech Republic.

Acknowledgments: The author would like to thank Alexander Kupčo and Jaroslav Antoš for reading the manuscript and providing useful comments.

Conflicts of Interest: The author is a member of the CDF and ATLAS collaboration. The funders had no role in the writing of the study.

References

1. Glashow, S. Partial Symmetries of Weak Interactions. *Nucl. Phys.* **1961**, *22*, 579–588. [CrossRef]
2. Weinberg, S. A Model of Leptons. *Phys. Rev. Lett.* **1967**, *19*, 1264–1266. [CrossRef]
3. Salam, A. Weak and electromagnetic interactions. In *Elementary Particle Physics: Relativistic Groups and Analyticity*; Proceedings of the eighth Nobel symposium; Svartholm, N., Ed.; Almqvist and Wiksell: Stockholm, Sweden, 1968; p. 367.
4. Freese, K. Review of Observational Evidence for Dark Matter in the Universe and in upcoming searches for Dark Stars. *EAS Publ. Ser.* **2009**, *36*, 113–126. [CrossRef]
5. Canetti, L.; Drewes, M.; Shaposhnikov, M. Matter and Antimatter in the Universe. *New J. Phys.* **2012**, *14*, 095012. [CrossRef]
6. Particle Data Group. Review of Particle Physics. *Phys. Rev. D* **2018**, *98*, 030001. [CrossRef]
7. Randall, L.; Sundrum, R. A Large mass hierarchy from a small extra dimension. *Phys. Rev. Lett.* **1999**, *83*, 3370–3373. [CrossRef]
8. Pomarol, A.; Serra, J. Top Quark Compositeness: Feasibility and Implications. *Phys. Rev. D* **2008**, *78*, 074026. [CrossRef]
9. CDF Collaboration. Observation of top quark production in $\bar{p}p$ collisions. *Phys. Rev. Lett.* **1995**, *74*, 2626–2631. [CrossRef]
10. D0 Collaboration. Observation of the top quark. *Phys. Rev. Lett.* **1995**, *74*, 2632–2637. [CrossRef]

11. Halzen, F.; Hoyer, P.; Kim, C. Forward - Backward Asymmetry of Hadroproduced Heavy Quarks in QCD. *Phys. Lett. B* **1987**, *195*, 74–77. [CrossRef]
12. Kuhn, J.H.; Rodrigo, G. Charge asymmetry in hadroproduction of heavy quarks. *Phys. Rev. Lett.* **1998**, *81*, 49–52. [CrossRef]
13. Kuhn, J.H.; Rodrigo, G. Charge asymmetry of heavy quarks at hadron colliders. *Phys. Rev. D* **1999**, *59*, 054017. [CrossRef]
14. Dittmaier, S.; Uwer, P.; Weinzierl, S. NLO QCD corrections to t anti-t + jet production at hadron colliders. *Phys. Rev. Lett.* **2007**, *98*, 262002. [CrossRef] [PubMed]
15. Dittmaier, S.; Uwer, P.; Weinzierl, S. Hadronic top-quark pair production in association with a hard jet at next-to-leading order QCD: Phenomenological studies for the Tevatron and the LHC. *Eur. Phys. J. C* **2009**, *59*, 625–646. [CrossRef]
16. Almeida, L.G.; Sterman, G.F.; Vogelsang, W. Threshold Resummation for the Top Quark Charge Asymmetry. *Phys. Rev. D* **2008**, *78*, 014008. [CrossRef]
17. Melnikov, K.; Scharf, A.; Schulze, M. Top quark pair production in association with a jet: QCD corrections and jet radiation in top quark decays. *Phys. Rev. D* **2012**, *85*, 054002. [CrossRef]
18. Kidonakis, N. The top quark rapidity distribution and forward-backward asymmetry. *Phys. Rev. D* **2011**, *84*, 011504. [CrossRef]
19. Ahrens, V.; Ferroglia, A.; Neubert, M.; Pecjak, B.D.; Yang, L.L. The top-pair forward-backward asymmetry beyond NLO. *Phys. Rev. D* **2011**, *84*, 074004. [CrossRef]
20. Hollik, W.; Pagani, D. The electroweak contribution to the top quark forward-backward asymmetry at the Tevatron. *Phys. Rev. D* **2011**, *84*, 093003. [CrossRef]
21. Kuhn, J.H.; Rodrigo, G. Charge asymmetries of top quarks at hadron colliders revisited. *JHEP* **2012**, *1*, 063. [CrossRef]
22. Manohar, A.V.; Trott, M. Electroweak Sudakov Corrections and the Top Quark Forward-Backward Asymmetry. *Phys. Lett. B* **2012**, *711*, 313–316. [CrossRef]
23. Bernreuther, W.; Si, Z.G. Distributions and correlations for top quark pair production and decay at the Tevatron and LHC. *Nucl. Phys. B* **2010**, *837*, 90–121. [CrossRef]
24. Bernreuther, W.; Si, Z.G. Top quark and leptonic charge asymmetries for the Tevatron and LHC. *Phys. Rev. D* **2012**, *86*, 034026. [CrossRef]
25. Bernreuther, W.; Heisler, D.; Si, Z.G. A set of top quark spin correlation and polarization observables for the LHC: Standard Model predictions and new physics contributions. *JHEP* **2015**, *12*, 026. [CrossRef]
26. Brodsky, S.J.; Wu, X.G. Application of the Principle of Maximum Conformality to the Top-Quark Forward-Backward Asymmetry at the Tevatron. *Phys. Rev. D* **2012**, *85*, 114040. [CrossRef]
27. Wang, S.Q.; Wu, X.G.; Si, Z.G.; Brodsky, S.J. Application of the Principle of Maximum Conformality to the Top-Quark Charge Asymmetry at the LHC. *Phys. Rev. D* **2014**, *90*, 114034. [CrossRef]
28. Wang, S.Q.; Wu, X.G.; Si, Z.G.; Brodsky, S.J. Predictions for the Top-Quark Forward-Backward Asymmetry at High Invariant Pair Mass Using the Principle of Maximum Conformality. *Phys. Rev. D* **2016**, *93*, 014004. [CrossRef]
29. Kidonakis, N. The top quark forward-backward asymmetry at approximate N^3LO. *Phys. Rev. D* **2015**, *91*, 071502, [CrossRef]
30. Czakon, M.; Fiedler, P.; Mitov, A. Resolving the Tevatron Top Quark Forward-Backward Asymmetry Puzzle: Fully Differential Next-to-Next-to-Leading-Order Calculation. *Phys. Rev. Lett.* **2015**, *115*, 052001. [CrossRef]
31. Czakon, M.; Fiedler, P.; Heymes, D.; Mitov, A. NNLO QCD predictions for fully-differential top-quark pair production at the Tevatron. *JHEP* **2016**, *05*, 034. [CrossRef]
32. Czakon, M.; Heymes, D.; Mitov, A.; Pagani, D.; Tsinikos, I.; Zaro, M. Top-quark charge asymmetry at the LHC and Tevatron through NNLO QCD and NLO EW. *Phys. Rev. D* **2018**, *98*, 014003. [CrossRef]
33. Aguilar-Saavedra, J.; Juste, A.; Rubbo, F. Boosting the $t\bar{t}$ charge asymmetry. *Phys. Lett. B* **2012**, *707*, 92–98. [CrossRef]
34. Aguilar-Saavedra, J.; Perez-Victoria, M. Asymmetries in $t\bar{t}$ production: LHC versus Tevatron. *Phys. Rev. D* **2011**, *84*, 115013. [CrossRef]
35. Aguilar-Saavedra, J.; Perez-Victoria, M. Simple models for the top asymmetry: Constraints and predictions. *JHEP* **2011**, *09*, 097. [CrossRef]

36. Drobnak, J.; Kamenik, J.F.; Zupan, J. Flipping t tbar Asymmetries at the Tevatron and the LHC. *Phys. Rev. D* **2012**, *86*, 054022. [CrossRef]
37. Drobnak, J.; Kagan, A.L.; Kamenik, J.F.; Perez, G.; Zupan, J. Forward Tevatron Tops and Backward LHC Tops with Associates. *Phys. Rev. D* **2012**, *86*, 094040. [CrossRef]
38. Aguilar-Saavedra, J.; Amidei, D.; Juste, A.; Perez-Victoria, M. Asymmetries in top quark pair production at hadron colliders. *Rev. Mod. Phys.* **2015**, *87*, 421–455. [CrossRef]
39. Aguilar-Saavedra, J. Portrait of a colour octet. *JHEP* **2014**, *8*, 172. [CrossRef]
40. D0 Collaboration. First measurement of the forward-backward charge asymmetry in top quark pair production. *Phys. Rev. Lett.* **2008**, *100*, 142002. [CrossRef]
41. CDF Collaboration. Forward-Backward Asymmetry in Top Quark Production in $p\bar{p}$ Collisions at $sqrts = 1.96$ TeV. *Phys. Rev. Lett.* **2008**, *101*, 202001. [CrossRef]
42. CDF Collaboration. Evidence for a Mass Dependent Forward-Backward Asymmetry in Top Quark Pair Production. *Phys. Rev. D* **2011**, *83*, 112003. [CrossRef]
43. D0 Collaboration. Forward-backward asymmetry in top quark-antiquark production. *Phys. Rev. D* **2011**, *84*, 112005. [CrossRef]
44. D0 Collaboration. Measurement of Leptonic Asymmetries and Top Quark Polarization in $t\bar{t}$ Production. *Phys. Rev. D* **2013**, *87*, 011103. [CrossRef]
45. Kamenik, J.F.; Shu, J.; Zupan, J. Review of New Physics Effects in t-tbar Production. *Eur. Phys. J. C* **2012**, *72*, 2102. [CrossRef]
46. Berger, E.L.; Cao, Q.H.; Chen, C.R.; Zhang, H. Interpretations and implications of the top quark rapidity asymmetries A_{FB}^t and A_{FB}^ℓ. *Phys. Rev. D* **2013**, *88*, 014033. [CrossRef]
47. Czakon, M.; Fiedler, P.; Mitov, A. Total Top-Quark Pair-Production Cross Section at Hadron Colliders Through $O(\alpha_S^4)$. *Phys. Rev. Lett.* **2013**, *110*, 252004. [CrossRef]
48. Czakon, M.; Mitov, A. Top++: A Program for the Calculation of the Top-Pair Cross-Section at Hadron Colliders. *Comput. Phys. Commun.* **2014**, *185*, 2930. [CrossRef]
49. CDF and D0 Collaboration. Combined Forward-Backward Asymmetry Measurements in Top-Antitop Quark Production at the Tevatron. *Phys. Rev. Lett.* **2018**, *120*, 042001. [CrossRef]
50. ATLAS Collaboration. Inclusive and Differential Measurement of the Charge Asymmetry in $t\bar{t}$ Events at 13 TeV with the ATLAS Detector. ATLAS-CONF-2019-026. Available online: http://cds.cern.ch/record/2682109 (accessed on 15 June 2020).
51. Kidonakis, N. NNNLO soft-gluon corrections for the top-antitop pair production cross section. *Phys. Rev. D* **2014**, *90*, 014006, [CrossRef]
52. Grazzini, M.; Kallweit, S.; Wiesemann, M. Fully differential NNLO computations with MATRIX. *Eur. Phys. J. C* **2018**, *78*, 537, [CrossRef]
53. Catani, S.; Devoto, S.; Grazzini, M.; Kallweit, S.; Mazzitelli, J.; Sargsyan, H. Top-quark pair hadroproduction at next-to-next-to-leading order in QCD. *Phys. Rev. D* **2019**, *99*, 051501, [CrossRef]
54. Catani, S.; Devoto, S.; Grazzini, M.; Kallweit, S.; Mazzitelli, J. Top-quark pair production at the LHC: Fully differential QCD predictions at NNLO. *JHEP* **2019**, *7*, 100, [CrossRef]
55. ATLAS Collaboration. *Top Working Group Cross-Section Summary Plots: Spring 2020*; ATL-PHYS-PUB-2020-012. Available online: http://cds.cern.ch/record/2718946 (accessed on 15 June 2020).
56. Berends, F.A.; Gaemers, K.; Gastmans, R. alpha**3 Contribution to the angular asymmetry in $e^+ e^- \to \mu^+ \mu^-$. *Nucl. Phys. B* **1973**, *63*, 381–397. [CrossRef]
57. Himel, T.; Richter, B.; Abrams, G. S.; Alam, M. S.; Boyarski, A.; Breidenbach, M.; Chinowsky, W.; Feldman, G.J.; Goldhaber, G.; Hanson, G.; et al. Limits on Strength of Neutral Currents From $e^+e^- \to \mu^+\mu^-$. *Phys. Rev. Lett.* **1978**, *41*, 449. [CrossRef]
58. Budny, R. Effects of neutral weak currents in annihilation. *Phys. Lett. B* **1973**, *45*, 340–344. [CrossRef]
59. ALEPH, DELPHI, L3, OPAL, SLD, LEP Electroweak Working Group, SLD Electroweak Group and SLD Heavy Flavour Group. Precision electroweak measurements on the Z resonance. *Phys. Rept.* **2006**, *427*, 257–454. [CrossRef]
60. Brown, R.; Sahdev, D.; Mikaelian, K. Probing Higher Order QCD: Charge Conjugation Asymmetries from Two Gluon Exchange. *Phys. Rev. Lett.* **1979**, *43*, 1069. [CrossRef]
61. LHCb Collaboration. First measurement of the charge asymmetry in beauty-quark pair production. *Phys. Rev. Lett.* **2014**, *113*, 082003. [CrossRef]

62. CDF Collaboration. First measurement of the forward-backward asymmetry in bottom-quark pair production at high mass. *Phys. Rev. D* **2015**, *92*, 032006. [CrossRef]
63. CDF Collaboration. Measurement of the forward-backward asymmetry in low-mass bottom-quark pairs produced in proton-antiproton collisions. *Phys. Rev. D* **2016**, *93*, 112003. [CrossRef]
64. D0 Collaboration. Measurement of the Forward-Backward Asymmetry in the Production of $B^\pm$ Mesons in $p\bar{p}$ Collisions at $\sqrt{s}$ = 1.96 TeV. *Phys. Rev. Lett.* **2015**, *114*, 051803. [CrossRef] [PubMed]
65. ATLAS Collaboration. Measurement of the charge asymmetry in top quark pair production in pp collisions at $\sqrt{s}$ = 7 TeV using the ATLAS detector. *Eur. Phys. J. C* **2012**, *72*, 2039. [CrossRef] [PubMed]
66. D0 Collaboration. Measurement of the top quark mass using dilepton events. *Phys. Rev. Lett.* **1998**, *80*, 2063–2068. [CrossRef]
67. CDF Collaboration. Measurement of the top quark mass using template methods on dilepton events in proton antiproton collisions at $\sqrt{s}$ = 1.96-TeV. *Phys. Rev. D* **2006**, *73*, 112006. [CrossRef]
68. CDF Collaboration. W boson polarization measurement in the $t\bar{t}$ dilepton channel using the CDF II Detector. *Phys. Lett. B* **2013**, *722*, 48–54. [CrossRef]
69. CMS Collaboration. Measurement of the charge asymmetry in top-quark pair production in proton-proton collisions at $\sqrt{s}$ = 7 TeV. *Phys. Lett. B* **2012**, *709*, 28–49. [CrossRef]
70. CMS Collaboration. Measurement of the $t\bar{t}$ production cross section and the top quark mass in the dilepton channel in pp collisions at $\sqrt{s}$ = 7 TeV. *JHEP* **2011**, *07*, 049. [CrossRef]
71. CDF Collaboration. Precision top quark mass measurement in the lepton + jets topology in p anti-p collisions at $\sqrt{s}$ = 1.96-TeV. *Phys. Rev. Lett.* **2006**, *96*, 022004. [CrossRef]
72. CDF Collaboration. Measurement of the forward–backward asymmetry of top-quark and antiquark pairs using the full CDF Run II data set. *Phys. Rev. D* **2016**, *93*, 112005. [CrossRef]
73. Betchart, B.A.; Demina, R.; Harel, A. Analytic solutions for neutrino momenta in decay of top quarks. *Nucl. Instrum. Meth. A* **2014**, *736*, 169–178, doi:10.1016/j.nima.2013.10.039. [CrossRef]
74. D0 Collaboration. Precise measurement of the top quark mass in the dilepton channel at D0. *Phys. Rev. Lett.* **2011**, *107*, 082004. [CrossRef] [PubMed]
75. D0 Collaboration. Experimental discrimination between charge 2e/3 top quark and charge 4e/3 exotic quark production scenarios. *Phys. Rev. Lett.* **2007**, *98*, 041801. [CrossRef] [PubMed]
76. CDF Collaboration. Exclusion of exotic top-like quarks with -4/3 electric charge using jet-charge tagging in single-lepton ttbar events at CDF. *Phys. Rev. D* **2013**, *88*, 032003. [CrossRef]
77. ATLAS Collaboration. Measurement of the top quark charge in pp collisions at $\sqrt{s}$ = 7 TeV with the ATLAS detector. *JHEP* **2013**, *11*, 031. [CrossRef]
78. Skands, P.; Webber, B.; Winter, J. QCD Coherence and the Top Quark Asymmetry. *JHEP* **2012**, *07*, 151. [CrossRef]
79. NNPDF Collaboration. Parton distributions from high-precision collider data. *Eur. Phys. J. C* **2017**, *77*, 663. [CrossRef]
80. Brodsky, S.J.; Lepage, G.; Mackenzie, P.B. On the Elimination of Scale Ambiguities in Perturbative Quantum Chromodynamics. *Phys. Rev. D* **1983**, *28*, 228. [CrossRef]
81. Brodsky, S.J.; Di Giustino, L. Setting the Renormalization Scale in QCD: The Principle of Maximum Conformality. *Phys. Rev. D* **2012**, *86*, 085026. [CrossRef]
82. Chawdhry, H.A.; Mitov, A. Ambiguities of the principle of maximum conformality procedure for hadron collider processes. *Phys. Rev. D* **2019**, *100*, 074013. [CrossRef]
83. Shen, J.M.; Wu, X.G.; Du, B.L.; Brodsky, S.J. Novel All-Orders Single-Scale Approach to QCD Renormalization Scale-Setting. *Phys. Rev. D* **2017**, *95*, 094006. [CrossRef]
84. Di Giustino, L.; Brodsky, S.J.; Wang, S.Q.; Wu, X.G. PMC$_\infty$: Infinite-Order Scale-Setting using the Principle of Maximum Conformality, A Remarkably Efficient Method for Eliminating Renormalization Scale Ambiguities for Perturbative QCD. *Phys. Rev. D* **2020**, *102*, 014015. [CrossRef]
85. Frixione, S.; Webber, B.R. Matching NLO QCD computations and parton shower simulations. *JHEP* **2002**, *6*, 029. [CrossRef]
86. Frixione, S.; Nason, P.; Ridolfi, G. A Positive-weight next-to-leading-order Monte Carlo for heavy flavour hadroproduction. *JHEP* **2007**, *09*, 126. [CrossRef]
87. Campbell, J.M.; Ellis, R. An Update on vector boson pair production at hadron colliders. *Phys. Rev. D* **1999**, *60*, 113006. [CrossRef]

88. Results—Centre for Precision Studies in Particle Physics. Available Online: http://www.precision.hep.phy.cam.ac.uk/results/ (accessed on 10 June 2020).
89. Berger, E.L.; Cao, Q.H.; Chen, C.R.; Yu, J.H.; Zhang, H. The Top Quark Production Asymmetries A_{FB}^t and A_{FB}^ℓ. *Phys. Rev. Lett.* **2012**, *108*, 072002. [CrossRef] [PubMed]
90. Westhoff, S. Top-Quark Asymmetry—A New Physics Overview. *PoS EPS-HEP2011* **2011**, 377. [CrossRef]
91. ATLAS and CMS Collaboration. Combination of inclusive and differential $t\bar{t}$ charge asymmetry measurements using ATLAS and CMS data at $\sqrt{s} = 7$ and 8 TeV. *JHEP* **2018**, *4*, 033. [CrossRef]
92. CDF Collaboration. Measurement of the J/ψ meson and b–hadron production cross sections in $p\bar{p}$ collisions at $\sqrt{s} = 1960$ GeV. *Phys. Rev. D* **2005**, *71*, 032001. [CrossRef]
93. D0 Collaboration. The Upgraded D0 detector. *Nucl. Instrum. Meth. A* **2006**, *565*, 463–537. [CrossRef]
94. ATLAS Collaboration. The ATLAS Experiment at the CERN Large Hadron Collider. *JINST* **2008**, *3*, S08003. [CrossRef]
95. CMS Collaboration. The CMS Experiment at the CERN LHC. *JINST* **2008**, *3*, S08004. doi:10.1088/1748-0221/3/08/S08004. [CrossRef]
96. LHCb Collaboration. First observation of top quark production in the forward region. *Phys. Rev. Lett.* **2015**, *115*, 112001, [CrossRef] [PubMed]
97. CDF Collaboration. Measurement of the top quark forward-backward production asymmetry and its dependence on event kinematic properties. *Phys. Rev. D* **2013**, *87*, 092002, [CrossRef]
98. CDF Collaboration. Measurement of the Differential Cross Section $d\sigma/d(\cos\theta t)$ for Top-Quark Pair Production in $p-\bar{p}$ Collisions at $\sqrt{s} = 1.96$ TeV. *Phys. Rev. Lett.* **2013**, *111*, 182002, [CrossRef]
99. CDF Collaboration. Measurement of the Leptonic Asymmetry in $t\bar{t}$ Events Produced in $p\bar{p}$ Collisions at $\sqrt{s} = 1.96$ TeV. *Phys. Rev. D* **2013**, *88*, 072003; Erratum in **2016**, *94*, 099901. [CrossRef]
100. CDF Collaboration. Measurement of the inclusive leptonic asymmetry in top-quark pairs that decay to two charged leptons at CDF. *Phys. Rev. Lett.* **2014**, *113*, 042001; Erratum in **2016**, *117*, 199901. [CrossRef]
101. D0 Collaboration. Measurement of the Asymmetry in Angular Distributions of Leptons Produced in Dilepton $t\bar{t}$ Final States in $p\bar{p}$ Collisions at $\sqrt{s}$=1.96 TeV. *Phys. Rev. D* **2013**, *88*, 112002. [CrossRef]
102. D0 Collaboration. Measurement of the Forward-Backward Asymmetry in Top Quark-Antiquark Production in $p\bar{p}$ Collisions using the Lepton+Jets Channel. *Phys. Rev. D* **2014**, *90*, 072011. [CrossRef]
103. D0 Collaboration. Simultaneous measurement of forward-backward asymmetry and top polarization in dilepton final states from $t\bar{t}$ production at the Tevatron. *Phys. Rev. D* **2015**, *92*, 052007. [CrossRef]
104. D0 Collaboration. Measurement of the forward-backward asymmetry in the distribution of leptons in $t\bar{t}$ events in the lepton+jets channel. *Phys. Rev. D* **2014**, *90*, 072001. [CrossRef]
105. ATLAS Collaboration. Measurement of the top quark pair production charge asymmetry in proton-proton collisions at $\sqrt{s}$ = 7 TeV using the ATLAS detector. *JHEP* **2014**, *02*, 107. [CrossRef]
106. ATLAS Collaboration. Measurement of the charge asymmetry in dileptonic decays of top quark pairs in pp collisions at $\sqrt{s}$ = 7 TeV using the ATLAS detector. *JHEP* **2015**, *05*, 061. [CrossRef]
107. CMS Collaboration. Inclusive and Differential Measurements of the $t\bar{t}$ Charge Asymmetry in Proton-Proton Collisions at $\sqrt{s}$ = 7 TeV. *Phys. Lett. B* **2012**, *717*, 129–150. [CrossRef]
108. CMS Collaboration. Measurements of the $t\bar{t}$ charge asymmetry using the dilepton decay channel in pp collisions at $\sqrt{s}$ = 7 TeV. *JHEP* **2014**, *04*, 191. [CrossRef]
109. ATLAS Collaboration. Measurement of the charge asymmetry in top-quark pair production in the lepton-plus-jets final state in pp collision data at $\sqrt{s}$ = 8 TeV with the ATLAS detector. *Eur. Phys. J. C* **2016**, *76*, 87; Erratum in **2017**, *77*, 564. [CrossRef]
110. ATLAS Collaboration. Measurement of the charge asymmetry in highly boosted top-quark pair production in $\sqrt{s}$ = 8 TeV pp collision data collected by the ATLAS experiment. *Phys. Lett. B* **2016**, *756*, 52–71. [CrossRef]
111. ATLAS Collaboration. Measurements of the charge asymmetry in top-quark pair production in the dilepton final state at $\sqrt{s}$ = 8 TeV with the ATLAS detector. *Phys. Rev. D* **2016**, *94*, 032006. [CrossRef]
112. CMS Collaboration. Inclusive and differential measurements of the $t\bar{t}$ charge asymmetry in pp collisions at $\sqrt{s}$ = 8 TeV. *Phys. Lett. B* **2016**, *757*, 154–179. [CrossRef]
113. CMS Collaboration. Measurement of the charge asymmetry in top quark pair production in pp collisions at $\sqrt{(s)}$ = 8 TeV using a template method. *Phys. Rev. D* **2016**, *93*, 034014. [CrossRef]
114. CMS Collaboration. Measurements of $t\bar{t}$ charge asymmetry using dilepton final states in pp collisions at $\sqrt{s}$ = 8 TeV. *Phys. Lett. B* **2016**, *760*, 365–386. [CrossRef]

115. CMS Collaboration. Measurement of the top quark forward-backward production asymmetry and the anomalous chromoelectric and chromomagnetic moments in pp collisions at $\sqrt{s} = 13$ TeV. *JHEP* **2020**, *6*, 146. [CrossRef]
116. CMS Collaboration. Measurements of $t\bar{t}$ differential cross sections in proton-proton collisions at $\sqrt{s} = 13$ TeV using events containing two leptons. *JHEP* **2019**, *2*, 149. [CrossRef]
117. Brooijmans, G.; Gripaios, B.; Moortgat, F.; Santiago, J.; Skands, P.; Albornoz Vásquez, D.; Allanach, B. C.; Alloul, A.; Arbey, A.; Azatov, A.; et al. Les Houches 2011: Physics at TeV Colliders New Physics Working Group Report. In Proceedings of the 7th Les Houches Workshop on Physics at TeV Colliders, Les Houches, France, 30 May–17 June 2011; pp. 221–463.
118. Aguilar-Saavedra, J.; Bernreuther, W.; Si, Z. Collider-independent top quark forward-backward asymmetries: standard model predictions. *Phys. Rev. D* **2012**, *86*, 115020. [CrossRef]
119. Grzadkowski, B.; Iskrzynski, M.; Misiak, M.; Rosiek, J. Dimension-Six Terms in the Standard Model Lagrangian. *JHEP* **2010**, *10*, 85. [CrossRef]
120. Zhang, C.; Willenbrock, S. Effective-Field-Theory Approach to Top-Quark Production and Decay. *Phys. Rev. D* **2011**, *83*, 034006. [CrossRef]
121. Azzi, P.; Farry, S.; Nason, P.; Tricoli, A.; Zeppenfeld, D.; Abdul Khalek, R.; Alimena, J.; Andari, N.; Aperio Bella, L.; Armbruster, A. J.; et al. Report from Working Group 1: Standard Model Physics at the HL-LHC and HE-LHC. *CERN Yellow Rep. Monogr.* **2019**, *7*, 1–220. [CrossRef]
122. Basan, A.; Berta, P.; Masetti, L.; Vryonidou, E.; Westhoff, S. Measuring the top energy asymmetry at the LHC: QCD and SMEFT interpretations. *JHEP* **2020**, *3*, 184. [CrossRef]
123. FCC Collaboration. FCC Physics Opportunities: Future Circular Collider Conceptual Design Report Volume 1. *Eur. Phys. J. C* **2019**, *79*, 474. [CrossRef]
124. Maltoni, F.; Mangano, M.; Tsinikos, I.; Zaro, M. Top-quark charge asymmetry and polarization in $t\bar{t}W^{\pm}$ production at the LHC. *Phys. Lett. B* **2014**, *736*, 252–260. [CrossRef]
125. Amjad, M. S.; Bilokin, S.; Boronat, M.; Doublet, P.; Frisson, T.; García, I. G.; Perelló, M.; Pöschl, R.; Richard, F.; Ros, E.; et al. A precise characterisation of the top quark electro-weak vertices at the ILC. *Eur. Phys. J. C* **2015**, *75*, 512. [CrossRef]
126. CLICdp Collaboration. Top-Quark Physics at the CLIC Electron-Positron Linear Collider. *JHEP* **2019**, *11*, 003. [CrossRef]

© 2020 by the author. Licensee MDPI, Basel, Switzerland. This article is an open access article distributed under the terms and conditions of the Creative Commons Attribution (CC BY) license (http://creativecommons.org/licenses/by/4.0/).

Article

Asymmetries in Processes of Electron–Positron Annihilation

Andrej Arbuzov [1,2,*,†], Serge Bondarenko [1,†] and Lidia Kalinovskaya [3,†]

1. Bogoliubov Laboratory of Theoretical Physics, JINR, Joliot-Curie str. 6, 141980 Dubna, Russia; bondarenko@jinr.ru
2. Dubna State University, Universitetskaya str. 19, 141982 Dubna, Russia
3. Dzhelepov Laboratory of Nuclear Problems, JINR, Joliot-Curie str. 6, 141980 Dubna, Russia; lidia.kalinovskaya@cern.ch
* Correspondence: arbuzov@theor.jinr.ru
† These authors contributed equally to this work.

Received: 14 June 2020; Accepted: 6 July 2020; Published: 7 July 2020

Abstract: Processes of electron–positron annihilation into a pair of fermions were considered. Forward–backward and left–right asymmetries were studied, taking into account polarization of initial and final particles. Complete 1-loop electroweak radiative corrections were included. A wide energy range including the Z boson peak and higher energies relevant for future e^+e^- colliders was covered. Sensitivity of observable asymmetries to the electroweak mixing angle and fermion weak coupling was discussed.

Keywords: high energy physics; electron–positron annihilation; forward–backward asymmetry; left–right asymmetry

PACS: 12.15.-y; 12.15.Lk; 13.66.Jn

1. Introduction

Symmetries play a key role in the construction of physical theories. In fact, they allow us to describe a huge variety of observables by means of compact formulae. We believe that the success of theoretical models based on symmetry principles is due to the presence of the corresponding properties in Nature. The Standard Model (SM) is the most successful physical theory ever. Its predictions are in excellent agreement with practically all experimental results in particle physics. The renormalizability of the model allows us to preserve unitarity and provide finite verifiable results. Both phenomenological achievements and nice theoretical features of the SM are mainly due to the extended usage of symmetries in its construction. The model is based on several symmetries of different type, including the Lorentz (Poincaré) symmetry, the gauge $SU(3)_C \times SU(2)_L \times U(1)_Y$ symmetries, the CPT symmetry, the spontaneously broken global $SU(2)_L \times SU(2)_R$ symmetry in the Higgs sector, etc. Some symmetries of the model are exact (or seem to be exact within the present precision) while others are spontaneously or explicitly broken. In particular, the nature of the symmetry among the three generations of fermions is one of the most serious puzzles in the SM and verification of the lepton universality hypothesis is on the task list of modern experiments.

Despite the great successes of the SM, we can hardly believe that it is the true fundamental theory of Nature. Most likely, it is an effective model with a limited applicability domain. The search for the upper energy limit of the SM applicability is the actual task at all high-energy colliders experiments. Up to now, all direct attempts to find elementary particles and interactions beyond the Standard Model have failed. The accent of experimental studies has shifted towards accurate verification of the SM features. Deep investigation of the SM symmetries is an important tool in this line of research.

Asymmetries form a special class of experimental observables. First of all, they explicitly access the breaking of a certain symmetry in Nature. Second, they are usually constructed as a ratio of observed quantities, in which the bulk of experimental and theoretical systematic uncertainties is canceled out. So the asymmetries provide independent additional information on particle interactions. They are especially sensitive to non-standard weak interactions including contributions of right currents and new intermediate Z' vector bosons, see e.g., [1].

The physical programs of future (super) high-energy electron–positron colliders such as CLIC [2], ILC [3–5], FCC-ee [6], and CEPC [7] necessarily include accurate tests of the SM. Studies of polarization effects and asymmetries will be important to probe of the fundamental properties of Higgs boson(s) and, in particular, in the process of annihilation into top quarks [8–10]. The future colliders plan to start operation in the so-called GigaZ mode at the Z peak and improve upon the LEP both in statistical and systematical uncertainties in tests of the SM [11] by at least one order of magnitude. Among these collider projects, the FCC-ee one has the most advanced program of high-precision measurements of SM processes at the Z peak. Such tests have been performed at LEP and SLC and they have confirmed the validity of the SM at the electroweak (EW) energy scale of about 100 GeV [12,13]. During the LEP era, extensive experimental and theoretical studies of asymmetries made an important contribution to the overall verification of the SM, see review [14] and references therein. The new precision level of future experiments motivates us to revisit the asymmetries and scrutinize the effects of radiative corrections (RCs) to them. In the analysis of LEP data, semi-analytic computer codes like ZFITTER [15] and TOPAZ0 [16] were extensively used. The forthcoming new generation of experiment requires more advanced programs, primarily Monte Carlo event generators.

The article is organized as follows. The next section contains preliminary remarks and the general notations. Section 3 is devoted to the left–right asymmetry. The forward–backward asymmetry is considered in Section 4. Discussion of the left–right forward–backward asymmetry is presented in Section 5. In Section 6, we provide results related to the final state fermion polarization. Section 7 contains a discussion and conclusions.

2. Preliminaries and Notations

In the recent paper [17] by the SANC group, high-precision theoretical predictions for the process $e^+e^- \to l^+l^-$ ($l = \mu$ or τ) were presented. With the help of computer system SANC [18], we calculated the complete 1-loop electroweak radiative corrections to these processes, taking into account possible longitudinal polarization of the initial beams. The calculations were performed within the helicity amplitude formalism, taking into account the initial and final state fermion masses. So, the SANC system provides a solid framework to access asymmetries in e^+e^- annihilation processes and to study various relevant effects. In particular, the system allows us to separate effects due to quantum electrodynamics (QED) and weak radiative corrections.

The focus of this article is on the description and assessment of the asymmetry family: the left–right asymmetry A_{LR}, the forward–backward asymmetry A_{FB}, the left–right forward–backward asymmetry A_{LRFB}, and the final state fermion polarization P_τ in collisions of high-energy polarized or unpolarized e^+e^- beams. The main aim was to verify the effect of radiative corrections on the extraction of the SM parameters from the asymmetries and to analyze the corresponding theoretical uncertainty.

We performed calculations for polarized initial and final state particles. Beam polarizations play an important role:

- They improve the sensitivity to CP-violating anomalous couplings or form factors, which are measurable even with unpolarized beams through the forward–backward asymmetry.
- With the polarization of both beams, the sensitivity to the new physics scale can be increased by a factor of up to 1.3 with respect to the case with only polarized electrons [1].
- A high-luminosity at the GigaZ stage of a collider running at the Z boson resonance with positron polarization allows us to improve the accuracy of the determination of $\sin^2 \vartheta_W$ (ϑ_W is

the electroweak mixing angle) by an order of magnitude, through studies of the left–right asymmetry [1].

Numerical illustrations for each asymmetry are given in two energy domains: the wide center-of-mass energy range $20 \leq \sqrt{s} \leq 500$ GeV and the narrow one around the Z resonance ($70 \leq \sqrt{s} \leq 100$ GeV), where a peculiar behavior of observables can be seen. All results were produced with the help of the e^+e^- branch [19] of the MCSANC Monte Carlo integrator [20].

Let us introduce the notation. First of all, we define quantities A_f ($f = e, \mu, \tau$) which are often used for description of asymmetries at the Z peak:

$$A_f \equiv 2 \frac{g_{V_f} g_{A_f}}{g_{V_f}^2 + g_{A_f}^2} = \frac{1 - (g_{R_f}/g_{L_f})^2}{1 + (g_{R_f}^2/g_{L_f}^2)^2}, \tag{1}$$

where the vector and axial-vector coupling constants of the weak neutral current of the fermion f with the electromagnetic charge q_f (in the units of the positron charge e) are

$$g_{V_f} \equiv I_f^3 - 2q_f \sin^2 \theta_W, \qquad g_{A_f} \equiv I_f^3. \tag{2}$$

The corresponding left and right fermion couplings are

$$g_{L_f} \equiv I_f^3 - q_f \sin^2 \theta_W, \qquad g_{R_f} \equiv -q_f \sin^2 \theta_W. \tag{3}$$

The neutral current couplings g_{L_f} and g_{R_f} quantify the strength of the interaction between the Z boson and the given chiral states of the fermion.

We claim that there are sizable corrections to all observable asymmetries due to radiative corrections which affect simple Born-level analytic formulae relating the asymmetries with electroweak parameters. It is especially interesting to consider the behavior of asymmetries in different EW schemes: $\alpha(0)$, $\alpha(M_Z^2)$, and G_μ, see their definitions below. We also will compare the results in the Born and 1-loop approximation. The latter means inclusion of 1-loop radiative corrections of one of the following types: pure QED photonic RCs (marked as "QED"), weak RCs (marked as "weak"), and the complete 1-loop electroweak RCs (marked as "EW"):

$$\sigma_{EW} = \sigma_{Born} + \sigma_{QED} + \sigma_{weak}.$$

The weak part in our notation includes 1-loop self-energy corrections to photon and Z boson propagators. In our notation, higher-order effects due to interference of pure QED and weak contributions are a part of σ_{weak}.

The cross section of a generic annihilation process of longitudinally polarized e^+ and e^- with polarization degrees P_{e^+} and P_{e^-} can be expressed as follows:

$$\begin{aligned}\sigma(P_{e^-}, P_{e^+}) &= (1 + P_{e^-})(1 + P_{e^+})\sigma_{RR} + (1 - P_{e^-})(1 + P_{e^+})\sigma_{LR} \\ &+ (1 + P_{e^-})(1 - P_{e^+})\sigma_{RL} + (1 - P_{e^-})(1 - P_{e^+})\sigma_{LL}.\end{aligned} \tag{4}$$

Here $\sigma_{ab} = \sum_{ij(k)} |\mathcal{H}_{abij(k)}|^2$ are the $2 \to 2(3)$ helicity amplitudes of the reaction, ($ab = RR, RL, LR, LL$) with right-handed R="+" or left-handed L="−" initial particles.

It is convenient to combine the electron P_{e^-} and positron P_{e^+} polarizations into the effective quantity

$$P_{eff} = \frac{P_{e^-} - P_{e^+}}{1 - P_{e^-} P_{e^+}}. \tag{5}$$

In the case when only the electron beam is polarized, the effective polarization coincides with the electron one.

To investigate theoretical uncertainties, we use the following three EW schemes:

1. the $\alpha(0)$ scheme in which the fine-structure constant $\alpha(0)$ is used as input. The contribution of RCs in this scheme is enhanced by the large logarithms of light fermion masses via $\alpha(0)\ln(s/m_f^2)$ terms.
2. The $\alpha(M_Z^2)$ scheme in which the effective electromagnetic constant $\alpha(M_Z^2)$ is used at Born level while virtual 1-loop and real photon bremsstrahlung contributions are proportional to $\alpha^2(M_Z^2)\alpha(0)$. In this scheme the virtual RCs receive contributions from the quantity $\Delta\alpha(M_Z^2)$ which describes the evolution of the electromagnetic coupling from the scale $Q^2 = 0$ to the $Q^2 = M_Z^2$ one and cancels the large terms with logarithms of light fermion masses.
3. the G_μ scheme in which the Fermi coupling constant G_μ, extracted from the muon life time, is used at the Born level while the virtual 1-loop and real photon bremsstrahlung contributions are proportional to $G_\mu^2\alpha(0)$. The virtual RCs receive contributions from the quantity Δr. Since the expression for Δr contains the $\Delta\alpha(M_Z^2)$, the large terms with logarithms of the light masses are also canceled. The quantity Δr rules the G_μ and $\alpha(0)$ relation in this scheme.

Results of fixed-order perturbative calculations in these schemes differ due to missing higher-order effects. In what follows, numerical calculations are performed in the $\alpha(0)$ EW scheme if another choice is not explicitly indicated.

3. Left–Right Asymmetry A_{LR}

A scheme to measure the A_{LR} polarization asymmetry at the Z peak was suggested in [21]. It was shown that this observable can be used as for extraction of electroweak couplings as well as for a polarimeter calibration.

If we neglect the initial electron masses, the polarized cross-section can be rewritten in the following form:

$$\sigma(P_{e^-}, P_{e^+}) = (1 - P_{e^-}P_{e^+})[1 - P_{eff}A_{LR}]\sigma_0, \quad (6)$$

where σ_0 is the unpolarized cross-section.

The left–right asymmetry in the presence of partially polarized ($|P_{eff}| < 1$) initial beams is defined as

$$A_{LR} = \frac{1}{P_{eff}} \frac{\sigma(-P_{eff}) - \sigma(P_{eff})}{\sigma(-P_{eff}) + \sigma(P_{eff})}, \quad (7)$$

where σ is the cross-section with polarization P_{eff}.

In the case of fully polarized initial particles ($|P_{e^\pm}| = 1$) the definition (7) becomes:

$$A_{LR} = \frac{\sigma_{L_e} - \sigma_{R_e}}{\sigma_{L_e} + \sigma_{R_e}}, \quad (8)$$

where L_e and R_e refer to the left and right helicity states of the incoming electron.

Equations (6) and (7) show that A_{LR} does not depend on the degree of the initial beam polarization. This type of asymmetry is sensitive to weak interaction effects in the initial vertex. In the Born approximation at energies close to the Z resonance, it is directly related to the electron coupling:

$$A_{LR} \approx A_e. \quad (9)$$

The left–right asymmetry A_{LR} as a function of the center-of-mass system (c.m.s.) energy in the ranges $20 \leq \sqrt{s} \leq 500$ GeV (Left) and $70 \leq \sqrt{s} \leq 110$ GeV (Right) is shown in Figure 1. We explore A_{LR} in different approximations and the corresponding shifts ΔA_{LR} between the Born level and 1-loop corrected approximations taking into account either pure QED, or weak, or complete EW effects: $\Delta A_{LR} = A_{LR}$(1-loop corrected)$-A_{LR}$(Born). The right figure shows the behavior of A_{LR} near the Z resonance, and the value A_e at $\sqrt{s} = M_Z$ is indicated by a black dot (see (9)).

One can notice that although the total 1-loop EW corrections to the process cross-section are equal to the sum of the pure QED and weak ones, the corresponding shifts ΔA_{LR} are not additive. That is because the asymmetry is defined as a ratio and the corrections affect both the numerator and denominator.

In Figure 2 we show A_{LR} for the Born and weak 1-loop corrected levels of accuracy in different EW schemes and the corresponding shifts $\Delta A_{LR}=A_{LR}$(weak, some EW scheme)-A_{LR}(Born). We see that the effects due to weak corrections in different EW schemes behave in a similar way. Nevertheless the scheme dependence is visible within the expected precision of future measurements. The deviations between the results in different schemes can be treated as a contribution into the theoretical uncertainty due to missing higher order corrections.

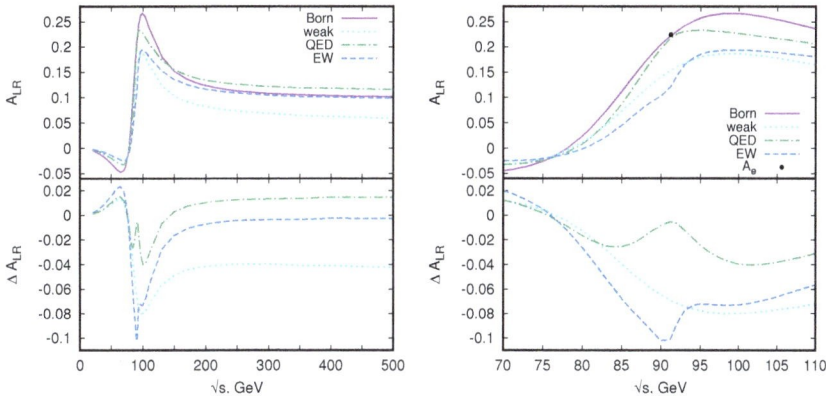

Figure 1. (**Left**) The A_{LR} asymmetry in the Born and 1-loop (weak, pure quantum electrodynamics (QED), and electroweak (EW)) approximations and ΔA_{LR} vs. center-of-mass system (c.m.s.) energy in a wide range; (**Right**) the same for the Z peak region.

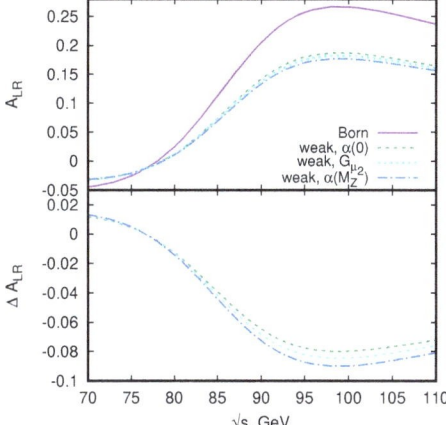

Figure 2. The A_{LR} asymmetry at the Born level and with 1-loop weak radiative corrections (RCs); the corresponding shifts ΔA_{LR} within $\alpha(0)$, G_μ, and $\alpha(M_Z^2)$ EW schemes vs. c.m.s. energy in the peak region.

The impact of 1-loop EW contributions to ΔA_{LR} is of the order -0.1 in the resonance region, but at energies above $\sqrt{s} = 200$ GeV there are considerable cancellations between weak and QED effects so that the combined EW corrections becomes small (but still numerically important for high-precision measurements).

Summary for A_{LR}

The left–right asymmetry A_{LR} is almost insensitive to the details of particle detection since the corresponding experimental uncertainties tend to cancel out in the ratio (7). It (almost) does not depend on the final state fermion couplings in the vicinity of the Z boson peak and can be measured for any final state with a large gain in statistics. For this reasons it is appropriate for extraction of the $\sin^2 \vartheta_W^{\text{eff}}$ value.

We observe that the values ΔA_{LR} due to weak and pure QED 1-loop corrections are very significant at high energies in general, but in the resonance region impact of QED is small, while the weak contribution to ΔA_{LR} reaches 0.07. Therefore, it is necessary to evaluate all possible radiative correction contributions to the weak parts of RCs carefully and thoroughly.

4. Forward–Backward Asymmetry A_{FB}

The forward–backward asymmetry is defined as

$$A_{FB} = \frac{\sigma_F - \sigma_B}{\sigma_F + \sigma_B},$$

$$\sigma_F = \int_0^1 \frac{d\sigma}{d\cos\vartheta_f} d\cos\vartheta_f, \quad \sigma_B = \int_{-1}^0 \frac{d\sigma}{d\cos\vartheta_f} d\cos\vartheta_f, \quad (10)$$

where ϑ_f is the angle between the momenta of the incoming electron and the outgoing negatively charged fermion. It can be measured in any $e^+e^- \to f\bar{f}$ channels but for precision test the most convenient channels are $f = e, \mu$. The channels with production of τ leptons, b or c quarks are very interesting as well.

At the Born level, this asymmetry is proportional to the product of initial and final state couplings and is caused by parity violation at both production and decay vertices:

$$A_{FB} \approx \frac{3}{4} A_e A_f. \quad (11)$$

In the case of partially polarized initial beams the condition (11) reduces to the following one

$$A_{FB} \approx \frac{3}{4} \frac{A_e - P_{\text{eff}}}{1 - A_e P_{\text{eff}}} A_f. \quad (12)$$

In Figure 3 we show the behavior of the A_{FB} asymmetry in the Born and 1-loop approximations (with weak, pure QED, or complete EW contributions) and the corresponding ΔA_{FB} for c.m.s. energy range $20 \leq \sqrt{s} \leq 500$ GeV in the left plot and for the Z peak region of c.m.s. energy $70 \leq \sqrt{s} \leq 110$ GeV in the right one. As in the previous case of A_{LR}, we indicate by a black dot the value $A_{FB} \approx 3/4 A_e A_\mu$ at the resonance. We observe that the weak contribution to A_{FB} is small and practically does not depend on energy. The shift ΔA_{FB} changes the sign at the resonance and tends to a constant value (~ -0.3) above 200 GeV. The huge magnitude of the shift ΔA_{FB} out of the Z resonance region is coming mainly from the pure QED corrections. In particular, above the peak the effect due to radiative return to the resonance is very important.

Figure 4 shows the dependence of A_{FB} for different levels of accuracy (Born and 1-loop weak) on the EW scheme choice: either $\alpha(0)$, or G_μ, or $\alpha(M_Z^2)$. The corresponding shifts ΔA_{FB} between the Born and the 1-loop weak corrected approximations are shown in the lower plot.

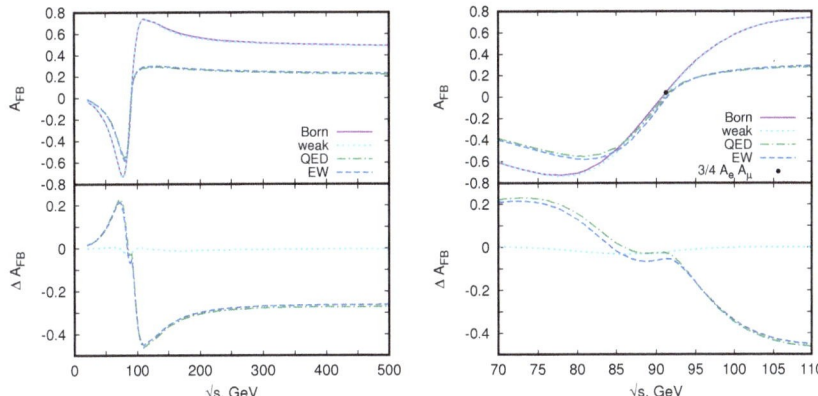

Figure 3. (**Left**) The A_{FB} asymmetry in the Born and 1-loop (weak, QED, EW) approximations and the corresponding shifts ΔA_{FB} for a wide c.m.s. energy range; (**Right**) the same for the Z peak region.

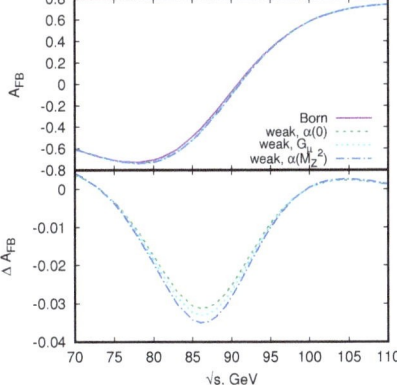

Figure 4. The A_{FB} asymmetry and ΔA_{FB} in the Born and complete 1-loop EW approximations within the $\alpha(0)$, G_μ, and $\alpha(M_Z^2)$ EW schemes vs. the c.m.s energy.

Below we investigate two sets of polarization degree $P_i = (P_{e^-}, P_{e^+})$:

$$P_1 = (-0.8, 0.3) \quad \text{and} \quad P_2 = (0.8, -0.3). \tag{13}$$

In Figure 5 we compare the values of A_{FB} asymmetry and the corresponding shifts due to EW corrections for the unpolarized case and two choices of polarized beams defined in the above equation. One can see that a combination of polarization degrees of initial particles can either increase or decrease the magnitude of the A_{FB} asymmetry with respect to the unpolarized case.

There is an interesting idea [22] to use the A_{FB} asymmetry at the FCC-ee in order to directly access the value of QED running coupling at M_Z. This idea was supported in [23] where it was demonstrated that higher-order QED radiative corrections to A_{FB} are under control. Our results show that higher-order effects due to weak interactions are not negligible in this observable; further studies are required.

At the Born level there are contributions suppressed by the small factor m_f^2/s with the fermion mass squared. It is interesting to note that in 1-loop radiative corrections there are contributions of the relative order $\alpha \cdot m_f/\sqrt{s}$ with the fermion mass to the first power [24], which are numerically relevant at high energies especially for the b quark channel.

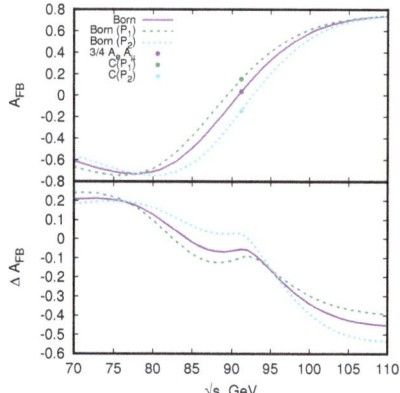

Figure 5. The A_{FB} asymmetry at the Born level (upper panel) and the corresponding ΔA_{FB} in the 1-loop EW approximation (bottom panel) for unpolarized and polarized cases with degrees of beam polarizations $P_{1,2}$ (13) vs. c.m.s. energy in the Z peak region. The constants $C(P_{1,2})$ stand for the expression (12) with polarization degrees (13).

Summary for A_{FB}

The weak 1-loop contribution ΔA_{FB} is rather small for the whole energy range, see Figure 3. Nevertheless in this asymmetry the difference between the pure QED and the complete 1-loop approximations near the resonance is numerically important. The dependence on the EW scheme choice, see Figure 4, is small but still relevant for high-precision measurements. The dependence of this asymmetry on polarization is very significant.

5. Left–Right Forward–Backward Asymmetry A_{LRFB}

In order to measure the weak couplings of the final state fermions, it was suggested to analyze the so-called left–right forward–backward asymmetry [25]:

$$A_{\text{LRFB}} = \frac{(\sigma_{L_e} - \sigma_{R_e})_F - (\sigma_{L_e} - \sigma_{R_e})_B}{(\sigma_{L_e} + \sigma_{R_e})_F + (\sigma_{L_e} + \sigma_{R_e})_B}, \tag{14}$$

where σ_L and σ_R are the cross sections with left and right handed helicities of the initial electrons.

From the definition (14) it follows that A_{LRFB} partially inherits the properties of the A_{LR} and, in particular, does not depend on the degree of the initial beam polarizations.

In the case of unpolarized beams on the Z resonance peak, the Born-level asymmetry is

$$A_{\text{LRFB}} \approx \frac{3}{4} A_f. \tag{15}$$

In Figure 6 we present the predictions for the A_{LRFB} asymmetry in several approximations, namely at the Born level and with 1-loop weak, pure QED, and complete EW contributions.

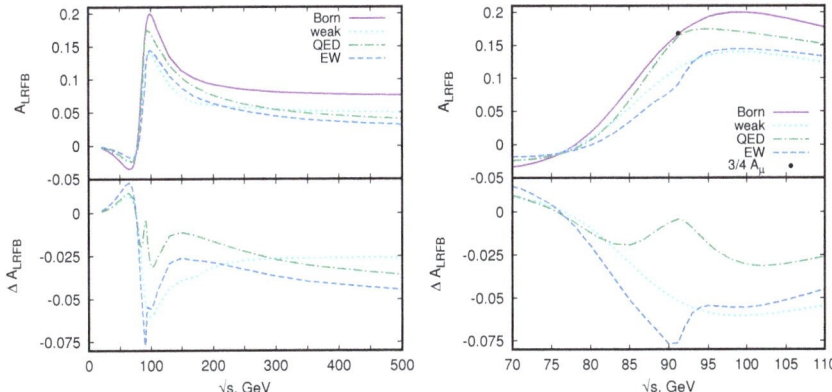

Figure 6. (**Left**) The A_{LRFB} asymmetry in the Born and 1-loop (weak, QED, EW) approximations and ΔA_{LRFB} for c.m.s. energy range; (**Right**) the same for the Z peak region.

Next, we repeat the study of the A_{LRFB} asymmetry behavior in different EW schemes. We have illustrated the energy dependence of the A_{LRFB} asymmetry in $\alpha(0)$, G_μ, and $\alpha(M_Z^2)$ schemes and the corresponding ΔA_{LRFB} in Figure 7. The impact of weak corrections on A_{LRFB} is large. For example, the Born-level value of A_{LRFB} at the Z peak is about 0.17, while accounting for the weak RCs contribution reduces the asymmetry value down to ~ 0.11.

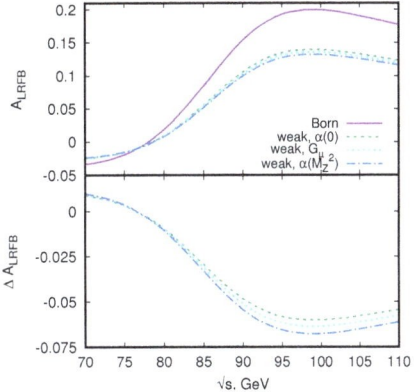

Figure 7. The A_{LRFB} asymmetry in the Born and 1-loop EW approximations and ΔA_{LRFB} within $\alpha(0)$, G_μ, and $\alpha(M_Z^2)$ EW schemes vs. c.m.s. energy in the Z peak region.

Summary for A_{LRFB}

We would like to emphasize that the above Formula (15) appears to be a rather rough approximation since radiative corrections shift the observable value of A_{LRFB} quite a lot. Apparently the A_{LRFB} asymmetry is more affected by weak corrections than A_{LR}. The shifts ΔA_{LRFB} only slightly depend on an EW scheme choice. The A_{LRFB} asymmetry at the Z boson peak depends on the final lepton coupling that could be used to measure the μ and τ weak couplings and their difference from the initial lepton (electron) one.

6. Final-State Fermion Polarization P_f

The polarization of a final-state fermion $P_{f=\mu,\tau}$ can be expressed as the ratio between the difference of the cross sections for right and left handed final state helicities and their sum

$$P_f = \frac{\sigma_{R_f} - \sigma_{L_f}}{\sigma_{R_f} + \sigma_{L_f}}. \qquad (16)$$

In an experiment, it can be measured for the $\tau^+\tau^-$ channel by reconstructing the τ polarization from the pion spectrum in the decay $\tau \to \pi\nu$. Details of the analysis of P_τ measurements at LEP are described in [13]. Computer programs TAOLA [26] and KORALZ [27,28] were applied for this analysis. Estimated improvement for P_τ and τ decay products over LEP time in ILC in the GigaZ program was done in [5].

In the case for unpolarized beams in the vicinity of the Z peak, the expression for channel $e^+e^- \to \tau^+\tau^-$ is simplified to

$$P_\tau(\cos\vartheta_\tau) \approx -\frac{A_\tau + \dfrac{2\cos\vartheta_\tau}{1+\cos^2\vartheta_\tau} A_e}{1 + \dfrac{2\cos\vartheta_\tau}{1+\cos^2\vartheta_\tau} A_e A_\tau}. \qquad (17)$$

From this observable, one can extract information on the couplings A_τ and A_e, simultaneously.

In Figure 8 (left) we show the distribution of P_τ in the cosine of the scattering angle at the Z peak in the Born and 1-loop (weak, QED, and EW) approximations. The same conventions as in previous sections are applied for the shifts ΔP_τ. The shift due to pure QED RCs is approximately a constant close to zero. But one can see that this observable is very sensitive to the presence of weak-interaction corrections.

In the presence of initial beams polarization the expression depends on P_{eff}:

$$P_\tau(\cos\vartheta) \approx -\frac{A_\tau(1 - A_e P_{\text{eff}}) + \dfrac{2\cos\vartheta_\tau}{(1+\cos^2\vartheta_\tau)}(A_e - P_{\text{eff}})}{(1 - A_e P_{\text{eff}}) + \dfrac{2\cos\vartheta_\tau}{(1+\cos^2\vartheta_\tau)} A_\tau(A_e - P_{\text{eff}})}. \qquad (18)$$

which can be reduced to the short form neglecting the $A_e A_\tau$ and $A_e P_{\text{eff}}$ terms:

$$P_\tau(\cos\vartheta_\tau) \approx -A_\tau - \frac{2\cos\vartheta_\tau}{(1+\cos^2\vartheta_\tau)}(A_e - P_{\text{eff}}). \qquad (19)$$

The influence of the initial particle polarization on P_τ at the Z peak is demonstrated in the Figure 8 (right). For comparison the unpolarized and two polarized cases (13) as functions of $\cos\vartheta_\tau$ are shown. It is seen that the behavior of P_τ depends on the polarization set choices very much, note that it even changes the sign for the P_2 case. The corresponding shifts ΔP_τ also strongly depend on the initial beam polarization degrees and change the shape accordingly (note the maximum for P_1).

In Figure 9 we show the dependence of P_τ on the c.m.s. energy in the Born and 1-loop approximations (weak, QED, and EW). We see that at energies above the Z resonance, both weak and QED radiative corrections to P_τ are large and considerable cancellations happen between their contributions. Note that theoretical uncertainties in weak and QED RCs are not correlated, so it is necessary to take into account higher-order effects to reduce the resulting uncertainty in the complete 1-loop result for P_τ at high energies.

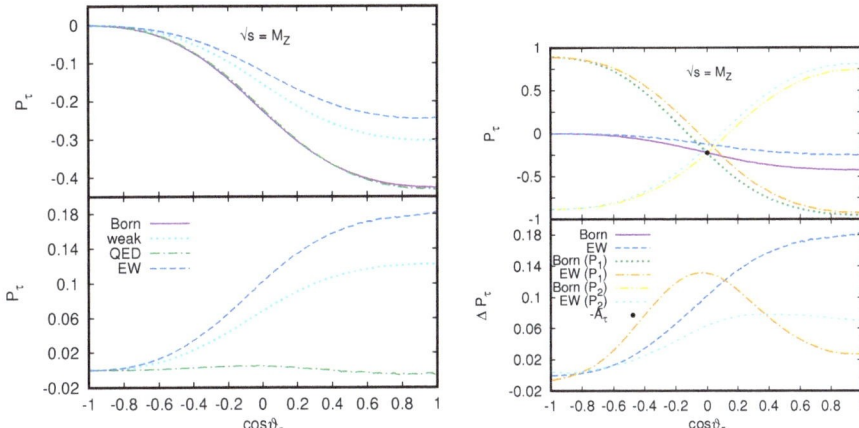

Figure 8. (**Left**) The P_τ polarization in the Born and 1-loop (weak, pure QED, and EW) approximations as a function of $\cos\vartheta_\tau$ at $\sqrt{s} = M_Z$. (**Right**) The P_τ polarization for unpolarized and polarized cases with (13) degrees of initial beam polarizations in the Born and EW 1-loop approximations vs. cosine of the final τ lepton scattering angle at the Z peak.

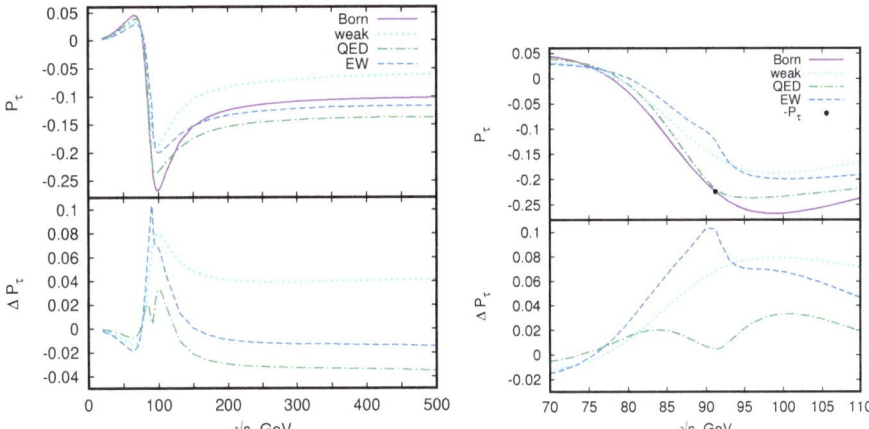

Figure 9. (**Left**) The P_τ polarization in the Born and 1-loop (weak, pure QED, and EW) approximations and ΔP_τ vs. c.m.s. energy in a wide range; (**Right**) the same for the Z peak region. The black dot indicates the value P_τ at the Z resonance.

In Figure 10 we show P_τ in the Born and 1-loop EW approximations for different sets of beam polarization degrees in a narrow bin around the Z resonance. The beam polarizations sets P_1 and P_2 are defined in Equation (13). One can see that the energy dependence of P_τ is strongly affected by a beam polarization choice outside the Z peak region. The same concerns the size of radiative corrections to P_τ, which are represented on the lower plot.

Summary for P_τ

The P_τ asymmetry is very sensitive to weak-interaction corrections and to the polarization degrees of the initial beams. Near the Z resonance the value of theoretical uncertainty of P_τ is determined by the interplay of uncertainties of rather large contributions pure QED and weak radiative corrections.

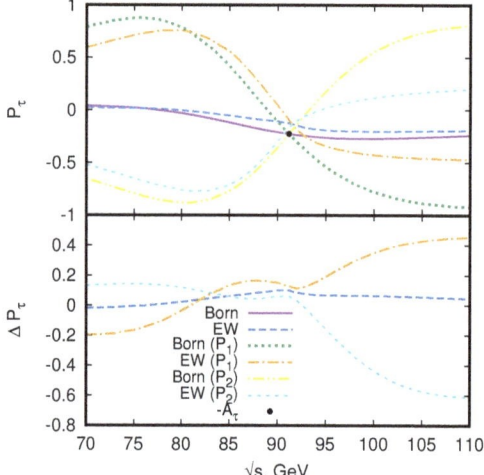

Figure 10. The P_τ polarization for (13) degrees of the initial beam polarizations in the Born and 1-loop EW approximations vs. c.m.s. energy in the Z peak region.

7. Conclusions

New opportunities of the future e^+e^- colliders: GigaZ options and new energy scale up to several TeV require modern tools for high-precision theoretical calculations of observables. We investigated A_{LR}, A_{FB} and A_{LRFB} for $e^+e^- \to \mu^+\mu^-$ channel and polarization P_τ for the final state in $e^+e^- \to \tau^+\tau^-$ channel on the Z resonance and in the high energy region up to 500 GeV by using MCSANC. We evaluated the resulting shifts of asymmetries at the Born and EW levels of accuracy in different EW schemes. The numerical results presented above for pure QED, weak, and complete EW radiative corrections show an interplay between the weak and QED contributions to asymmetries. This fact indicates the necessity to consider those contributions always in combined way.

Asymmetries in e^+e^- annihilation processes provide a powerful tool for investigation of symmetries between three fermion generations. By studying all available asymmetries, one can extract parameters of weak interactions in the neutral current for all three charged leptons. So, by comparing the parameters it will be possible to verify the lepton universality hypothesis at a new level of precision.

Hypothetical extra neutral Z' vector bosons [29] can contribute to the processes of e^+e^- annihilation. For example, effects of Kaluza–Klein excited vector bosons in the gauge Higgs unification on e^+e^- annihilation cross sections were considered in [30,31]. Since the new bosons can have couplings to left and right fermions being different from the SM ones, the asymmetries (especially with polarized beams) can help a lot in search for such Z' bosons.

At the FCC-ee we have experimental precision tag in the $\sin^2 \vartheta_W^{\text{eff}}$ measurement of the order of 5×10^{-6}, which means more than a thirty-fold improvement with respect to the current precision of 1.6×10^{-4}. This is due to a factor of several hundred improvement on statistical errors and because of a considerable improvement in particle identification and vertexing. In order to provide theoretical predictions for the considered asymmetries with sufficiently small uncertainties which would not spoil the precision of the future experiments besides the complete 1-loop EW radiative corrections presented here we need:

- higher order pure QED corrections preferably with resummation;
- higher order (electro)weak corrections;
- taking into account perturbative and nonperturbative quantum chromodynamics (QCD) effects in RCs;
- Monte Carlo event generators and integrators which ensure the required technical precision.

Challenges in calculations of higher order QED effects for FCC-ee were discussed in Ref. [32]. The complete two-loop electroweak corrections in the vicinity of the Z boson peak have been presented in [33]. More details on challenges for high-precision theoretical calculations for future e^+e^- colliders can be found in [34,35].

Author Contributions: Conceptualization, A.A., S.B. and L.K.; methodology, A.A., S.B. and L.K.; software, A.A., S.B. and L.K.; validation, A.A., S.B. and L.K.; formal analysis, A.A., S.B. and L.K.; investigation, A.A., S.B. and L.K.; resources, A.A., S.B. and L.K.; data curation, A.A., S.B. and L.K.; writing—original draft preparation, A.A., S.B. and L.K.; writing—review and editing, A.A., S.B. and L.K. The authors claim to have contributed equally and significantly in this paper. All authors have read and agreed to the published version of the manuscript.

Funding: This research was funded by RFBR grant 20-02-00441.

Acknowledgments: The authors are grateful to Ya. Dydyshka, R. Sadykov, V. Yermolchyk, and A. Sapronov for fruitful discussions and numerical cross checks, and to A. Kalinovskaya for the help with preparation of the manuscript.

Conflicts of Interest: The authors declare no conflict of interest.

Abbreviations

The following abbreviations are used in this manuscript:

SM	Standard Model
QED	quantum electrodynamics
QCD	quantum chromodynamics
EW	electroweak
RCs	radiative corrections
FB	forward–backward
LR	left–right
c.m.s.	center-of-mass system

References

1. Fujii, K.; Grojean, C.; Peskin, M.E.; Barklow, T.; Gao, Y.; Kanemura, S.; Kim, H.; List, J.; Nojiri, M.; Perelstein, M.; et al. Tests of the Standard Model at the International Linear Collider. *arXiv* **2019**, arXiv:1908.11299.
2. *A Multi-TeV Linear Collider Based on CLIC Technology: CLIC Conceptual Design Report*; SLAC National Accelerator Lab.: Menlo Park, CA, USA, 2012. [CrossRef]
3. Moortgat-Pick, G. The Role of polarized positrons and electrons in revealing fundamental interactions at the linear collider. *Phys. Rep.* **2008**, *460*, 131–243. [CrossRef]
4. The International Linear Collider Technical Design Report—Volume 2: Physics. *arXiv* **2013**, arXiv:1306.6352.
5. Bambade, P.; Barklow, T.; Behnke, T.; Berggren, M.; Brau, J.; Burrows, P.; Denisov, D.; Faus-Golfe, A.; Foster, B.; Fujii, K.; et al. The International Linear Collider: A Global Project. *arXiv* **2019**, arXiv:1903.01629.
6. Abada, A. FCC-ee: The Lepton Collider: Future Circular Collider Conceptual Design Report Volume 2. *Eur. Phys. J. ST* **2019**, *228*, 261–623. [CrossRef]
7. Ahmad, M.; Alves, D.; An, H.; An, Q.; Arhrib, A.; Arkani-Hamed, N.; Ahmed, I.; Bai, Y.; Ferroli, R.B.; Ban, Y.; et al. CEPC-SPPC Preliminary Conceptual Design Report. 1. Physics and Detector. Preprint IHEP-CEPC-DR-2015-01. 2015. Available online: https://inspirehep.net/literature/1395734 (accessed on 1 June 2020).
8. Bhupal Dev, P.; Djouadi, A.; Godbole, R.; Muhlleitner, M.; Rindani, S. Determining the CP properties of the Higgs boson. *Phys. Rev. Lett.* **2008**, *100*, 051801. [CrossRef]
9. Hagiwara, K.; Ma, K.; Yokoya, H. Probing CP violation in e^+e^- production of the Higgs boson and toponia. *JHEP* **2016**, *6*, 48. [CrossRef]

10. Ma, K. Enhancing CP Measurement of the Yukawa Interactions of Top-Quark at e^-e^+ Collider. *Phys. Lett. B* **2019**, *797*, 134928. [CrossRef]
11. Erler, J.; Heinemeyer, S.; Hollik, W.; Weiglein, G.; Zerwas, P. Physics impact of GigaZ. *Phys. Lett. B* **2000**, *486*, 1389–1402. [CrossRef]
12. Grunewald, M. Experimental tests of the electroweak standard model at high-energies. *Phys. Rep.* **1999**, *322*, 125–346. [CrossRef]
13. Schael, S. Precision electroweak measurements on the Z resonance. *Phys. Rep.* **2006**, *427*, 257–454. [CrossRef]
14. Mnich, J. Experimental tests of the standard model in $e^+e^- \to$ f anti-f at the Z resonance. *Phys. Rep.* **1996**, *271*, 181–266. [CrossRef]
15. Arbuzov, A.; Awramik, M.; Czakon, M.; Freitas, A.; Grunewald, M.; Monig, K.; Riemann, S.; Riemann, T. ZFITTER: A Semi-analytical program for fermion pair production in e^+e^- annihilation, from version 6.21 to version 6.42. *Comput. Phys. Commun.* **2006**, *174*, 728–758. [CrossRef]
16. Montagna, G.; Nicrosini, O.; Piccinini, F.; Passarino, G. TOPAZ0 4.0: A New version of a computer program for evaluation of deconvoluted and realistic observables at LEP-1 and LEP-2. *Comput. Phys. Commun.* **1999**, *117*, 278–289. [CrossRef]
17. Bondarenko, S.; Dydyshka, Y.; Kalinovskaya, L.; Sadykov, R.; Yermolchyk, V. One-loop electroweak radiative corrections to lepton pair production in polarized electron-positron collisions. *arXiv* **2020**, arXiv:2005.04748.
18. Andonov, A.; Arbuzov, A.; Bardin, D.; Bondarenko, S.; Christova, P.; Kalinovskaya, L.; Nanava, G.; von Schlippe, W. SANCscope—v.1.00. *Comput. Phys. Commun.* **2006**, *174*, 481–517. [CrossRef]
19. Arbuzov, A.; Bondarenko, S.; Dydyshka, Y.; Kalinovskaya, L.; Rumyantsev, L.; Sadykov, R.; Yermolchyk, V. Electron-positron annihilation processes in MCSANCee. *CERN Yellow Rep. Monogr.* **2020**, *3*, 213–216. [CrossRef]
20. Arbuzov, A.; Bardin, D.; Bondarenko, S.; Christova, P.; Kalinovskaya, L.; Klein, U.; Kolesnikov, V.; Rumyantsev, L.; Sadykov, R.; Sapronov, A. Update of the MCSANC Monte Carlo integrator, v. 1.20. *JETP Lett.* **2016**, *103*, 131–136. [CrossRef]
21. Blondel, A. A Scheme to Measure the Polarization Asymmetry at the Z Pole in LEP. *Phys. Lett. B* **1988**, *202*, 145. [CrossRef]
22. Janot, P. Direct measurement of $\alpha_{QED}(m_Z^2)$ at the FCC-ee. *JHEP* **2016**, *2*, 53. [CrossRef]
23. Jadach, S.; Yost, S. QED Interference in Charge Asymmetry Near the Z Resonance at Future Electron-Positron Colliders. *Phys. Rev. D* **2019**, *100*, 013002. [CrossRef]
24. Arbuzov, A.; Bardin, D.; Leike, A. Analytic final state corrections with cut for e+ e− —> massive fermions. *Mod. Phys. Lett. A* **1992**, *7*, 2029–2038. [CrossRef]
25. Blondel, A.; Lynn, B.; Renard, F.; Verzegnassi, C. Precision Measurements of Final State Weak Coupling from Polarized Electron—Positron Annihilation. *Nucl. Phys. B* **1988**, *304*, 438–450. [CrossRef]
26. Jadach, S.; Was, Z.; Decker, R.; Kuhn, J.H. The tau decay library TAUOLA: Version 2.4. *Comput. Phys. Commun.* **1993**, *76*, 361–380. [CrossRef]
27. Jadach, S.; Ward, B.; Was, Z. The Monte Carlo program KORALZ, version 4.0, for the lepton or quark pair production at LEP/SLC energies. *Comput. Phys. Commun.* **1994**, *79*, 503–522. [CrossRef]
28. Jadach, S.; Ward, B.; Was, Z. The Monte Carlo program KORALZ, for the lepton or quark pair production at LEP/SLC energies: From version 4.0 to version 4.04. *Comput. Phys. Commun.* **2000**, *124*, 233–237. [CrossRef]
29. Langacker, P. The Physics of Heavy Z' Gauge Bosons. *Rev. Mod. Phys.* **2009**, *81*, 1199–1228. [CrossRef]
30. Funatsu, S.; Hatanaka, H.; Hosotani, Y.; Orikasa, Y. Distinct signals of the gauge-Higgs unification in e^+e^- collider experiments. *Phys. Lett. B* **2017**, *775*, 297–302. [CrossRef]
31. Funatsu, S. Forward-backward asymmetry in the gauge-Higgs unification at the International Linear Collider. *Eur. Phys. J. C* **2019**, *79*, 854. [CrossRef]
32. Jadach, S.; Skrzypek, M. QED challenges at FCC-ee precision measurements. *Eur. Phys. J. C* **2019**, *79*, 756. [CrossRef]
33. Dubovyk, I.; Freitas, A.; Gluza, J.; Riemann, T.; Usovitsch, J. Electroweak pseudo-observables and Z-boson form factors at two-loop accuracy. *JHEP* **2019**, *8*, 113. [CrossRef]

34. Blondel, A.; Freitas, A.; Gluza, J.; Riemann, T.; Heinemeyer, S.; Jadach, S.; Janot, P. Theory Requirements and Possibilities for the FCC-ee and other Future High Energy and Precision Frontier Lepton Colliders. *arXiv* **2019**, arXiv:1901.02648.
35. Blondel, A.; Gluza, J.; Jadach, S.; Janot, P.; Riemann, T. (Eds.) *Theory for the FCC-ee: Report on the 11th FCC-ee Workshop Theory and Experiments*; Vol. 3/2020; CERN Yellow Reports: Monographs; CERN: Geneva, Switzerland, 2019. [CrossRef]

© 2020 by the authors. Licensee MDPI, Basel, Switzerland. This article is an open access article distributed under the terms and conditions of the Creative Commons Attribution (CC BY) license (http://creativecommons.org/licenses/by/4.0/).

Article

Long-Range Correlations between Observables in a Model with Translational Invariance in Rapidity

Svetlana Belokurova and Vladimir Vechernin *

Department of High Energy Physics and Elementary Particles, Saint-Petersburg State University, 7-9 Universitetskaya emb., 199034 St. Petersburg, Russia; sveta.1596@mail.ru
* Correspondence: v.vechernin@spbu.ru

Received: 30 May 2020; Accepted: 24 June 2020; Published: 2 July 2020

Abstract: We estimate the impact of the fixation of the total number of sources (quark–gluon strings) on the long-range rapidity correlations between different observables. In our approach this condition models the fixation of the collision centrality class, what is the usual practice in modern collider experiments, like Relativistic Heavy Ion Collider (RHIC), Large Hadron Collider (LHC) and so on. The estimates are obtained under the assumption of the translational invariance in rapidity, which is usually assumed in mid-rapidity region at high energies. Based on these assumptions, we are developing a technique for the analytical calculation of various average values of extensive and intense variables at high string densities on the transverse lattice, taking into account the effects of string fusion, leading to the formation of string clusters. Using this technique we calculate the asymptotes of the correlations coefficients both between the multiplicities and between the multiplicity and the event-mean transverse momentum of particles in two separated rapidity intervals. As a result, we found that fixing the total number of strings has a significant effect on the behavior of both types of correlations, especially in the case of a uniform distribution of strings in the transverse plane.

Keywords: strong interaction; high energy; multiparticle production; multiplicity; transverse momentum; forward-backward correlations; long-range rapidity correlations; translation invariance in rapidity; quark-gluon strings; string fusion

1. Introduction

In modern particle physics, one of the urgent tasks is to extract information about the initial stage of high-energy hadronic interactions. The valuable source of such information can be the experimental and theoretical studies of the long-range correlations (LRCs) between observables in two separated rapidity windows usually refereed as forward (F) and backward (B) [1]. In this approach one suggests that at the initial stage of the strong interaction at high energy the formation of boost invariant flux tubes of color fields take place between colliding hadrons. It is important that the long-range rapidity correlations originating to the formation of the color flux tubes persist during the evolution of the Quark Gluon Plasma formed later in the collision and hence can be observed experimentally as the LRC between produced particles.

In the framework of the similar approach, rather long ago, in paper [2] the study of the long-range forward–backward correlations between multiplicities (n) in two separated rapidity intervals has been proposed with the aim to find signatures of the string fusion and percolation phenomenon [3–6] in ultrarelativistic heavy ion collisions. It was found later that the investigations of the FB correlations involving along with extensive, n, also intense observables e.g., such as the event-mean transverse momentum [7–18],

$$p_t = \frac{1}{n} \sum_{i=1}^{n} |\mathbf{p}_t^i|,\qquad(1)$$

or going to more sophisticated correlation variables, e.g., to the so-called strongly intensive quantities [19–23], enable us to obtain a clearer signal about the initial stage of hadronic interaction, including the process of string fusion, compared to usual FB multiplicity correlations. In the present paper we estimate the impact of the fixing of the total number of sources (quark-gluon strings) on the value of LRC both between the multiplicities in the forward (n_F) and backward (n_B) rapidity windows and between the multiplicity (n_F) in the forward window and the event-mean transverse momentum (p_{tB}) in the backward one. In our approach the fixing of the total number of strings models the fixation of the event centrality class, what is currently a widespread practice in analyzing experimental data in modern collider experiments (Relativistic Heavy Ion Collider—RHIC, Large Hadron Collider—LHC and so on).

The estimates are obtained under the assumption of validity of the translational (boost) invariance in rapidity, which is usually assumed in the central rapidity region for symmetric reactions at high energies. This assumption implies the uniform rapidity distribution of multiplicity and the dependence of the two-particle correlation function $C_2(y_1, y_2)$ only on the difference $\Delta y = y_1 - y_2$ of the particle rapidities y_1 and y_2 [24].

For symmetric reactions the uniformity of the rapidity distribution of multiplicity is approximately fulfilled at $|y| < 1$ and $|y| < 2$ for the RHIC and LHC energies correspondingly (see e.g., [25–27]). Referring to the ALICE data, [26,27]), one must to take into account that in these papers the distributions are presented in pseudorapidity not in rapidity, what leads to the characteristic kinematical drop of the spectra in the vicinity of $y = 0$, which is absent in the rapidity distributions. The dependence of the two-particle correlation function $C_2(y_1, y_2)$ only on the difference $\Delta y = y_1 - y_2$ of the particle rapidities in the central region is also commonly used when extracting this correlation function from experimental data both at RHIC and LHC energies (see e.g., [28–30]).

Note that for asymmetric reactions, like the proton-nucleus and deuteron-nucleus interactions, the boost invariance in rapidity is absent even in the central region. The rapidity distribution of multiplicity is not uniform at mid-rapidities [31,32] and basically one must to take into account the dependence of the two-particle correlation function $C_2(y_1, y_2)$ both on y_1 and y_2 in this case (see e.g., discussion in paper [33]).

Calculations of the asymptotes of the correlation coefficients are carried out by introducing a lattice (grid) in the impact parameter plane, which enables effectively to take into account the influence of the color string fusion processes, leading to the formation of string clusters in ultra relativistic heavy ion collisions. We present in details the developed methods for the analytical calculation of various average values of extensive and intensive variables at high string densities on the transverse lattice, what was announced in our short note [34], published as the proceedings of the WPCF Conference.

Basing on the averages found with high accuracy we calculate the strength of the LRC between the multiplicities (n_B-n_F) and between the multiplicity and the event-mean transverse momentum (p_{tB}-n_F) in the FB observation windows. It turns out that the fixation of the total number of strings, has a significant impact on the behavior of the both type of the correlations.

The paper organized as follows. In Section 2 we introduce the definitions of the n-n and p_t-n FB correlation coefficients and briefly describe our model with a lattice in transverse plane, which enables to take into account the string fusion effects on the correlation strength. Section 3 presents the developed method for the analytical calculation of various averages at high string densities on a transverse lattice. In Section 4 basing on the calculated averages we found the covariances $\text{cov}(n_B, n_F)$ and $\text{cov}(p_{tB}, n_F)$ determining the LRC coefficients, b_{nn} and $b_{p_t n}$, for the cases of uniform and non-uniform distribution of strings in the transverse plane. In Section 5 we summarize the influence of the fixation of the total number of strings (imitating in our approach the fixation of the collision centrality) on the behavior of the asymptotes of the LRC n-n and p_t-n coefficients at high string density.

2. The Model with the String Fusion on a Transverse Lattice

To quantify the strength of the FB correlations between observables measured in two separated rapidity intervals δy_F and δy_B it is convenient to use the correlation coefficients defined between the so-called relative variables $n/\langle n \rangle$ and $p_t/\langle p_t \rangle$ (see [24,35]). Therefore for the n-n correlation between multiplicities n_F and n_B in forward and backward rapidity intervals we will use the following definition:

$$b_{nn} \equiv \frac{\langle n_F \rangle}{\langle n_B \rangle} \frac{\text{cov}(n_B, n_F)}{D_{n_F}} = \frac{\langle n_F \rangle}{\langle n_B \rangle} \frac{\langle n_B n_F \rangle - \langle n_B \rangle \langle n_F \rangle}{\langle n_F^2 \rangle - \langle n_F \rangle^2}, \qquad (2)$$

where D_{n_F} is the variance of the n_F. Correspondingly, for the p_t-n correlation between the multiplicity, n_F, in the forward window and the event-mean transverse momentum of the particles observed and in the backward window, p_{tB}, we will use the similar definition:

$$b_{p_t n} \equiv \frac{\langle n_F \rangle}{\langle p_{tB} \rangle} \frac{\text{cov}(p_{tB}, n_F)}{D_{n_F}} = \frac{\langle n_F \rangle}{\langle p_{tB} \rangle} \frac{\langle p_{tB} n_F \rangle - \langle p_{tB} \rangle \langle n_F \rangle}{\langle n_F^2 \rangle - \langle n_F \rangle^2}. \qquad (3)$$

One can find the value of these correlations considering the effects from the interaction between strings in the framework of the models with string fusion and percolation [3–6]. In the present paper we will take these effects into account in simplified form, by introducing the finite lattice (the grid) in the transverse plane of the collision. This approach was suggested in paper [8]. Later, it was used for a description of various phenomena in ultra relativistic nuclear collisions (azimuthal flows, correlations, the ridge) [9–18,36–40].

In this approach we split the transverse plane into M cells. The area of each cell is equal to the transverse area of a single string. Then we consider that all initial strings, which centers occur in a given cell, merge into one string cluster. In this simplified model each string configuration is completely specified by the set of integers: $C_\eta = \{\eta_1, ..., \eta_M\}$, where η_i is a number of initial strings merged in a given i-th cell.

In fact, in this approach, the transverse plane is divided into cells with different, fluctuating values of the color field inside them. That is similar to the considering the color field density variation in the impact parameter plane in models based on the Color Glass Condensate (CGC) approach [1,41].

So we will suppose that in each cell the η_i fluctuates around some average, $\bar\eta_i$, with a scaled variance ω. Physically the $\bar\eta_i$ are determined by the geometry of a hadronic collision at given value of the impact parameter. Then in accordance with the string fusion prescriptions [5,6] the mean number of charged particles in given observation rapidity window δy, produced from the fragmentation of a string cluster in the i-th cell, and their mean transverse momentum are given by the expressions:

$$\bar n_i = \mu \sqrt{\eta_i}, \qquad \bar p_t^i = p_0 \sqrt[4]{\eta_i}, \qquad (4)$$

where μ and p_0 is the average multiplicity and the average transverse momentum for particles formed in the decay of a single string.

Note that we assume the translational invariance of the string picture in rapidity, originating from the locality of the strong interaction in the rapidity space, which is usually assumed in the central rapidity region at LHC energies. This translational invariance in rapidity corresponds to the boost invariance of the flux tubes in CGC models [1]. It leads to the independence of the $\bar n_i$ and $\bar p_t^i$ on a rapidity for a given string configuration C_η in our model.

We will also assume independent fragmentation of each string cluster into acceptances of the forward and backward windows, because in the present work we are only interested in studying long-range correlations, i.e., we will suppose that the n_i^F and n_i^B fluctuates independently around their mean values with some scaled variance ω_μ. One can find a more detailed description of the model in [8,14,16].

In the present paper for simplicity we will restrict ourselves to the case of a symmetric reaction and symmetric windows, $\delta y_F = \delta y_B \equiv \delta y$. For this case in definitions (2) and (3) of the FB correlation coefficients we have $\mu_F = \mu_B \equiv \mu$, $\langle n_F \rangle = \langle n_B \rangle \equiv \langle n \rangle$, $D_{n_F} = D_{n_B} \equiv D_n$, $\langle p_{tF} \rangle = \langle p_{tB} \rangle \equiv \langle p_t \rangle$, and so on. Due to mentioned translation invariance in rapidity these quantities do not depend on rapidity. In this case one can also show, [16], that for the LRC:

$$\langle n_F^2 \rangle = \langle n_B^2 \rangle = \langle n^2 \rangle = \sum_{i=1}^M d_{n_i} + \langle n_F n_B \rangle = \omega_\mu \langle n \rangle + \langle n_F n_B \rangle \,. \tag{5}$$

In the last transition we have used the assumption that the variance $d_{n_i} \equiv \langle n_i^2 \rangle - \langle n_i \rangle^2$ of the number of particles, n_i, produced in any rapidity window from string cluster decay in i-th lattice cell, is proportional to their mean multiplicity, $d_{n_i} = \omega_\mu \langle n_i \rangle$, with the same factor ω_μ. Then

$$D_{n_F} = D_{n_B} = D_n = \langle n^2 \rangle - \langle n \rangle^2 = \omega_\mu \langle n \rangle + \mathrm{cov}(n_F, n_B) \,. \tag{6}$$

Hence we can find the LRC coefficients using the formula

$$b_{nn} = \frac{\mathrm{cov}(n_B, n_F)}{\omega_\mu \langle n \rangle + \mathrm{cov}(n_B, n_F)}, \qquad b_{p_t n} = \frac{\langle n \rangle}{\langle p_t \rangle} \frac{\mathrm{cov}(p_{tB}, n_F)}{\omega_\mu \langle n \rangle + \mathrm{cov}(n_B, n_F)} \,. \tag{7}$$

instead of the general definitions (2) and (3). So to find these correlation coefficients we need to calculate only the following averages: $\langle n \rangle$, $\langle p_t \rangle$, $\langle n_B n_F \rangle$ and $\langle p_{tB} n_F \rangle$.

As was shown in [8,9,11,12,14,16] in this model with Gaussian distributions one can find the asymptotes of the long-range FB correlation coefficients at large string density in an explicit analytical form. In all these papers we supposed that the number of strings, η_i, in each cell fluctuates around their mean values $\bar\eta_i$ independently. In the present work we impose the additional condition fixing the total number of initial strings, N, in each event and study its impact on the LRC coefficients in the asymptotic regime of high string density.

For this purpose in present analysis we use the following event-by-event string distribution:

$$P(\eta_1, ..., \eta_M) = \sqrt{2\pi \omega \overline{N}}\, \delta(N - \overline{N}) \prod_{i=1}^M \frac{1}{\sqrt{2\pi \omega \bar\eta_i}} e^{-\frac{(\eta_i - \bar\eta_i)^2}{2\omega \bar\eta_i}} \,, \tag{8}$$

where $N = \sum_{i=1}^M \eta_i$ and $\overline{N} = \sum_{i=1}^M \bar\eta_i$. At high string density we can consider η_i, as continuous variables [8,14,16]. So the distribution (8) is normalized as follows

$$\int P(\eta_1, ..., \eta_M)\, d\eta_1 ... d\eta_M = 1 \,.$$

Below we will denote by $\langle ... \rangle$ the averages over string configurations, calculated with this distribution. One can easy check that the mean number of strings in each cell $\langle \eta_i \rangle$ is equal to the parameter $\bar\eta_i$, $\langle \eta_i \rangle = \bar\eta_i$. In the following consideration the important role will play the variables

$$v_i = \eta_i - \bar\eta_i \,, \tag{9}$$

characterizing the deviation of η_i from $\bar\eta_i$. One can easily verify that for the string distribution (8) we have the following *exact* relations:

$$\langle 1 \rangle = 1, \quad \langle v_k \rangle = 0, \quad \langle v_k^2 \rangle = \bar\eta_k \omega \left(1 - \frac{\bar\eta_k}{\overline{N}}\right), \quad \langle v_k^4 \rangle = 3\bar\eta_k^2 \omega^2 \left(1 - \frac{\bar\eta_k}{\overline{N}}\right)^2, \tag{10}$$

$$\langle v_k v_m \rangle = -\bar\eta_k \bar\eta_m \frac{\omega}{\overline{N}}, \quad \langle v_k v_m^3 \rangle = -3\bar\eta_k \bar\eta_m^2 \frac{\omega^2}{\overline{N}} \left(1 - \frac{\bar\eta_m}{\overline{N}}\right), \quad \langle v_i v_k v_m^2 \rangle = -\bar\eta_i \bar\eta_k \bar\eta_m \frac{\omega^2}{\overline{N}} \left(1 - 3\frac{\bar\eta_m}{\overline{N}}\right),$$

$$\langle v_k^2 v_m^2 \rangle = \bar{\eta}_k \bar{\eta}_m \omega^2 \left(1 - \frac{\bar{\eta}_k}{N} - \frac{\bar{\eta}_m}{N} + 3\frac{\bar{\eta}_k \bar{\eta}_m}{N^2}\right), \qquad \langle v_i v_j v_k v_m \rangle = 3\bar{\eta}_i \bar{\eta}_j \bar{\eta}_k \bar{\eta}_m \frac{\omega^2}{N^2}.$$

If $\alpha \equiv \sum_{i=1}^n \alpha_i$ is odd, then we have

$$\langle v_i^{\alpha_1} v_j^{\alpha_2} v_k^{\alpha_3} \ldots \rangle = 0, \quad \text{at} \quad \alpha \equiv \sum_{i=1}^n \alpha_i = 2l+1. \tag{11}$$

We note that the relations (10) are valid only if $i \neq j \neq k \neq m$. Really, we see that $\langle v_k v_m \rangle|_{k=m} \neq \langle v_k^2 \rangle$, $\langle v_k v_m^3 \rangle|_{k=m} \neq \langle v_k^4 \rangle$ and so on. That is a consequence of the correlations between fluctuations of the η_i in different cells arising due to the conservation of the total number of strings (see [34]).

Use of the relations (10) and (11) enables drastically simplify the calculation of various averages in this model, because all integrations over the η_i come down to using these simple rules.

We will calculate the asymptotes of the LRC coefficients (7) at high string density supposing that all $\bar{\eta}_i \gg 1$. We will also suppose that $M \gg 1$, because as it was discussed in [15,16] with a realistic string radius $r_{str} = 0.2 \div 0.3$ fm we need lattices with a large number of cells $M \sim 10^2$ and 10^4 for a description of pp and AA collisions correspondingly.

3. Averaging over String Configurations

We will demonstrate the technique of the analytical calculation of the different lattice averages with the distribution (8) by using as example the most complicated calculation of the mean value of the intensive variable—the mean transverse momentum of the produced particles, $\langle p_t \rangle$.

Regarding the accuracy of the calculation, we need to take into account the terms of the order $1/\bar{\eta}$, $1/\bar{\eta}^2$, $1/(M\bar{\eta})$ and $1/(M\bar{\eta}^2)$. Because, as we will see later, the terms of the leading order in M (the $1/\bar{\eta}$ and $1/\bar{\eta}^2$ in the case of $\langle p_t \rangle$ calculation) are cancelled when calculating the covariances entering expressions (7) for the LRC coefficients b_{nn} and $b_{p_t n}$. Moreover, in the case of homogeneous string spreading in transverse plane, when all $\bar{\eta}_i = \bar{\eta}$, we have additional mutual cancellation of the contributions of the order of $1/(M\bar{\eta})$ to the LRC coefficients calculated with the distribution (8) corresponding to a fixed total number of initial strings. In last case the only contribution to the LRC coefficient originates from the terms of the order of $1/(M\bar{\eta}^2)$.

As was shown in [16] with the prescriptions (4) we can find $\langle p_t \rangle$ by calculating the following average over string configurations:

$$\frac{\langle p_{tB} \rangle}{p_0} = \left\langle \frac{\sum_{i=1}^M \eta_i^{\frac{3}{4}}}{\sum_{k=1}^M \eta_i^{\frac{1}{2}}} \right\rangle \equiv \langle YZ \rangle. \tag{12}$$

Here we introduce the following notations

$$Y \equiv \sum_{i=1}^M \eta_i^{\frac{3}{4}}, \qquad Z \equiv \left(\sum_{i=1}^M \eta_i^{\frac{1}{2}}\right)^{-1}. \tag{13}$$

Taking into account the definition (9) we can present the Y with the accuracy up to v_i^4 as follows

$$Y = M S_{3/4} \left[1 + \frac{1}{M S_{3/4}} \sum_{i=1}^M \bar{\eta}_i^{\frac{3}{4}} \left(\frac{3v_i}{4\bar{\eta}_i} - \frac{3v_i^2}{32\bar{\eta}_i^2} + \frac{15v_i^3}{384\bar{\eta}_i^3} - \frac{45v_i^4}{2048\bar{\eta}_i^4}\right)\right], \tag{14}$$

where we have introduced the following convenient notation

$$S_\beta \equiv \frac{1}{M} \sum_{i=1}^M \bar{\eta}_i^\beta. \tag{15}$$

To calculate the Z with the same accuracy we at first have to use the expansion

$$\sum_{i=1}^{M}\eta_i^{\frac{1}{2}} = MS_{1/2}\left[1+\frac{1}{MS_{1/2}}\sum_{i=1}^{M}\eta_i^{\frac{1}{2}}\left(\frac{v_i}{2\overline{\eta}_i}-\frac{v_i^2}{8\overline{\eta}_i^2}+\frac{v_i^3}{16\overline{\eta}_i^3}-\frac{5v_i^4}{128\overline{\eta}_i^4}\right)\right] \equiv MS_{1/2}[1+a], \qquad (16)$$

where

$$a = \frac{1}{MS_{1/2}}\sum_{i=1}^{M}\eta_i^{\frac{1}{2}}\left(\frac{v_i}{2\overline{\eta}_i}-\frac{v_i^2}{8\overline{\eta}_i^2}+\frac{v_i^3}{16\overline{\eta}_i^3}-\frac{5v_i^4}{128\overline{\eta}_i^4}\right). \qquad (17)$$

Then we can write

$$Z \equiv \frac{1}{MS_{1/2}[1+a]} = \frac{1}{MS_{1/2}}[1-a+a^2-a^3+a^4] \qquad (18)$$

Multiplying Y by Z and taking into account only the terms $v_i^{\alpha_1}v_j^{\alpha_2}v_k^{\alpha_3}v_m^{\alpha_4}$, satisfying the conditions $\alpha = \sum_i \alpha_i = 2$ or 4 (see Formulas (10) and (11)), we find

$$YZ = \frac{S_{3/4}}{S_{1/2}}[1 + A_1 + A_2 + B_1 + B_2 + C]. \qquad (19)$$

Here the A_1 and A_2 collects four terms with $\alpha = 2$:

$$A_1 = \frac{1}{8MS_{1/2}}\sum_{i=1}^{M}\frac{v_i^2}{\overline{\eta}_i^{3/2}} - \frac{3}{32MS_{3/4}}\sum_{i=1}^{M}\frac{v_i^2}{\overline{\eta}_i^{5/4}}, \qquad (20)$$

$$A_2 = \frac{1}{4M^2 S_{1/2}^2}\sum_{i,j}^{M}\frac{v_i v_j}{\overline{\eta}_i^{1/2}\overline{\eta}_j^{1/2}} - \frac{3}{8M^2 S_{1/2} S_{3/4}}\sum_{i,j}^{M}\frac{v_i v_j}{\overline{\eta}_i^{1/2}\overline{\eta}_j^{1/4}} \qquad (21)$$

and the B_1, B_2 and the C collect 12 terms with $\alpha = 4$:

$$B_1 = \frac{5}{128 MS_{1/2}}\sum_{i=1}^{M}\frac{v_i^4}{\overline{\eta}_i^{7/2}} - \frac{45}{2048 MS_{3/4}}\sum_{i=1}^{M}\frac{v_i^4}{\overline{\eta}_i^{13/4}} \qquad (22)$$

$$+ \frac{1}{64 M^2 S_{1/2}^2}\sum_{i,j}^{M}\frac{v_i^2 v_j^2}{\overline{\eta}_i^{3/2}\overline{\eta}_j^{3/2}} - \frac{3}{256 M^2 S_{1/2} S_{3/4}}\sum_{i,j}^{M}\frac{v_i^2 v_j^2}{\overline{\eta}_i^{3/2}\overline{\eta}_j^{5/4}},$$

$$B_2 = \frac{1}{16 M^2 S_{1/2}^2}\sum_{i,j}^{M}\frac{v_i v_j^3}{\overline{\eta}_i^{1/2}\overline{\eta}_j^{5/2}} - \frac{5}{256 M^2 S_{1/2} S_{3/4}}\sum_{i,j}^{M}\frac{v_i v_j^3}{\overline{\eta}_i^{1/2}\overline{\eta}_j^{9/4}} - \frac{3}{64 M^2 S_{1/2} S_{3/4}}\sum_{i,j}^{M}\frac{v_i v_j^3}{\overline{\eta}_i^{1/4}\overline{\eta}_j^{5/2}} \qquad (23)$$

$$- \frac{3}{32 M^3 S_{1/2}^3}\sum_{i,j,k}^{M}\frac{v_i v_j v_k^2}{\overline{\eta}_i^{1/2}\overline{\eta}_j^{1/2}\overline{\eta}_k^{3/2}} - \frac{3}{128 M^3 S_{1/2}^2 S_{3/4}}\sum_{i,j,k}^{M}\frac{v_i v_j v_k^2}{\overline{\eta}_i^{1/2}\overline{\eta}_j^{1/2}\overline{\eta}_k^{5/4}} - \frac{3}{32 M^3 S_{1/2}^2 S_{3/4}}\sum_{i,j,k}^{M}\frac{v_i v_j v_k^2}{\overline{\eta}_i^{1/2}\overline{\eta}_j^{1/4}\overline{\eta}_k^{3/2}},$$

$$C = \frac{1}{16 M^4 S_{1/2}^4}\sum_{i,j,k,m}^{M}\frac{v_i v_j v_k v_m}{\overline{\eta}_i^{1/2}\overline{\eta}_j^{1/2}\overline{\eta}_k^{1/2}\overline{\eta}_m^{1/2}} - \frac{3}{32 M^4 S_{1/2}^3 S_{3/4}}\sum_{i,j,k,m}^{M}\frac{v_i v_j v_k v_m}{\overline{\eta}_i^{1/2}\overline{\eta}_j^{1/2}\overline{\eta}_k^{1/2}\overline{\eta}_m^{1/4}}. \qquad (24)$$

We will see below that the leading contributions to $\langle p_t \rangle$ originating from the terms A_1, A_2, B_1, B_2 and C are of the following order:

$$A_1 \sim \frac{1}{\overline{\eta}}, \quad A_2 \sim \frac{1}{M\overline{\eta}}, \quad B_1 \sim \frac{1}{\overline{\eta}^2}, \quad B_2 \sim \frac{1}{M\overline{\eta}^2}, \quad C \sim \frac{1}{M^2 \overline{\eta}^2}. \qquad (25)$$

So, taking into account the remark in the beginning of the present Section in the leading approximation we can do not take the C contribution into consideration.

Now to calculate the $\langle p_t \rangle$ by (12) we need to average the expression (19) over string fluctuations, given by the distribution (8):

$$\frac{\langle p_t \rangle}{p_0} = \langle YZ \rangle = \frac{S_{3/4}}{S_{1/2}}[1 + \langle A_1 \rangle + \langle A_2 \rangle + \langle B_1 \rangle + \langle B_2 \rangle + \langle C \rangle]. \tag{26}$$

We can do this using the rules (10) and (11) obtained above. At that we have to take into account that these rules are valid only for non coinciding arguments (see the remark after the Formula (11)). So, at first we must express all sums entering the Formulas (20)–(24) through the sums with non coinciding arguments. We can do it easily using the following obvious relations:

$$\sum_{i,j} = \sum_{i\neq j} + \sum_{i=j}, \qquad \sum_{i,j,k} = \sum_{i\neq j\neq k} + \sum_{i=j\neq k} + \sum_{i\neq j=k} + \sum_{i=k\neq j} + \sum_{i=j=k}, \tag{27}$$

and so on. Then for terms of the general form we have

$$\sum_{i,j} \frac{\langle v_i^{\alpha_1} v_j^{\alpha_2} \rangle}{\bar{\eta}_i^{\beta_1} \bar{\eta}_j^{\beta_2}} = \sum_{i\neq j} \frac{\langle v_i^{\alpha_1} v_j^{\alpha_2} \rangle}{\bar{\eta}_i^{\beta_1} \bar{\eta}_j^{\beta_2}} + \sum_i \frac{\langle v_i^{\alpha_1+\alpha_2} \rangle}{\bar{\eta}_i^{\beta_1+\beta_2}}, \qquad \sum_{i,j,k} \frac{\langle v_i^{\alpha_1} v_j^{\alpha_2} v_k^{\alpha_3} \rangle}{\bar{\eta}_i^{\beta_1} \bar{\eta}_j^{\beta_2} \bar{\eta}_k^{\beta_3}} \tag{28}$$

$$= \sum_{i\neq j\neq k} \frac{\langle v_i^{\alpha_1} v_j^{\alpha_2} v_k^{\alpha_3} \rangle}{\bar{\eta}_i^{\beta_1} \bar{\eta}_j^{\beta_2} \bar{\eta}_k^{\beta_3}} + \sum_{i\neq k} \frac{\langle v_i^{\alpha_1+\alpha_2} v_k^{\alpha_3} \rangle}{\bar{\eta}_i^{\beta_1+\beta_2} \bar{\eta}_k^{\beta_3}} + \sum_{i\neq j} \frac{\langle v_i^{\alpha_1} v_j^{\alpha_2+\alpha_3} \rangle}{\bar{\eta}_i^{\beta_1} \bar{\eta}_j^{\beta_2+\beta_3}} + \sum_{i\neq j} \frac{\langle v_i^{\alpha_1+\alpha_3} v_j^{\alpha_2} \rangle}{\bar{\eta}_i^{\beta_1+\beta_3} \bar{\eta}_j^{\beta_2}} + \sum_i \frac{\langle v_i^{\alpha_1+\alpha_2+\alpha_3} \rangle}{\bar{\eta}_i^{\beta_1+\beta_2+\beta_3}}.$$

After that, using the rules (10) and (11) and taking also into account that $\bar{N} = M S_1$, we find the answer for $\langle p_t \rangle$ as the linear combination of the sums of the following type:

$$\sum_i \bar{\eta}_i^\beta = M S_\beta, \qquad \sum_{i\neq j} \bar{\eta}_i^\beta \bar{\eta}_j^\gamma, \qquad \sum_{i\neq j\neq k} \bar{\eta}_i^\beta \bar{\eta}_j^\gamma \bar{\eta}_k^\delta, \tag{29}$$

and so on. Now to express all these sums through the S_β, defined by (15), we have to use the relations inverse to (27):

$$\sum_{i\neq j} \bar{\eta}_i^\beta \bar{\eta}_j^\gamma = \sum_{i,j} \bar{\eta}_i^\beta \bar{\eta}_j^\gamma - \sum_{i=j} \bar{\eta}_i^\beta \bar{\eta}_j^\gamma = \left(\sum_i \bar{\eta}_i^\beta\right)\left(\sum_j \bar{\eta}_j^\gamma\right) - \sum_i \bar{\eta}_i^{\beta+\gamma} = M^2 S_\beta S_\gamma - M S_{\beta+\gamma}, \tag{30}$$

$$\sum_{i\neq j\neq k} \bar{\eta}_i^\beta \bar{\eta}_j^\gamma \bar{\eta}_k^\delta = \sum_{i,j,k} \bar{\eta}_i^\beta \bar{\eta}_j^\gamma \bar{\eta}_k^\delta - \sum_{i=j,k} \bar{\eta}_i^\beta \bar{\eta}_j^\gamma \bar{\eta}_k^\delta - \sum_{i,j=k} \bar{\eta}_i^\beta \bar{\eta}_j^\gamma \bar{\eta}_k^\delta - \sum_{i=k,j} \bar{\eta}_i^\beta \bar{\eta}_j^\gamma \bar{\eta}_k^\delta + 2\sum_{i=j=k} \bar{\eta}_i^\beta \bar{\eta}_j^\gamma \bar{\eta}_k^\delta$$

$$= M^3 S_\beta S_\gamma S_\delta - M^2 S_\beta S_{\gamma+\delta} - M^2 S_\gamma S_{\beta+\delta} - M^2 S_\delta S_{\beta+\gamma} + 2M S_{\beta+\gamma+\delta}.$$

Using this technique we can easily check that the leading orders of the terms in the A_1, A_2, B_1, B_2 and the C are given by (25). Then, applying this approach and taking into account only the terms of the order $1/\bar{\eta}$, $1/\bar{\eta}^2$, $1/(M\bar{\eta})$ and $1/(M\bar{\eta}^2)$ in the contributions A_1, A_2, B_1, B_2 (see the remark in the beginning of the present Section) we find

$$\frac{\langle p_t \rangle}{p_0} = \langle YZ \rangle = \frac{S_{3/4}}{S_{1/2}} \left\{ 1 + \omega \left(\frac{S_{-1/2}}{8 S_{1/2}} - \frac{3 S_{-1/4}}{32 S_{3/4}} \right) + \omega^2 \left(\frac{S_{-1/2}^2}{64 S_{1/2}^2} - \frac{3 S_{-1/2} S_{-1/4}}{256 S_{1/2} S_{3/4}} \right. \right. \tag{31}$$

$$+ \frac{15 S_{-3/2}}{128 S_{1/2}} - \frac{135 S_{-5/4}}{2048 S_{3/4}} \right) + \frac{1}{M} \left[\omega \left(\frac{3}{32 S_1} + \frac{1}{4 S_{1/2}^2} - \frac{3 S_{1/4}}{8 S_{1/2} S_{3/4}} \right) + \omega^2 \left(\frac{7 S_{-1}}{32 S_{1/2}^2} - \frac{3 S_{-1/2}}{32 S_{1/2}^3} \right. $$

$$\left. \left. - \frac{29 S_{-1/2}}{256 S_1 S_{1/2}} - \frac{3 S_{1/4} S_{-1/2}}{32 S_{1/2}^2 S_{3/4}} + \frac{231 S_{-1/4}}{1024 S_1 S_{3/4}} - \frac{3 S_{-1/4}}{128 S_{1/2}^2 S_{3/4}} - \frac{57 S_{-3/4}}{256 S_{1/2} S_{3/4}} \right) \right] \right\}$$

4. Calculation of the Long-Range Correlation Coefficients

Using the methods developed in Section 3 with the example of the $\langle p_t \rangle$ calculation, we can now easily find all other averages entering the correlation coefficients b_{nn} and $b_{p_t n}$ defined by (7) with necessary accuracy.

In accordance with the prescriptions (4), taking into account (16) and applying the developed technique we find

$$\frac{\langle n \rangle}{\mu} = \langle \sum_{i=1}^{M} \eta_i^{\frac{1}{2}} \rangle = M S_{1/2} \left\{ 1 - \omega \frac{S_{-1/2}}{8 S_{1/2}} - \omega^2 \frac{15 S_{-3/2}}{128 S_{1/2}} + \frac{1}{M} \left[\omega \frac{1}{8 S_1} + \omega^2 \frac{15 S_{-1/2}}{64 S_1 S_{1/2}} \right] \right\}. \tag{32}$$

Using the general expression for $\langle p_{tB} n_F \rangle$:

$$\frac{\langle p_{tB} n_F \rangle}{p_0 \mu} = \langle \sum_{i=1}^{M} \eta_i^{\frac{3}{4}} \rangle, \tag{33}$$

obtained in [16] for the LRC, and the Formula (14) we also find that in the framework of the developed approach:

$$\frac{\langle p_{tB} n_F \rangle}{p_0 \mu} = M S_{3/4} \left\{ 1 - \omega \frac{3 S_{-1/4}}{32 S_{3/4}} - \omega^2 \frac{135 S_{-5/4}}{2048 S_{3/4}} + \frac{1}{M} \left[\omega \frac{3}{32 S_1} + \omega^2 \frac{135 S_{-1/4}}{1024 S_1 S_{3/4}} \right] \right\}. \tag{34}$$

Finally by this technique taking into account the Formula (16) we find for the contribution of the LRC to $\langle n_B n_F \rangle$ the following expression:

$$\frac{\langle n_B n_F \rangle}{\mu^2} = \langle \sum_{i=1}^{M} \eta_i^{\frac{1}{2}} \sum_{j=1}^{M} \eta_j^{\frac{1}{2}} \rangle = M^2 S_{1/2}^2 \left\{ 1 - \frac{\omega S_{-1/2}}{4 S_{1/2}} - \frac{\omega^2}{64 S_{1/2}} \left(\frac{S_{-1/2}^2}{S_{1/2}} - 15 S_{-3/2} \right) \right. $$
$$\left. + \frac{1}{M} \left[\frac{\omega}{4 S_{1/2}^2} + \frac{\omega^2}{4 S_{1/2}} \left(\frac{7 S_{-1}}{8 S_{1/2}} + \frac{S_{-1/2}}{S_1} \right) \right] \right\}. \tag{35}$$

Now we can calculate the covariances (the correlators) entering the correlation coefficients b_{nn} and $b_{p_t n}$ (see the Formula (7)):

$$\frac{\mathrm{cov}(n_B, n_F)}{\mu^2} = \frac{\langle n_B n_F \rangle - \langle n \rangle^2}{\mu^2} = M \left[\frac{\omega}{4} \left(1 - \frac{S_{1/2}^2}{S_1} \right) + \frac{\omega^2}{32} \left(7 S_{-1} - 6 \frac{S_{1/2} S_{-1/2}}{S_1} \right) \right], \tag{36}$$

$$\frac{\mathrm{cov}(p_{tB}, n_F)}{p_0 \mu} = \frac{\langle p_{tB} n_F \rangle - \langle p_t \rangle \langle n \rangle}{p_0 \mu} = \frac{\omega}{4} \left(\frac{3 S_{1/4}}{2 S_{1/2}} - \frac{S_{3/4}}{S_{1/2}^2} - \frac{S_{3/4}}{2 S_1} \right) \tag{37}$$

$$+ \frac{\omega^2}{8} \left(\frac{57 S_{-3/4}}{32 S_{1/2}} - \frac{21 S_{-1/4}}{32 S_1} + \frac{3 S_{-1/4}}{16 S_{1/2}^2} - \frac{7 S_{3/4} S_{-1}}{4 S_{1/2}^2} + \frac{3 S_{1/4} S_{-1/2}}{8 S_{1/2}^2} + \frac{S_{3/4} S_{-1/2}}{S_{1/2}^3} - \frac{S_{3/4} S_{-1/2}}{S_1 S_{1/2}} \right).$$

We really see that all terms proportional M^2 in Formula (35) for $\langle n_B n_F \rangle$ are cancelled by the terms of this order in $\langle n \rangle^2$. similarly, all terms proportional M in Formula (34) for $\langle p_{tB} n_F \rangle$ are cancelled by the terms of this order in the product $\langle p_t \rangle \langle n \rangle$, given by the Formulas (31) and (32).

Moreover if we will go to the case with a homogenous string distribution in the transverse plane with some mean string density, corresponding to the same mean number, $\bar{\eta}$, of initial strings in a lattice (grid) cell, when all $\bar{\eta}_i = \bar{\eta}$, then we will have

$$S_\beta \equiv \frac{1}{M} \sum_{i=1}^{M} \bar{\eta}_i^\beta = \bar{\eta}^\beta. \tag{38}$$

In this case all contributions proportional to ω in Formulas (35) and (34) for the correlators (covariances) $\text{cov}(n_B, n_F)$ and $\text{cov}(p_{tB}, n_F)$ are also mutually cancelled and only the contributions of the terms of the order ω^2 survive. In this simple case the formula for the correlators (covariances) reduce to

$$\frac{\text{cov}(n_B, n_F)}{\mu^2} = M \frac{\omega^2}{32\,\overline{\eta}}, \qquad (39)$$

$$\frac{\text{cov}(p_{tB}, n_F)}{p_0\,\mu} = -\frac{\omega^2}{128\,\overline{\eta}^2}. \qquad (40)$$

This leads to the proportionality of the the b_{nn} and $b_{p_t n}$ correlation coefficients (7) at large string density to $1/\overline{\eta}^{3/2}$ in this case, instead of $1/\sqrt{\overline{\eta}}$ that took place in the case without the fixation of the total string number [8,9,11,14].

At that by (39) and (40) we see that in this case the $b_{p_t n}$ correlation coefficient is negative, whereas the b_{nn} correlation coefficient is positive. Note that without imposing this additional condition, fixing the total number of strings, both b_{nn} and $b_{p_t n}$ correlation coefficients were always positive for a homogeneous string distribution in the transverse plane [8,9,11,14].

5. Summary

We present the developed technique for the analytical calculation of various average values of extensive and intensive variables at high string densities on the transverse lattice with taking into account the string fusion effects leading to the formation of string clusters. Using this technique we calculate the asymptotes of the LRC coefficients between the multiplicities, b_{nn}, and between the multiplicity and the event-mean transverse momentum, $b_{p_t n}$, in two separated rapidity intervals at high string density and with the fixation of the total number of initial strings. This last condition models in our approach the fixation of the collision centrality class, which is the usual practice of analyzing experimental data in modern collider experiments, like RHIC, LHC and so on.

As a result we found that the fixation of the total number of strings has a significant impact on the behavior of the both type of the correlations, especially in the case of uniform string distribution in transverse plane. In this case at large string density the b_{nn} and $b_{p_t n}$ LRC coefficients become proportional to $1/\overline{\eta}^{3/2}$ instead of $1/\sqrt{\overline{\eta}}$ that took place without the fixation of the total number of strings [8,9,11,14].

We also found that in this case the correlation coefficient $b_{p_t n}$ always has a negative value, while the correlation coefficient b_{nn} is positive. Whereas without fixing the total number of strings both correlation coefficients b_{nn} and $b_{p_t n}$ were always positive for a homogeneous distribution of the strings in the transverse plane [8,9,11,14].

In general, the proposed lattice approach to the analysis of correlations between various extensive and intense observables can be useful for modeling the magnitude of these correlations under developing of various detecting systems aimed to study these effects, in particular, in the design and construction of vertex detectors for the NICA accelerator complex.

Author Contributions: Both authors contributed equally to this work. All authors have read and agreed to the published version of the manuscript.

Funding: This research was funded by the Russian Foundation for Basic Research, project number 18-02-40075.

Acknowledgments: V.V. acknowledges the support given by the Saint-Petersburg State University grant for outgoing academic mobility, id. 41159705.

Conflicts of Interest: The authors declare no conflict of interest.

Abbreviations

The following abbreviations are used in this manuscript:

LRC   Long-Range Correlation
FB   Forward-Backward
CGC   Color Glass Condensate
RHIC   Relativistic Heavy Ion Collider
LHC   Large Hadron Collider

References

1. Dumitru, A.; Gelis, F.; McLerran, L.; Venugopalan, R. Glasma flux tubes and the near side ridge phenomenon at RHIC. *Nucl. Phys. A* **2008**, *810*, 91–108. [CrossRef]
2. Amelin, N.S.; Armesto, N.; Braun, M.A.; Ferreiro, E.G.; Pajares, C. Long and short range correlations and the search of the quark gluon plasma. *Phys. Rev. Lett.* **1994**, *73*, 2813. [CrossRef] [PubMed]
3. Biro, T.S.; Nielsen, H.B.; Knoll, J. Colour rope model for extreme relativistic heavy ion collisions. *Nucl. Phys. B* **1984**, *245*, 449–468. [CrossRef]
4. Bialas, A.; Czyz, W. Conversion of color field into $q\bar{q}$ matter in the central region of high-energy heavy ion collisions. *Nucl. Phys. B* **1986**, *267*, 242–252. [CrossRef]
5. Braun, M.A.; Pajares, C. Particle production in nuclear collisions and string interactions. *Phys. Lett. B* **1992**, *287*, 154–158. [CrossRef]
6. Braun, M.A.; Pajares, C. A probabilistic model of interacting strings. *Nucl. Phys. B* **1993**, *390*, 542–558. [CrossRef]
7. Braun, M.A.; Pajares, C. Transverse momentum distributions and their forward-backward correlations in the percolating color string approach. *Phys. Rev. Lett.* **2000**, *85*, 4864. [CrossRef]
8. Vechernin, V.V.; Kolevatov, R.S. Cellular approach to long-range p_T and multiplicity correlations in the string fusion model. *Vestn. SPbU Ser.* **2004**, *4*, 11–27.
9. Braun, M.A.; Kolevatov, R.S.; Pajares, C.; Vechernin, V.V. Correlations between multiplicities and average transverse momentum in the percolating color strings approach. *Eur. Phys. J. C* **2004**, *32*, 535–546. [CrossRef]
10. Alessandro, B.; Antinori, F.; Belikov, J.A.; Blume, C.; Dainese, A.; Foka, P.; Giubellino, P.; Hippolyte, B.; Kuhn, C.; Martinez, G.; et al. [ALICE Collaboration]. ALICE: Physics Performance Report, Volume II. *Phys. J. G* **2006**, *32*, 1295–2040.
11. Vechernin, V.V.; Kolevatov, R.S. On multiplicity and transverse-momentum correlations in collisions of ultrarelativistic ions. *Phys. Atom. Nucl.* **2007**, *70*, 1797–1808. [CrossRef]
12. Vechernin, V.V.; Kolevatov, R.S. Long-range correlations between transverse momenta of charged particles produced in relativistic nucleus-nucleus collisions. *Phys. Atom. Nucl.* **2007**, *70*, 1809–1818. [CrossRef]
13. Kovalenko, V.; Vechernin, V. Long-range rapidity correlations in high energy AA collisions in Monte Carlo model with string fusion. In Proceedings of the XXVth International Nuclear Physics Conference (INPC 2013), Firenze, Italy, 2–7 June 2013; Volume 66, p. 04015.
14. Vechernin, V.V. Correlation Between Transverse Momenta in the String Fusion Model. *Theor. Math. Phys.* **2015**, *184*, 1271–1280. [CrossRef]
15. Vechernin, V. Long-range rapidity correlations between mean transverse momenta in the model with string fusion. In Proceedings of the 19-th International Seminar on High Energy Physics (QUARKS-2016), Pushkin, Russia, 29 May–4 June 2016; Volume 125, p. 04022.
16. Vechernin, V.V. Asymptotic behavior of the correlation coefficients of transverse momenta in the model with string fusion. *Theor. Math. Phys.* **2017**, *190*, 251–267. [CrossRef]
17. Belokurova, S. Asymptotes of multiplicity and transverse momentum correlation coefficients at large string density. In Proceedings of the XXth International Seminar on High Energy Physics (QUARKS-2018), Valday, Russia, 27 May–2 June 2018; Volume 191, p. 04010.
18. Belokurova, S.N.; Vechernin, V.V. Strongly intensive variables and the long-range correlations in the model with a lattice in transverse plane. *Theor. Math. Phys.* **2019**, *200*, 1094–1109. [CrossRef]
19. Gorenstein, M.I.; Gazdzicki, M. Strongly intensive quantities. *Phys. Rev. C* **2011**, *84*, 014904. [CrossRef]
20. Andronov, E.V. Influence of the quark-gluon string fusion mechanism on long-range rapidity correlations and fluctuations. *Theor. Math. Phys.* **2015**, *185*, 1383–1390. [CrossRef]

21. Vechernin, V. Short- and long-range rapidity correlations in the model with a lattice in transverse plane. In Proceedings of the XXth International Seminar on High Energy Physics (QUARKS-2018), Valday, Russia, 27 May–2 June 2018; Volume 191, p. 04011.
22. Andronov, E.; Vechernin, V. Strongly intensive observable between multiplicities in two acceptance windows in a string model. *Eur. Phys. J. A* **2019**, *55*, 14. [CrossRef]
23. Vechernin, V.; Andronov, E. Strongly intensive observables in the model with string fusion. *Universe* **2019**, *5*, 15. [CrossRef]
24. Vechernin, V. Forward-backward correlations between multiplicities in windows separated in azimuth and rapidity. *Nucl. Phys. A* **2015**, *939*, 21–45. [CrossRef]
25. Bearden, I.G.; Beavis, D.; Besliu, C.; Budick, B.; Boggild, H.; Chasman, C.; Christensen, C.H.; Christiansen, P.; Cibor, J.; Debbe, R.; et al. Charged meson rapidity distributions in central Au+Au collisions at $\sqrt{s_{NN}}$ = 200 GeV. *Phys. Rev. Lett.* **2005**, *94*, 162301. [CrossRef] [PubMed]
26. Abbas, E.; Abelev, B.; Adam, J.; Adamova, D.; Adare, A.M.; Aggarwal, M.M.; Aglieri Rinella, G.; Agnello, M.; Agocs, A.G.; Agostinelli, A.; et al. Centrality dependence of the pseudorapidity density distribution for charged particles in Pb–Pb collisions at $\sqrt{s_{NN}}$ = 2.76 TeV. *Phys. Lett. B* **2013**, *726*, 610–622. [CrossRef]
27. Adam, J.; Adamova, D.; Aggarwal, M.M.; Aglieri Rinella, G.; Agnello, M.; Agrawal, N.; Ahammed, Z.; Ahmad, S.; Ahn, S.U.; Aiola, S.; et al. Centrality dependence of the pseudorapidity density distribution for charged particles in Pb–Pb collisions at $\sqrt{s_{NN}}$ = 5.02 TeV. *Phys. Lett. B* **2017**, *772*, 567–577. [CrossRef]
28. Abelev, B.I.; Aggarwal, M.M.; Ahammed, Z.; Alakhverdyants, A.V.; Anderson, B.D.; Arkhipkin, D.; Averichev, G.S.; Balewski, J.; Barannikova, O.; Barnby, L.S.; et al. Long range rapidity correlations and jet production in high energy nuclear collisions. *Phys. Rev. C* **2009**, *80*, 064912. [CrossRef]
29. Chatrchyan, S.; Khachatryan, V.; Sirunyan, A.M.; Tumasyan, A.; Adam, W.; Bergauer, T.; Dragicevic, M.; Ero, J.; Fabjan, C.; Friedl, M.; et al. Long-range and short-range dihadron angular correlations in central PbPb collisions at $\sqrt{s_{NN}}$ = 2.76 TeV. *JHEP* **2011**, *7*, 76. [CrossRef]
30. Adam, J.; Adamova, D.; Aggarwal, M.M.; Aglieri Rinella, G.; Agnello, M.; Agrawal, N.; Ahammed, Z.; Ahmad, S.; Ahn, S.U.; Aiola, S.; et al. Insight into particle production mechanisms via angular correlations of identified particles in pp collisions at $\sqrt{s}$ = 7 TeV. *Eur. Phys. J. C* **2017**, *77*, 569. [CrossRef]
31. Back, B.B.; Baker, M.D.; Ballintijn, M.; Barton, D.S.; Becker, B.; Betts, R.R.; Bickley, A.A.; Bindel, R.; Busza, W.; Carroll, A.; et al. Pseudorapidity Distribution of Charged Particles in d+Au Collisions at $\sqrt{s_{NN}}$ = 200 GeV. *Phys. Rev. Lett.* **2004**, *93*, 082301. [CrossRef]
32. Acharya, S.; Acosta, F.-T.; Adamova, D.; Adhya, S.P.; Adler, A.; Adolfsson, J.; Aggarwal, M.M.; Aglieri Rinella, G.; Agnello, M.; Ahammed, Z.; et al. Charged-particle pseudorapidity density at mid-rapidity in p–Pb collisions at $\sqrt{s_{NN}}$ = 8.16 TeV. *Eur. Phys. J. C* **2019**, *79*, 307. [CrossRef]
33. Vechernin, V.V.; Ivanov, K.O.; Neverov, D.I. Two-particle correlation function and dihadron correlation approach. *Phys. At. Nucl.* **2016**, *79*, 798–806. [CrossRef]
34. Belokurova, S.N.; Vechernin, V.V. Calculation of long-range rapidity correlations in the model with string fusion on a transverse lattice. *Phys. Part. Nucl.* **2020**, *51*, 319–322. [CrossRef]
35. Adam, J.; Adamova, D.; Aggarwal, M.M.; Aglieri Rinella, G.; Agnello, M.; Agrawal, N.; Ahammed, Z.; Ahmed, S.; Ahn, S.U.; Aimo, I.; et al. Forward-backward multiplicity correlations in pp collisions at $\sqrt{s}$ = 0.9, 2.76 and 7 TeV. *JHEP* **2015**, *5*, 097. [CrossRef]
36. Braun, M.A.; Pajares, C. Elliptic flow from colour strings. *Eur. Phys. J. C* **2011**, *71*, 1558. [CrossRef]
37. Kovalenko, V.N. Modelling of exclusive parton distributions and long-range rapidity correlations for pp collisions at the LHC energy. *Phys. Atom. Nucl.* **2013**, *76*, 1189–1195. [CrossRef]
38. Kovalenko, V.; Vechernin, V. Model of pp and AA collisions for the description of long-range correlations. *Proc. Sci.* **2012**, *Baldin-ISHEPP-XXI*, 077.
39. Braun, M.A.; Pajares, C.; Vechernin, V.V. Anisotropic flows from colour strings: Monte-Carlo simulations. *Nucl. Phys. A* **2013**, *906*, 14–27. [CrossRef]
40. Braun, M.A.; Pajares, C.; Vechernin, V.V. Ridge from strings. *Eur. Phys. J. A* **2015**, *51*, 44. [CrossRef]
41. Kovner, A.; Lublinsky, M. Angular correlations and high energy evolution. *Phys. Rev. D* **2011**, *83*, 034017. [CrossRef]

 © 2020 by the authors. Licensee MDPI, Basel, Switzerland. This article is an open access article distributed under the terms and conditions of the Creative Commons Attribution (CC BY) license (http://creativecommons.org/licenses/by/4.0/).

Article

Three Flavor Quasi-Dirac Neutrino Mixing, Oscillations and Neutrinoless Double Beta Decay

Amina Khatun [1], Adam Smetana [2] and Fedor Šimkovic [1,2,3,*]

[1] Department of Nuclear Physics and Biophysics, Faculty of Mathematics, Physics and Informatics, Comenius University, Mlynská dolina F1, SK842 48 Bratislava, Slovakia; amina.khatun@fmph.uniba.sk
[2] Institute of Experimental and Applied Physics, Czech Technical University in Prague, 166 36 Prague, Czech Republic; adam.smetana@cvut.cz
[3] BLTP, JINR, 141980 Dubna, Russia
* Correspondence: fedor.simkovic@fmph.uniba.sk

Received: 22 June 2020; Accepted: 31 July 2020; Published: 5 August 2020

Abstract: The extension of the Standard model by three right-handed neutrino fields exhibit appealing symmetry between left-handed and right-handed sectors, which is only violated by interactions. It can accommodate three flavor quasi-Dirac neutrino mixing scheme, which allows processes with violation of both lepton flavor and total lepton number symmetries. We propose a 6×6 unitary matrix for parameterizing the mixing among three flavors of quasi-Dirac neutrino. This mixing matrix is constructed by two 3×3 unitary matrices that diagonalizes the Dirac mass term in the Lagrangian. By only assuming the Standard Model $V - A$ weak interaction, it is found that probabilities of neutrino oscillations among active flavor states and effective masses measured by single beta decay, by neutrinoless double-beta decay and by cosmology only depend on single 3×3 unitary matrix relevant to mixing of active neutrino flavors. Further, by considering 1σ and 3σ uncertainties in the measured oscillation probability of electron antineutrino from reactor, derivation of the constraint on the Majorana neutrino mass component is demonstrated. The consequence for effective Majorana neutrino mass governing the neutrinoless double-beta decay is discussed.

Keywords: quasi-Dirac; neutrino oscillation; Majorana neutrino mass; neutrinoless double beta decay

1. Introduction

The discovery of neutrino oscillations in experiments with atmospheric, solar, reactor, and accelerator neutrinos have provided compelling evidence that flavor neutrinos oscillate from one flavor (electron-, muon-, and tau-) to another due to neutrino mixing and that neutrinos possess nonzero masses [1], which offer an insight on new physics beyond the Standard Model (SM) [2].

The data from all neutrino oscillation experiments are well described by the three-neutrino mixing:

$$\nu_{\alpha L} = \sum_{j=1}^{3} U_{\alpha j} \nu_{jL} \quad (\alpha = e, \mu, \tau), \qquad (1)$$

where ν_j is the field of the neutrino with mass m_j and $U_{\alpha j}$ are the elements of the Pontecorvo-Maki-Nakagawa-Sakata unitary neutrino matrix [3,4].

The observation of neutrino oscillations implies that the flavor lepton numbers L_e, L_μ, and L_τ are not conserved, which follows from the presence of flavor-mixing neutrino mass term in Lagrangian of the theory. If the total lepton number $L = L_e + L_\mu + L_\tau$ is conserved, neutrinos with definite masses ν_j are Dirac particles (i.e., different from their antiparticle). The theoretical expectation is that L is not conserved and, consequently, neutrinos are Majorana particles (i.e., identical to its own antiparticle). The fundamental problem of nature of neutrinos, which is directly related with the origin of neutrino masses and mixing, can be experimentally solved by the observation of the L violating processes, e.g., neutrinoless double-beta ($0\nu\beta\beta$) decay [5,6].

In the case of the most general Dirac-Majorana mass term, fields of neutrinos with definite masses are of Majorana nature and their number depends on number of sterile fields (not entering in the gauge interaction Lagrangian of the SM) and is larger than three [7]. The Dirac-Majorana mass term can accommodate the seesaw scenario [8–11], which helps to understand the smallness of the neutrino masses constrained by laboratory and cosmological measurements. In the classical realization of the seesaw scenario with three right-handed neutrino fields, the Dirac-Majorana mass term is dominated by the lepton-number-violating right-handed neutrino Majorana masses giving rise to three light and active neutrinos, and three very heavy sterile neutrinos. Out of these, only the three active neutrinos participate in solar, atmospheric, and terrestrial neutrino flavor oscillations.

The goal of this paper is to discuss an opposite scenario, in which the Dirac-Majorana mass term is dominated by the Dirac masses. Such a scenario, in general, leads to six Majorana neutrino states with pairwise quasi-degenerate masses, referred to as quasi-Dirac neutrinos, see, e.g., [12–15] and references therein. Here, all six states participate in neutrino flavor oscillations providing much richer oscillation phenomenology. The tiny neutrino masses can be ascribed to the smallness of neutrino Yukawa couplings with difficulty to explain why the fermion masses span twelve orders of magnitude. The solution to this problem can be inspired by extra-dimensional models [16] or can be due a radiative mechanism for neutrino mass generation [17]. Once the right-handed neutrino fields are accepted in the theory, it is mandatory to also investigate the quasi-Dirac neutrino regime of the Dirac-Majorana mass term. The quasi-Dirac neutrinos are distinct from, so called, pseudo-Dirac neutrinos [7,18], which also exhibit quasi-degenerate mass spectrum, however they are composed exclusively of active neutrino flavors.

In this paper, a special form of mixing matrix corresponding to this case, which is constructed with two 3×3 unitary matrices, will be presented and motivated. By assuming single small Majorana component of neutrino masses, the oscillation probabilities and quantities measured in single and $0\nu\beta\beta$-decay experiments and in cosmology will be determined. Further, restriction on this parameter coming from oscillations of electron antineutrinos will be studied and consequences for observation of the $0\nu\beta\beta$-decay will be given.

2. Theory

The quasi-Dirac (QD) neutrino scenario requires a number N_R of right-handed neutrino fields added into the SM Lagrangian, which mix with the SM left-handed neutrino fields via both the Dirac mass matrix M_D and the Majorana mass matrices $M_{L,R}$, as opposed to the older idea of pseudo-Dirac neutrinos [7]. In this work, we limit ourselves to the natural case of $N_R = 3$. In that case, the neutrino Dirac–Majorana mass term in Lagrangian is given as

$$\mathcal{L}_m = \frac{1}{2} \begin{pmatrix} \overline{\nu_L} & \overline{\nu_R^c} \end{pmatrix} \mathcal{M} \begin{pmatrix} \nu_L^c \\ \nu_R \end{pmatrix} + \text{h.c.}, \qquad (2)$$

where the mass matrix $\mathcal{M}$ is a 6×6 symmetric matrix

$$\mathcal{M} = \begin{pmatrix} M_L & M_D \\ M_D^T & M_R \end{pmatrix}, \qquad (3)$$

where M_D is a 3×3 complex matrix parametrized by 18 real numbers, and $M_{L,R}$ are 3×3 complex symmetric matrices parametrized by 12 real numbers each. Altogether, it makes 42 real parameters, out of which six are mass eigenvalues and six are phases absorbed by three left-handed and three right-handed neutrino fields. One is left with 15 angles and 15 phases of the 6×6 unitary mixing matrix $\mathcal{U}$, which diagonalizes the mass matrix $\mathcal{M}$ according to

$$\mathcal{U}^T \tilde{\mathcal{M}} \mathcal{U} = \mathcal{M}, \tag{4}$$

where $\tilde{\mathcal{M}}$ is a diagonal mass matrix, which, in general, is given by three Dirac masses m_i and 3 mass splittings ϵ_i, so that six neutrino mass eigenvalues $m_i^\pm$ are parametrized as

$$m_i^\pm = \pm m_i + \epsilon_i. \tag{5}$$

The QD neutrino regime is defined by hierarchy $m_i \gg \epsilon_i$. This can be achieved in two different regimes: (A) $|M_D| \gg |M_{L,R}|$, or (B) $|M_L| \approx |M_R|$. In order to get the QD mass spectrum within the regime (B), extreme fine-tuning of elements of M_L and M_R is required, so that large contributions to the mass splitting cancel out. Clearly, some symmetry would be needed to make it natural. In what follows, the subject of our interest is regime (A).

A general parametrization of the 6×6 unitary diagonalization matrix can be introduced according to Xing [19] as a product of three unitary matrices (Here, we are using just slightly different parametrization from that used in [19], $\hat{\mathcal{U}} = \mathcal{A} \cdot \mathcal{X} \cdot \mathcal{S}$. The difference is in the ordering of matrices in the product).

$$\hat{\mathcal{U}} = \mathcal{X} \cdot \mathcal{A} \cdot \mathcal{S}, \tag{6}$$

where $\mathcal{A}$ and $\mathcal{S}$ mix exclusively active or sterile neutrino flavors, ν_L or ν_R, respectively,

$$\mathcal{A} \equiv \begin{pmatrix} U^T & 0 \\ 0 & \mathbb{1} \end{pmatrix} \quad \text{and} \quad \mathcal{S} \equiv \begin{pmatrix} \mathbb{1} & 0 \\ 0 & V^\dagger \end{pmatrix}, \tag{7}$$

each containing three angles and three phases. The remaining nine angles and nine phases are included in the matrix $\mathcal{X}$, for which the perturbative expansion up to the linear order in the mixing angles gathered in the general 3×3 matrix X is

$$\mathcal{X} = \begin{pmatrix} \mathbb{1} & X^\dagger \\ -X & \mathbb{1} \end{pmatrix} + \mathcal{O}(X^2). \tag{8}$$

For the purpose of the QD scenario, it is useful to reproduce the pure Dirac case for $X = 0$. This can be done by inserting a constant unitary matrix $\mathcal{T}$ into the definition of the QD unitary diagonalization matrix $\mathcal{U}$

$$\mathcal{U} = \mathcal{X} \cdot \mathcal{T} \cdot \mathcal{A} \cdot \mathcal{S}, \tag{9}$$

where

$$\mathcal{T} \equiv \frac{1}{\sqrt{2}} \begin{pmatrix} \mathbb{1} & -\mathbb{1} \\ \mathbb{1} & \mathbb{1} \end{pmatrix}. \tag{10}$$

The expansion of the matrix $\mathcal{U}$ up to first order in X is then given as

$$\mathcal{U} = \frac{1}{\sqrt{2}} \begin{pmatrix} (\mathbb{1} + X^\dagger) U^T & -(\mathbb{1} - X^\dagger) V^\dagger \\ (\mathbb{1} - X) U^T & (\mathbb{1} + X) V^\dagger \end{pmatrix} + \mathcal{O}(X^2). \tag{11}$$

From that, it can be seen that for $X = 0$ the unitary matrix diagonalizing the pure Dirac mass term is reproduced,

$$\tilde{\mathcal{M}}_{(\epsilon_i = 0)} = \mathcal{U}^*_{(X=0)} \mathcal{M}_{(M_{L,R}=0)} \mathcal{U}^\dagger_{(X=0)}, \tag{12}$$

provided that the Dirac mass matrix M_D is diagonalized by a bi-unitary transformation

$$\tilde{M} = U^\dagger M_D V \equiv \begin{pmatrix} m_1 & 0 & 0 \\ 0 & m_2 & 0 \\ 0 & 0 & m_3 \end{pmatrix}, \qquad (13)$$

$$U^\dagger U = \mathbf{1} = V^\dagger V, \qquad (14)$$

where the tilde denotes that the 3×3 matrix $\tilde{M}$ is diagonalized with the eigenvalues of M_D on its diagonal.

The elements of the matrix X are actually calculable perturbatively from the known entries of the neutrino mass matrix, M_D and $M_{L,R}$, under the assumption that the X is just small perturbation of the Dirac diagonalization matrix at the same level as the Majorana masses $M_{L,R}$ are small perturbations of the purely Dirac mass matrix $\mathcal{M}_{(M_{L,R}=0)}$, i.e., under the assumption

$$|X| \sim |M_{L,R}|/|M_D|. \qquad (15)$$

The perturbative diagonalization of the Dirac–Majorana mass matrix (3) to the first order gives the relation between X and $M_{L,R}$, dependent on M_D and its bi-unitary diagonalization,

$$X^*\tilde{M} + \tilde{M}X = -\frac{1}{2}\left[U^\dagger M_L U^* - V^T M_R V\right], \qquad (16)$$

which is obtained by the requirement that the off-diagonal blocks of $\mathcal{U}^*\mathcal{M}\mathcal{U}^\dagger$ vanish to the first order in X and $M_{L,R}$. Simultaneously, the perturbative expressions for diagonal blocks $M^\pm$ of QD mass matrix after the block-diagonalization to the first order in X and $M_{L,R}$ is given as

$$M^\pm = \pm\tilde{M} + \frac{1}{2}\left[U^\dagger M_L U^* + V^T M_R V\right] + \ldots, \qquad (17)$$

which should be further diagonalized in order to come to the mass eigenvalues (5). Interestingly, the X does not enter the first-order expression for the block-diagonalized masses.

The Equations (11) and (17) are the three-flavor generalization of the toy one-flavor QD neutrino case discussed in [14,15]. It exhibits the same feature that the mass splitting and mixing angles X are two independent sets of beyond-Dirac parameters. In the special case, when

$$U^\dagger M_L U^* = -V^T M_R V \qquad (18)$$

we encounter the analogous situation to the pseudo-Dirac neutrinos that are described in [18]. In that case, the neutrino mass matrix provides three pairs of eigenvalues $\pm\tilde{M}_{ii}$ degenerate in magnitude, which correspond to three Dirac neutrinos. The lepton number violating masses M_{LR} are, however, non-zero, as well as the beyond-Dirac mixing angles X. These new Dirac neutrinos carry a new lepton number $\hat{L}$, which is, however, explicitly broken by weak interactions. Weak interactions generate tiny mass splitting [18] (One should be careful here, as the condition (18) is derived from the perturbative expressions linear in M_L and M_R. It is expected that higher-order terms also lift the degeneracy).

In the analysis within the present work, we will use just simplified model, which is exclusively focused on studying the effects of neutrino mass splitting. Therefore, we set all beyond-Dirac mixing angles to zero, i.e.,

$$X = 0. \qquad (19)$$

As a consequence, the first order expression for the block-diagonalized masses (17) becomes exact. On top of that, again for simplicity, we choose M_L and M_R in such way that

$$\frac{1}{2}\left[U^\dagger M_L U^* + V^T M_R V\right] = \epsilon\mathbf{1}, \qquad (20)$$

leading to the simplified QD neutrino mass spectrum with a universal Majorana mass splitting ϵ

$$m_i^\pm = \pm m_i + \epsilon, \quad \epsilon > 0. \tag{21}$$

As a result of these assumptions, it is the matrix

$$\mathcal{U}_{QD} = \frac{1}{\sqrt{2}} \begin{pmatrix} U & U \\ -V^* & V^* \end{pmatrix}, \tag{22}$$

which plays the role of the QD 6×6 generalization of the PMNS mixing matrix.

3. Consequences of Our Specific QD Scenario for Processes Measuring Neutrino Masses

The general formula for probabilities of neutrino oscillations from flavor α to flavor β for our specific scenario that is given by (21) and (22) can be written as

$$P_{\alpha\beta} = \left| \sum_{i=1}^{6} \mathcal{U}_{QD,\alpha i}^* \mathcal{U}_{QD,\beta i} e^{-i \tilde{M}_{ii}^2 L/2E} \right|^2. \tag{23}$$

From here, it can be clearly seen that, if α and β takes value only for active neutrino flavors, i.e., $\alpha, \beta = 1, 2, 3$, only the matrix U, and not V, is entering the oscillation probabilities. For the matrix U, we take the standard parmetrization

$$U = \begin{pmatrix} c_{12}c_{13} & s_{12}c_{13} & e^{-i\delta}s_{13} \\ -s_{12}c_{23} - e^{i\delta}c_{12}s_{13}s_{23} & c_{12}c_{23} - e^{i\delta}s_{12}s_{13}s_{23} & s_{23}c_{13} \\ s_{12}s_{23} - e^{i\delta}c_{12}s_{13}c_{23} & -c_{12}s_{23} - e^{i\delta}s_{12}s_{13}c_{23} & c_{13}c_{23} \end{pmatrix} \begin{pmatrix} 1 & 0 & 0 \\ 0 & e^{i\alpha_{21}} & 0 \\ 0 & 0 & e^{i(\alpha_{31}+\delta)} \end{pmatrix}, \tag{24}$$

where

$$c_{ij} \equiv \cos\theta_{ij}, \quad s_{ij} \equiv \sin\theta_{ij}, \tag{25}$$

θ_{12}, θ_{13} and θ_{23} are three mixing angles, δ is the CP violating Dirac phase and α_{21} and α_{31} are two CP violating Majorana phases. In terms of U, the oscillation probabilities among active neutrinos under our assumptions (19) and (20) are given by,

$$P_{\alpha\beta} = \delta_{\alpha\beta} - \sum_{i=1}^{3} |U_{\alpha i}|^2 |U_{\beta i}|^2 \sin^2 \frac{m_i \epsilon}{E} L - \sum_{i>j=1}^{3} \text{Re}(U_{\alpha i}^* U_{\beta i} U_{\alpha j} U_{\beta j}^*) \left(\sin^2 \frac{\Delta m_{ij}^2 + 2\epsilon \Delta m_{ij}}{4E} L \right.$$
$$+ \sin^2 \frac{\Delta m_{ij}^2 - 2\epsilon \Sigma m_{ij}}{4E} L + \sin^2 \frac{\Delta m_{ij}^2 + 2\epsilon \Sigma m_{ij}}{4E} L + \sin^2 \frac{\Delta m_{ij}^2 - 2\epsilon \Delta m_{ij}}{4E} L \right)$$
$$+ \frac{1}{2} \sum_{i>j=1}^{3} \text{Im}(U_{\alpha i}^* U_{\beta i} U_{\alpha j} U_{\beta j}^*) \left(\sin \frac{\Delta m_{ij}^2 + 2\epsilon \Delta m_{ij}}{2E} L + \sin \frac{\Delta m_{ij}^2 - 2\epsilon \Sigma m_{ij}}{2E} L \right.$$
$$+ \sin \frac{\Delta m_{ij}^2 + 2\epsilon \Sigma m_{ij}}{2E} L + \sin \frac{\Delta m_{ij}^2 - 2\epsilon \Delta m_{ij}}{2E} L \right) \tag{26}$$

The matrix V enters the probability for oscillations, in which sterile neutrino flavors, ν_R, are involved. With $\epsilon = 0$, Equation (26) reproduces the well known expression of oscillation probability for three-neutrino mixing.

The oscillation probabilities are functions of 15 mass-squared differences. Among them, just five are independent and are expressed in terms of 6 parameters, either $m_i^\pm$, or m_i and ϵ_i. Within our constrained neutrino mass spectrum (21), $\epsilon_i = \epsilon$, they are explicitly given as

$$(m_i^\pm)^2 - (m_j^\pm)^2 = \Delta m_{ij}^2 \pm 2\epsilon \Delta m_{ij}, \qquad (m_i^\pm)^2 - (m_j^\mp)^2 = \Delta m_{ij}^2 \pm 2\epsilon \Sigma m_{ij} \tag{27}$$

for $i < j = 1, 2, 3$, and for $i = j = 1, 2, 3$

$$(m_i^+)^2 - (m_i^-)^2 = 4\epsilon \Delta m_{ij}. \tag{28}$$

There are just four parameters, m_i and ϵ. These can be traded for another set of four parameters, Δm_{21}^2, Δm_{31}^2, m_1 and ϵ, as we can write

$$\Delta m_{32}^2 = \Delta m_{31}^2 - \Delta m_{21}^2, \quad \Delta m_{ij} = m_i - m_j, \quad \Sigma m_{ij} = m_i + m_j,$$

$$m_2 = \sqrt{\Delta m_{21}^2 + m_1^2}, \quad m_3 = \sqrt{\Delta m_{31}^2 + m_1^2}. \tag{29}$$

As a result, due to the additional assumption (21), $\epsilon_i = \epsilon$, we can completely fix the neutrino mass spectrum, including its absolute mass scale by fitting all oscillation frequencies given by five independent mass-squared differences. This is, of course, not possible in general case with three independent mass splittings ϵ_i, as fixing of the five independent mass-squared differences is not enough to determine six mass parameters, m_i and ϵ_i.

The amplitude for the $0\nu\beta\beta$ decay is given by the effective Majorana neutrino mass defined as

$$m_{\beta\beta} = [M_L]_{ee} = [U_{QD} \tilde{\mathcal{M}} U_{QD}^T]_{ee}. \tag{30}$$

For our constrained case $\epsilon_i = \epsilon$ and $X = 0$, it reduces to the expression

$$m_{\beta\beta} = \left| \sum_{i=1}^{3} U_{ei}^2 \epsilon \right| = \epsilon \left| c_{12}^2 c_{13}^2 + e^{2i\alpha_{21}} c_{13}^2 s_{12}^2 + e^{2i\alpha_{31}} s_{13}^2 \right|. \tag{31}$$

It means that the effective neutrino mass for $0\nu\beta\beta$ decay is, in our scenario, directly proportional to the mass splitting ϵ with the factor of proportionality of the order of $\sim \mathcal{O}(1)$ given the best fit values for θ_{12} and θ_{13}, and for marginalized values of the Majorana phases α_{21} and α_{31}.

The effective electron neutrino mass for single beta decay is in our case

$$m_\beta = \sqrt{\sum_{i=1}^{6} |U_{QD,ei}|^2 \tilde{\mathcal{M}}_{ii}^2} = \sqrt{\sum_{i=1}^{3} |U_{ei}|^2 (m_i^2 + \epsilon^2)}$$

$$= \sqrt{m_1^2 c_{12}^2 c_{13}^2 + m_2^2 c_{13}^2 s_{12}^2 + m_3^2 s_{13}^2 + \epsilon^2} = m_\beta^{(0)} \left(1 + \frac{1}{2} \left(\epsilon / m_\beta^{(0)} \right)^2 + \ldots \right), \tag{32}$$

where $m_\beta^{(0)} \equiv m_\beta|_{\epsilon=0}$ is the effective neutrino mass for standard three neutrino mixing case.

The sum of the six QD neutrino mass eigenvalues is the parameter relevant for cosmology. It turns out trivially that the cosmology is insensitive to the universal mass splitting ϵ used in our simplified model as long as $m_i > \epsilon$:

$$\frac{1}{2} \sum_{i=1}^{6} |\tilde{\mathcal{M}}_{ii}| = \sum_{i=1}^{3} m_i. \tag{33}$$

The factor of $\frac{1}{2}$ reflects the QD (or Dirac) nature of neutrinos, in which case, effectively, only two out of their four states are kept in equilibrium with cosmological plasma of the early Universe by the $V - A$ interactions of the SM.

4. The Survival Probabilities of Electron Antineutrino

In this section, we discuss the survival probability ($P_{\bar{\nu}_e \to \bar{\nu}_e}$) of electron antineutrino produced at the reactor with energy E and detected at the detector after traversing a baseline L. In three-flavor model of neutrino oscillation, the well-known expression of $P_{\bar{\nu}_e \to \bar{\nu}_e}$, which can be obtained by putting $\epsilon = 0$ in Equation (26), is given by

$$P_{\bar{\nu}_e \to \bar{\nu}_e}(\epsilon = 0) = 1 - 4c_{13}^4 s_{12}^2 c_{12}^2 \sin^2 \frac{\Delta m_{21}^2 L}{4E} - 4s_{13}^2 c_{13}^2 \left(c_{12}^2 \sin^2 \frac{\Delta m_{31}^2 L}{4E} + s_{12}^2 \sin^2 \frac{\Delta m_{32}^2 L}{4E}\right). \quad (34)$$

In above equation, Δm_{21}^2 and Δm_{31}^2 are solar and atmospheric mass-squared differences, respectively. The parameter θ_{12} and θ_{13} are mixing angles that are related to our matrix U [3,4]. One can see that the $\bar{\nu}_e$ survival probability is a function of mass-squared differences, and does not depend on the absolute mass scale of neutrino. In case of quasi-Dirac nature of neutrino, $\epsilon \neq 0$, and the mixing among active and sterile neutrinos modifies the neutrino oscillation probabilities. For $\epsilon \neq 0$, we obtain the expression of $\bar{\nu}_e$ survival probability from Equation (26), as follows,

$$P_{\bar{\nu}_e \to \bar{\nu}_e}(\epsilon \neq 0) = 1 - P_1 - 4c_{13}^4 c_{12}^2 s_{12}^2 F_{21} - 4s_{13}^2 c_{13}^2 (c_{12}^2 F_{31} + s_{12}^2 F_{32}), \quad (35)$$

with

$$P_1 = c_{13}^4 c_{12}^4 \sin^2 \frac{m_1 \epsilon L}{E} + c_{13}^4 s_{12}^4 \sin^2 \frac{m_2 \epsilon L}{E} + s_{13}^4 \sin^2 \frac{m_3 \epsilon L}{E}, \quad (36)$$

$$F_{ij} = \frac{1}{4}\left(\sin^2 \frac{\Delta m_{ij}^2 + 2\epsilon \Delta m_{ij}}{4E}L + \sin^2 \frac{\Delta m_{ij}^2 - 2\epsilon \Delta m_{ij}}{4E}L \right.$$
$$\left. + \sin^2 \frac{\Delta m_{ij}^2 + 2\epsilon \sum m_{ij}}{4E}L + \sin^2 \frac{\Delta m_{ij}^2 - 2\epsilon \sum m_{ij}}{4E}L\right). \quad (37)$$

For very small ϵ, with an approximation of $\epsilon \Delta m_{ij} \ll \Delta m_{ij}^2$, Equation (37) boils down to the following simplified expression of survival probability of $\bar{\nu}_e$,

$$P_{\bar{\nu}_e \to \bar{\nu}_e}(\epsilon \neq 0) = P_{\bar{\nu}_e \to \bar{\nu}_e}(\epsilon = 0) - \frac{\epsilon^2 L^2}{E^2}\left[c_{13}^4 c_{12}^4 m_1^2 + c_{13}^4 s_{12}^4 m_2^2 + s_{13}^4 m_3^2\right]$$
$$- \frac{\epsilon^2 L^2}{4E^2}\left[4 c_{13}^4 s_{12}^2 c_{12}^2 \Sigma m_{21}^2 \cos \frac{\Delta m_{21}^2 L}{2E} + 4 s_{13}^2 c_{13}^2 c_{12}^2 \Sigma m_{31}^2 \cos \frac{\Delta m_{31}^2 L}{2E}\right.$$
$$\left. + 4 s_{13}^2 c_{13}^2 s_{12}^2 \Sigma m_{32}^2 \cos \frac{\Delta m_{32}^2 L}{2E}\right] + \mathcal{O}(\epsilon^4), \quad (38)$$

where $\Sigma m_{ij}^2 \equiv m_i^2 + m_j^2$. Using this simple expression, we can explain the following features of $\bar{\nu}_e$ survival probability with a small value of ϵ.

- We see that $P_{\bar{\nu}_e \to \bar{\nu}_e}(\epsilon \neq 0) - P_{\bar{\nu}_e \to \bar{\nu}_e}(\epsilon = 0)$ is directly proportional to L/E ratio multiplied with ϵ. Thus, as we go to higher L/E, the effect of ϵ becomes larger.
- Since θ_{13} is very small when compared to θ_{12}, the term $m_3^2 \sin^4 \theta_{13}$ in second part of Equation (38) becomes negligible as compared to two other terms containing m_1 and m_2, respectively. As a result, the second part of Equation (38) in the case of NO ($m_1 < m_2 < m_3$) is always smaller than that in IO scenario ($m_3 < m_1 < m_2$). Therefore, we expect the modification in survival probability due to non-zero ϵ to be smaller when the mass pattern is NO than that for IO.

In Figure 1, we show the survival probabilities of electron antineutrino as a function of energy for 1.5 km (top panels), 53 km (middle panels), and 180 km (bottom panels) baselines—relevant for short, medium, and long-baseline reactor neutrino oscillation experiments, respectively. We show the probabilities for three cases: (i) $\epsilon = 0$, 3ν mixing case, (ii) $\epsilon = 10^{-4}$ eV, and (iii) $\epsilon = 2 \times 10^{-4}$ with black, green, and red lines, respectively. The plots shown in left (right panels) are with normal (inverted) ordering for which m_1 (m_3) is lightest Dirac mass. Here, for cases (ii) and (iii), lightest Dirac mass is assumed to be 0.01 eV. The value of oscillation parameters that we use in this study are given in Table 1. These values are similar as obtained in the global fit to neutrino oscillation data [20–22]. Note that, if the neutrino oscillation data are fitted in the current framework with six quasi-Dirac neutrinos, the best-fit value of mixing angles and mass-squared difference may be slightly different than that we use. However, we expect that these values will not be beyond the current 3σ allowed range, as obtained in the global fit of neutrino data in three-flavor Dirac neutrino mixing framework [20–22].

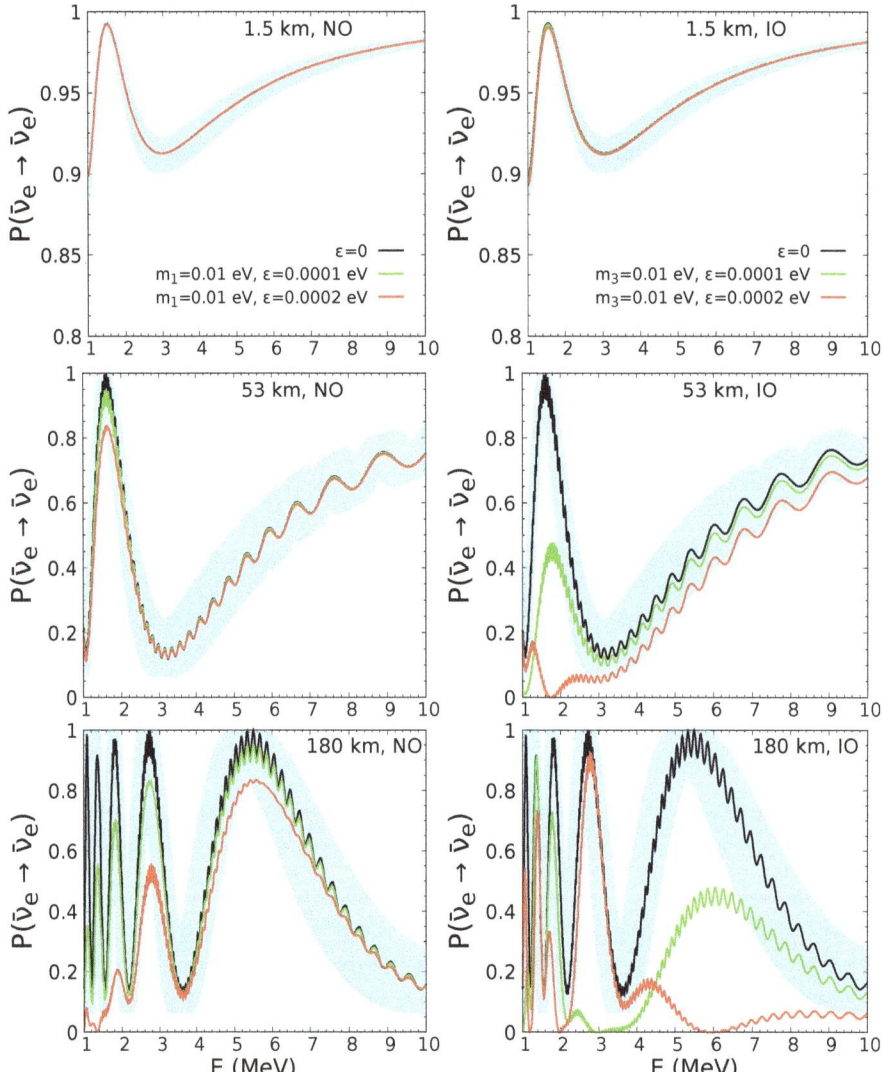

Figure 1. The survival probabilities of $\bar{\nu}_e$ as the functions of energy E for 1.5 km (upper panel), 53 km (middle panel), and 180 km (lower panel) baselines. The plots in left and right panels are with NO and IO as mass orderings, respectively. The probabilities are shown for three cases: (i) $\epsilon = 0$, 3ν mixing case, (ii) $\epsilon = 10^{-4}$ eV, and (iii) $\epsilon = 2 \times 10^{-4}$ eV, with black, green, and red lines, respectively. In cases (ii) and (iii), the lightest Dirac mass is 0.01 eV. The cyan bands represent $\bar{\nu}_e$ survival probabilities in three-flavor neutrino oscillation framework with 3σ allowed range of oscillation parameters. The benchmark values of oscillation parameters along and their 3σ allowed ranges, as used in this study, are given in Table 1.

The common feature that emerges from all of the panels of Figure 1 is the larger effect of non-zero ϵ in $\bar{\nu}_e$ survival probabilities for IO than NO. The reason behind this is already discussed while using

Equation (37). In case of $L = 1.5$ km baseline, the survival probability of $\bar{\nu}_e$ shown in red and green lines (cases *ii* and *iii*) are exactly same as black line (with $\epsilon = 0$), as can be seen from top panels of Figure 1. This is true for both the mass orderings, NO and IO. From this observation, one can infer that the short-baseline experiments will not be able to see the signal for the quasi-Dirac nature of neutrino if ϵ is of the order of 10^{-4} eV. For baseline $L = 53$ km (see middle panels of Figure 1), $\bar{\nu}_e$ survival probabilities with cases (ii) and (iii) are similar to case (i), except a small difference at $E \approx 1.8$ MeV if mass ordering is NO. However, with IO, $\bar{\nu}_e$ survival probabilities get modified by a large amount in the whole range of E (1 MeV to 10 MeV) due to non-zero ϵ considered here. This proves that medium baseline neutrino oscillation experiment, like JUNO, will be able to see the signal for quasi-Dirac nature of neutrino with $\epsilon \sim 10^{-4}$ eV only if mass ordering is IO. As we go to higher L, for an example $L = 180$ km (see bottom panels of Figure 1), $\bar{\nu}_e$ survival probabilities with $\epsilon \sim 10^{-4}$ eV (green and red lines) are significantly different than that of the three-flavor. Thus, we expect that long-baseline reactor neutrino experiments are suitable for providing better constraint on ϵ.

Table 1. The benchmark values of the oscillation parameters and their 1σ and 3σ allowed ranges that we use in this paper. These values are similar, as obtained in the global fit to neutrino oscillation data [20–22].

Parameters	Best Fit Values	1σ Range	3σ Range
θ_{12} (°)	34	[33.1, 34.6]	[31, 37]
θ_{13} (°)	8.5	[8.48, 8.74]	[40, 53]
$\|\Delta m_{31}^2/10^{-3}\|$ (eV^2)	2.5	[2.49, 2.55]	[2.4, 2.6]
$\Delta m_{21}^2/10^{-5}$ (eV^2)	7.5	[7.2, 7.6]	[6.7, 8.0]

5. Constraints on Majorana Component of Neutrino Masses

A preliminary idea about the allowed values of lightest Dirac mass and ϵ can be achieved from $\bar{\nu}_e$ survival probabilities for a fixed neutrino energy and baseline. Keeping the oscillation parameters fixed at the benchmark values, we scan the lightest Dirac mass (m_1 for NO and m_3 for IO) and ϵ in the range of 10^{-5} eV to 0.1 eV to reproduce the $\bar{\nu}_e$ survival probabilities in the range that is allowed by the three-flavor neutrino oscillation framework and the current 1σ or 3σ uncertainties of oscillation parameters.

Figure 2 presents the allowed region in the plane of lightest Dirac mass and ϵ which we obtain following the above mentioned method for 1.5 km baseline and two fixed energies 4 MeV (top panels) and 8 MeV (bottom panels) for demonstration purpose. We present these limits with benchmark values (green line), 1σ (pink and red lines) and 3σ (cyan and blue lines) allowed range of oscillation parameters. The limits in the case of IO (right panels) is more stringent than that for NO (left panels) for both of the energies due to a larger effect of ϵ in $\bar{\nu}_e$ survival probabilities for IO, which is explained in Section 4.

If we ignore the small features above $\epsilon > 0.01$ eV, which are expected to have just a small chance to survive after the full oscillation data analysis of the QD scenario, one may say that the region below the cyan line in all of the panels Figure 2 are allowed by 3σ uncertainty of oscillation parameters. Here, we do not demand that these limits are final since the detailed statistical analysis with spectral information of events with detector properties would give the concrete results. Our attempt here is to demonstrate the validity of the theory that we propose based on the oscillation probabilities in a simplified manner. The study of quasi-Dirac neutrino with detailed analysis of events at the neutrino oscillation experiments to constraint the lightest Dirac mass and ϵ in this framework would be interesting for future study.

Obviously, Figure 2 shows larger region of ϵ-$m_1(m_3)$ parametric space than corresponding to QD scenario defined by $m_i \gg \epsilon$. Namely, the additional regions lie around and above the diagonal axis, where the lightest Dirac mass gets comparable and smaller to ϵ, respectively. The allowed region, below the cyan line, however still guarantees that at least the heaviest pair of neutrinos is of QD nature.

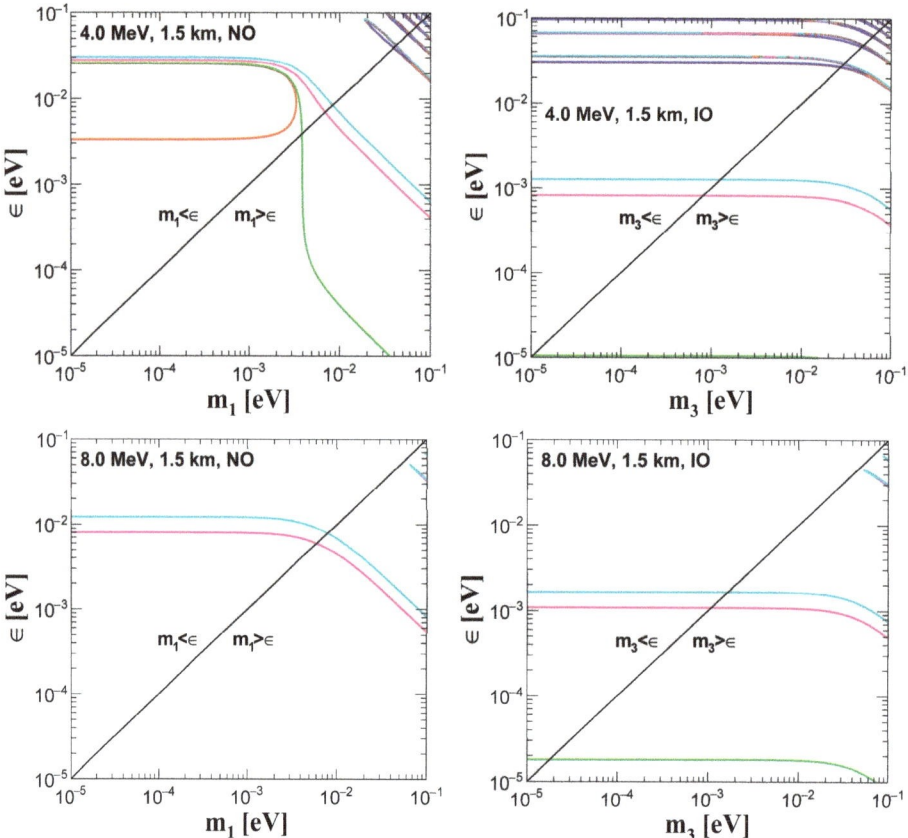

Figure 2. The top (bottom) panels show the allowed ranges of lightest Dirac mass and ϵ that are obtained using 1.5 km baseline and 4 MeV (8 MeV) energy. The left and right panels are obtained with NO and IO, respectively. In all of the panels, pink (cyan) and red (blue) lines correspond to minimum and maximum $\bar{\nu}_e$ survival probabilities, respectively, allowed in three-flavor neutrino oscillation picture with 1σ (3σ) uncertainty of oscillation parameters. The green line corresponds to $\bar{\nu}_e$ survival probability in 3ν framework with the benchmark value of oscillation parameters. For details, see text. The benchmark values of oscillation parameters along with their 1σ and 3σ allowed ranges, used in this study, are given in Table 1.

If the limits that are given by the cyan lines (3σ) in the Figure 2 will be approved by the future full analysis, the conservative limits on the effective neutrino Majorana mass within our simplified QD model will be

$$m_{\beta\beta} \lesssim 30 \text{ meV} \quad \text{for} \quad \text{NO},$$
$$\lesssim 1 \text{ meV} \quad \text{for} \quad \text{IO}. \qquad (39)$$

As seen from Equation (31), the $m_{\beta\beta}$ represents linear effect of ϵ, in contrast to the correction to the effective electron mass for single-beta decay m_β given in Equation (32), which is quadratic in ϵ. Therefore, the effect of ϵ on the single-beta decay within our simplified QD model is expected to be, at most, at the level of few percent, as long as $m_1 > \epsilon$ in NO or $m_3 > \epsilon$ in IO. The cosmology is in this regime even completely insensitive to ϵ, see Equation (33).

The above conclusions are valid for the universal Majorana mass contribution ϵ to neutrino masses. More comprehensive analysis with non-universal ϵ_i ($i = 1$ to 3) and mixing angles as free parameters could modify this conclusion about $m_{\beta\beta}$ and its dependence on Dirac component of lightest neutrino mass via neutrino oscillation analysis. Recall that the oscillation probabilities of neutrinos depend on both Dirac m_i and Majorana ϵ_i masses.

6. Conclusions

The quasi-Dirac neutrino mixing scheme incorporating three pairs of Majorana neutrinos with quasi-degenerate masses was analyzed. The diagonalization of the Dirac-Majorana mass term with 6×6 unitary neutrino mixing matrix, which is generally parametrized with 15 mixing angles and 15 CP-violating phases, was discussed. By exploiting the limiting case of three Dirac neutrinos and assuming a small Majorana component in neutrino masses, the quasi-Dirac 6×6 neutrino mixing matrix $\mathcal{U}_{QD}$ constructed with two 3×3 unitary mixing matrices, each of them incorporating three mixing angles and three phases, was proposed. For the sake of simplicity, only left-handed weak interaction of the SM was assumed. It was found that probabilities for oscillations of 3 flavor neutrinos (ν_e, ν_μ, ν_τ) can be described with eight parameters, namely three angles (θ_{12}, θ_{13}, and θ_{23}) and one Dirac CP phase δ having origin in a single unitary matrix, two mass squared differences (Δm^2_{21} and Δm^2_{31}), the lightest neutrino mass (m_1 for normal ordering and m_3 for inverted ordering of Dirac neutrino masses), and a small Majorana neutrino mass parameter ϵ. Recall that, within the commonly considered mixing scheme of three neutrinos, the required number of parameters is only two less (three mixing angles, one phase, and two mass squared differences). Further, it was established that the effective Majorana mass entering the $0\nu\beta\beta$-decay rate is proportional to ϵ, the sum of neutrino masses measured by cosmology only contains contributions of Dirac masses m_1, m_2, and m_3, and that the effective neutrino mass measured in tritium β-decay is practically not affected by ϵ as long as it is a small quantity when compared to the Dirac masses, $\epsilon < m_i$.

The first simplified analysis of parameters of 3+3 quasi-Dirac neutrino mixing scheme was performed by exploiting the 1σ and 3σ uncertainty of measured probability of $\bar{\nu}_e$ oscillations from a reactor. The mixing angles θ_{12}, θ_{13}, and mass squared differences Δm^2_{21} and Δm^2_{31} were considered to be those that were determined within the PMNS unitary mixing scheme with three neutrinos. The lightest Dirac neutrino mass and ϵ were considered to be free parameters. It was manifested that a tiny value of ϵ below 1 MeV is not affecting the current phenomenology representing, e.g., by the Daya–Bay experiment, but they can have significant impact on the JUNO experiment registering oscillations of antineutrinos at significantly larger distance. A detailed study on restriction of ϵ, which depends on the considered value of lightest neutrino mass and ordering of neutrinos (normal or inverted) was presented. We keep our study simplified to have better insight. For this, in this paper, we study the consequences of quasi-Dirac scenario in neutrino oscillation when only considering the reactor experiments, since the oscillation channel that governs the disappearance of reactor $\bar{\nu}_e$ is independent of Dirac CP phase as well as same as vacuum oscillation probabilities. It goes without saying that a more comprehensive analysis covering oscillations of atmospheric, solar, and terrestrial neutrinos in which all involved parameters are assumed to be free are a subject of interest. The three Majorana constituents of the neutrino mass would be considered as a free parameters and a comprehensive study of all parameters of this model would be performed by considering data of all types of neutrino oscillations experiments. It might be that due to a large number of degrees of freedom some unconventional solutions could be found, which will require additional neutrino oscillations experiments with different baselines, energy, and high statistics of data. Of course, this task is beyond the scope of the present article.

Author Contributions: A.K., A.S. and F.Š. contributed equally to the manuscript. All authors have read and agreed to the published version of the manuscript.

Funding: This work was funded by the VEGA Grant Agency of the Slovak Republic under Contract No. 1/0607/20 and by the Ministry of Education, Youth and Sports of the Czech Republic under the INAFYM Grant No. CZ.02.1.01/0.0/0.0/16_019/0000766.

Conflicts of Interest: The authors declare no conflict of interest.

References

1. Tanabashi, M.; Hagiwara, K.; Hikasa, K.; Nakamura, K.; Sumino, Y.; Takahashi, F.; Tanaka, J.; Agashe, K.; Aielli, G.; Amsler, C.; et al. Review of Particle Physics. *Phys. Rev. D* **2018**, *98*, 030001. [CrossRef]
2. Bilenky, S. Neutrino oscillations: From a historical perspective to the present status. *Nucl. Phys. B* **2016**, *908*, 2–13. [CrossRef]
3. Pontecorvo, B. Neutrino Experiments and the Problem of Conservation of Leptonic Charge. *Sov. Phys. JETP* **1968**, *26*, 984–988.
4. Maki, Z.; Nakagawa, M.; Sakata, S. Remarks on the unified model of elementary particles. *Prog. Theor. Phys.* **1962**, *28*, 870–880. [CrossRef]
5. Vergados, J.D.; Ejiri, H.; Šimkovic, F. Neutrinoless double beta decay and neutrino mass. *Int. J. Mod. Phys. E* **2016**, *25*, 1630007. [CrossRef]
6. Dolinski, M.J.; Poon, A.W.; Rodejohann, W. Neutrinoless Double-Beta Decay: Status and Prospects. *Ann. Rev. Nucl. Part. Sci.* **2019**, *69*, 219–251. [CrossRef]
7. Bilenky, S.M.; Petcov, S. Massive Neutrinos and Neutrino Oscillations. *Rev. Mod. Phys.* **1987**, *59*, 671; Erratum in **1989**, *61*, 169; Erratum in **1988**, *60*, 575–575. [CrossRef]
8. Gell-Mann, M.; Ramond, P.; Slansky, R. Complex Spinors and Unified Theories. *Conf. Proc. C* **1979**, *790927*, 315–321.
9. Yanagida, T. Horizontal gauge symmetry and masses of neutrinos. *Conf. Proc. C* **1979**, *7902131*, 95–99.
10. Mohapatra, R.N.; Senjanović, G. Neutrino Mass and Spontaneous Parity Nonconservation. *Phys. Rev. Lett.* **1980**, *44*, 912–915. [CrossRef]
11. Schechter, J.; Valle, J.W.F. Neutrino masses in SU(2) ⊗ U(1) theories. *Phys. Rev. D* **1980**, *22*, 2227–2235. [CrossRef]
12. Valle, J.W.F. Neutrinoless double-β decay with quasi-Dirac neutrinos. *Phys. Rev. D* **1983**, *27*, 1672–1674. [CrossRef]
13. de Gouvêa, A.; Huang, W.C.; Jenkins, J. Pseudo-Dirac neutrinos in the new standard model. *Phys. Rev. D* **2009**, *80*, 073007. [CrossRef]
14. Anamiati, G.; Fonseca, R.M.; Hirsch, M. Quasi-Dirac neutrino oscillations. *Phys. Rev. D* **2018**, *97*, 095008. [CrossRef]
15. Anamiati, G.; De Romeri, V.; Hirsch, M.; Ternes, C.A.; Tórtola, M. Quasi-Dirac neutrino oscillations at DUNE and JUNO. *Phys. Rev. D* **2019**, *100*, 035032. [CrossRef]
16. Dienes, K.R.; Dudas, E.; Gherghetta, T. Neutrino oscillations without neutrino masses or heavy mass scales: A Higher dimensional seesaw mechanism. *Nucl. Phys. B* **1999**, *557*, 25–59. [CrossRef]
17. Cai, Y.; Herrero-García, J.; Schmidt, M.A.; Vicente, A.; Volkas, R.R. From the trees to the forest: A review of radiative neutrino mass models. *Front. Phys.* **2017**, *5*, 63. [CrossRef]
18. Wolfenstein, L. Different Varieties of Massive Dirac Neutrinos. *Nucl. Phys. B* **1981**, *186*, 147–152. [CrossRef]
19. Xing, Z.Z. A full parametrization of the 6 X 6 flavor mixing matrix in the presence of three light or heavy sterile neutrinos. *Phys. Rev. D* **2012**, *85*, 013008. [CrossRef]
20. Capozzi, F.; Di Valentino, E.; Lisi, E.; Marrone, A.; Melchiorri, A.; Palazzo, A. Addendum to: Global constraints on absolute neutrino masses and their ordering. *arXiv* **2020**, arXiv:2003.08511.
21. De Salas, P.; Gariazzo, S.; Mena, O.; Ternes, C.; Tórtola, M. Neutrino Mass Ordering from Oscillations and Beyond: 2018 Status and Future Prospects. *Front. Astron. Space Sci.* **2018**, *5*, 36-1–36-50. [CrossRef]
22. Esteban, I.; Gonzalez-Garcia, M.C.; Hernandez-Cabezudo, A.; Maltoni, M.; Schwetz, T. Global analysis of three-flavour neutrino oscillations: synergies and tensions in the determination of θ_{23}, δ_{CP}, and the mass ordering. *JHEP* **2019**, *1*, 106. [CrossRef]

© 2020 by the authors. Licensee MDPI, Basel, Switzerland. This article is an open access article distributed under the terms and conditions of the Creative Commons Attribution (CC BY) license (http://creativecommons.org/licenses/by/4.0/).

Article

Rephasing Invariant for Three-Neutrino Oscillations Governed by a Non-Hermitian Hamiltonian

Dmitry V. Naumov, Vadim A. Naumov * and Dmitry S. Shkirmanov

JINR, Dubna 141980, Russia; dnaumov@jinr.ru (D.V.N.); shkirmanov@theor.jinr.ru (D.S.S.)
* Correspondence: vnaumov@theor.jinr.ru

Received: 18 June 2020; Accepted: 13 July 2020; Published: 3 August 2020

Abstract: Time-reversal symmetry is broken for mixed and possibly unstable Dirac neutrino propagation through absorbing media. This implies that interplay between the neutrino mixing, refraction, absorption and/or decay can be described by non-Hermitian quantum dynamics. We derive an identity which sets up direct connection between the fundamental neutrino parameters (mixing angles, CP-violating phase, mass-squared splittings) in vacuum and their effective counterparts in matter.

Keywords: neutrino oscillations in matter; rephasing invariant; neutrino absorption

1. Introduction

High-energy neutrinos, unique messengers of the most violent processes that occurred during the evolution of the Universe, are under extensive study by the modern neutrino telescopes (see reference [1] for a comprehensive recent review and further references). The propagation of these particles through dense matter requires a theoretical consideration accounting for two major phenomena. (i) The quantum coherence and decoherence, most clearly manifested in the neutrino oscillation phenomenon, firmly established experimentally [2–10]. The corresponding theoretical approaches rely on either quantum mechanical [11–13] or quantum field theory [14–20] considerations. (ii) Neutrino production, inelastic interactions, and possible decays, typically considered by the classical transport theory [21–23].

In this paper we consider a more particular aspect of the full problem—propagation of high-energy neutrinos in dense environment with accounting for neutrino masses, mixing, CP violation, refraction, and absorption. We do not consider neutrino energy loss through neutral-current (NC) interactions and charged-current (CC) induced reaction chains, but of course we take into account disappearance of the neutrinos due to all these processes. In other words, the formalism does not predict the energy spectrum transformation due to the energy losses. This is acceptable in the case of sufficiently narrow boundary energy spectrum or nearly-monochromatic neutrino source, when we are interested in the flavor evolution at the same energy as on the boundary or in the source (e.g., annihilating non-relativistic WIMS). Since, in this statement of the problem, neutrinos simply disappear with time (due to both CC and NC interactions), the time-reversal symmetry is broken and the neutrino flavor evolution can be described within a non-Hermitian formulation of quantum mechanics. The inclusion of the neutrino energy loss effects is of course very important in more general and practically interesting conditions, but it will also require the more universal formalism, like quantum kinetic equations or a hybrid (approximate) technique based on the non-Hermitian dynamics and classical transport theory. One of the simplest realization of the hybrid approach is in the replacement of the mean free paths Λ_α (see Section 2) to effective functions $\widetilde{\Lambda}_\alpha$ derived from solution of the classical transport equations [22] for the given initial spectrum/source and given profiles of density and composition of the medium along the neutrino beam direction. Such a method conserves the generic results of the present study. The approach based on the the non-Hermitian dynamics has been considered earlier in reference [24], for a generic three-level system and latter in reference [25], for a simplified two-flavor mixing model, which included either the mixing between the active (standard) or active and sterile neutrinos; see also references [26,27] for recent developments and further references. Here we follow these studies and consider the Standard Model's three-neutrino species.

Consideration of the neutrino oscillation phenomenon in the simplest adiabatic regime usually requires a diagonalization of the corresponding Hamiltonian. The instantaneous eigenstates are defined not uniquely but up to certain ("rephasing") transformations, keeping the observable transition and survival probabilities $P_{\alpha\beta} \equiv P_{\nu_\alpha \to \nu_\beta}$ invariant. This is discussed in greater details in Section 3. An important class of observables invariant under the same transformations is known as flavor-symmetric Jarlskog invariants, introduced by Jarlskog [28] for quarks. In the three-generation case, the nine Jarlskog invariants are equal and uniquely determine the amount of CP violation in the quark sector of the Standard Electroweak Model. Similar rephasing invariants determine the amount of CP violation in the lepton sector. In present work, we found an extension of the Jarlskog invariants for the dissipating three-neutrino system; this is one of the results of this study.

As was first pointed out by Wolfenstein [29], the neutrino mixing is modified when neutrinos propagate through normal C-asymmetric matter, owing to the CC forward scattering of electron neutrinos on electrons in matter. In some circumstances, these ghostly interactions may drastically modify the neutrino oscillation pattern [30,31]. It is however interesting that a nontrivial observable proportional to the Jarlskog invariant, J, is also a "matter invariant". More precisely, in references [32,33] (see also reference [34] for a relevant result), an identity has been found which relates the products of J and neutrino squared-mass splittings, $\Delta m_{ij} = m_i^2 - m_j^2$, in vacuum and in matter:

$$J \Delta m_{23}^2 \Delta m_{31}^2 \Delta m_{12}^2 = \tilde{J} \Delta \tilde{m}_{23}^2 \Delta \tilde{m}_{31}^2 \Delta \tilde{m}_{12}^2. \tag{1}$$

Here tilde marks the quantities perturbed by the matter. This identity has proved to be useful in various phenomenological and mathematical aspects of the oscillating neutrino propagation in matter [35–80]. The main result of the present work is a generalization of the identity (1) to the case of the neutrino propagation in absorbing media, which can be described by non-Hermitian dynamics. Since the quantities in the RHS of Equation (1) are defined as instantaneous functions, the adiabaticity conditions are not formally essential (so we do not study the corresponding constraints). However, the actual usage of the generalized identity is mainly reasonable in the environments where the neutrino flavors evolve adiabatically or quasi-adiabatically. It is also pertinent to note that the adiabatic solution can be adapted to form the basis of a numerical method: by dividing the medium into a number of layers with slowly varying densities, the solution is obtained as chronological product of the (non-unitary) evolution operators for each layer [25]. Though, our primary interest is motivated by the neutrino oscillation phenomenon, the obtained identity has a much wider range of applicability relevant to arbitrary quantum three-level system governed by a non-Hermitian Hamiltonian.

The paper is organized as follows. The master equation and appropriate theoretical framework are considered in Section 2. In Section 3 we introduce two "mixing matrices" for a generic three-level quantum dissipative system, describing, in particular, the neutrino mixing, refraction, decay, and absorption due to standard or nonstandard inelastic neutrino-matter interactions. We show that these matrices are not uniquely defined. In Section 4 we study the generalized "rephasing" and "dynamic" invariants constructed from the elements of the mixing matrices and of the Hamiltonian matrix, respectively. Then we put forward the relation generalizing the identity (1). The proof of this relation is delivered in Appendix A. Finally, we draw the summary in Section 5. Some auxiliary information is summarized in Appendix B.

2. Master Equation

The Schrödinger equation

$$i\frac{d}{dt}|\nu_f(t)\rangle = \mathbf{H}(t)|\nu_f(t)\rangle \tag{2}$$

describes the time evolution of the three-neutrino state

$$|v_f(t)\rangle = \left(|v_e(t)\rangle, |v_\mu(t)\rangle, |v_\tau(t)\rangle\right)^T \qquad (3)$$

governed by a Hamiltonian $\mathbf{H}(t)$. The bold face is used for matrices in what follows. The flavor v_α ($\alpha = e, \mu, \tau$) and mass v_i ($i = 1, 2, 3$) eigenstates are related to each other as

$$|v_\alpha\rangle = \sum_{i=1}^{3} V_{\alpha i} |v_i\rangle. \qquad (4)$$

This definition differs from that used in quantum filed theoretical (QFT) description. Their relationship is given by $V_{\text{QFT}} = V_{\text{QM}}^*$. Since the observables are flavor changing probabilities $P_{\alpha\beta}(t) = |\langle v_\beta | v_\alpha(t)\rangle|^2$, it is convenient to rewrite Equation (2) as one for the corresponding amplitudes

$$S_{\beta\alpha}(t) = \langle v_\beta | v_\alpha(t)\rangle \qquad (5)$$

as follows

$$i\frac{d}{dt}\mathbf{S}(t) = \mathbf{H}(t)\mathbf{S}(t) = \left[\mathbf{V}\mathbf{H}_0\mathbf{V}^\dagger + \mathbf{W}(t)\right]\mathbf{S}(t), \quad (\mathbf{S}(0) = \mathbf{1}), \qquad (6)$$

where $\mathbf{S}(t)$ is a matrix with elements $S_{\alpha\beta}(t)$ (evolution operator), $\mathbf{V}$ is the Pontecorvo–Maki–Nakagawa–Sakata (PMNS) mixing matrix with elements $V_{\alpha i}$, $\mathbf{H}_0$, and $\mathbf{W}(t)$ are the free and neutrino-matter interaction Hamiltonians, respectively,

$$\mathbf{H}_0 = \begin{pmatrix} E_1 & 0 & 0 \\ 0 & E_2 & 0 \\ 0 & 0 & E_3 \end{pmatrix}, \quad \mathbf{W}(t) = -p_v \begin{pmatrix} n_e(t) - 1 & 0 & 0 \\ 0 & n_\mu(t) - 1 & 0 \\ 0 & 0 & n_\tau(t) - 1 \end{pmatrix}, \qquad (7)$$

$E_i = \sqrt{p_v^2 + m_i^2} \simeq p_v + m_i^2/2p_v$ and m_i are, respectively, the total energies and masses of the neutrino mass eigenstates, and $n_\alpha(t)$ are the complex indices of refraction; where we assume, as usual, that neutrinos are ultrarelativistic, $p_v^2 \simeq E_v^2 \gg \max(m_i^2)$. In normal matter, the functions n_α are linear with respect to the densities of scatterers. The same is also true for hot media under the assumption that introduction of a finite temperature does not break the coherent condition [81]. With these assumptions

$$n_\alpha(t) = 1 + \frac{2\pi N_0 \rho(t)}{p_v^2} \sum_k Y_k(t) f_{v_\alpha k}(0), \qquad (8)$$

where $N_0 = 6.022 \times 10^{23}$ cm^{-3}, $f_{v_\alpha k}(0)$ is the amplitude for the v_α zero-angle scattering from particle k ($k = e, p, n, \ldots$), $\rho(t)$ is the density of the matter (in g/cm^3) and $Y_k(t)$ is the number of particles k per AMU in the point t of the medium. The optical theorem says (see, e.g., reference [82]):

$$\text{Im}\left[f_{v_\alpha k}(0)\right] = \frac{p_v}{4\pi} \sigma_{v_\alpha k}^{\text{tot}}(p_v), \qquad (9)$$

where $\sigma_{v_\alpha k}^{\text{tot}}(p_v)$ is the total cross section for $v_\alpha k$ scattering due to both CC and NC interactions. This implies that

$$p_v \text{Im}\left[n_\alpha(t)\right] = \frac{N_0 \rho(t)}{2} \sum_k Y_k(t) \sigma_{v_\alpha k}^{\text{tot}}(p_v) = \frac{1}{2\Lambda_\alpha(t)}, \qquad (10)$$

where $\Lambda_\alpha(t)$ is the (energy dependent) mean free path of neutrino v_α in the point t of the medium.

It is convenient to transform Equation (6) into the one with a traceless Hamiltonian. For this purpose we define the matrix

$$\tilde{\mathbf{S}}(t) = \exp\left\{\frac{i}{3}\int_0^t \text{Tr}\left[\mathbf{H}_0 + \mathbf{W}(t')\right] dt'\right\} \mathbf{S}(t). \qquad (11)$$

After substituting Equation (11) into Equation (6), we have

$$i\frac{d}{dt}\tilde{\mathbf{S}}(t) = \mathbf{H}(t)\tilde{\mathbf{S}}(t), \quad \tilde{\mathbf{S}}(0) = 1, \tag{12}$$

where

$$\mathbf{H}(t) = \begin{pmatrix} \mathcal{W}_e - q_e & \mathcal{H}_\tau & \mathcal{H}_\mu^* \\ \mathcal{H}_\tau^* & \mathcal{W}_\mu - q_\mu & \mathcal{H}_e \\ \mathcal{H}_\mu & \mathcal{H}_e^* & \mathcal{W}_\tau - q_\tau \end{pmatrix}. \tag{13}$$

The constants $\mathcal{W}_\alpha$ and $\mathcal{H}_\alpha$ are determined by the elements of the PMNS matrix, $\mathbf{V} = \|V_{\alpha i}\|$, and by the neutrino masses m_i:

$$\mathcal{W}_\alpha = \sum_i |V_{\alpha i}|^2 \Delta_i, \quad \mathcal{H}_\alpha = \sum_i \eta_\alpha^{\beta\gamma} V_{\beta i} V_{\gamma i}^* \Delta_i,$$

$$\Delta_i = \frac{m_i^2 - \langle m^2 \rangle}{2p_\nu}, \quad \langle m^2 \rangle = \frac{1}{3}\sum_i m_i^2. \tag{14}$$

The PMNS matrix is usually parameterized in terms of three mixing angles and the CP-violating (Dirac) phase (see Appendix B); the two additional phases present in the Majorana case do not affect the neutrino oscillation pattern in matter. Here and below, the symbol $\eta_\alpha^{\beta\gamma}$ is defined to be 1 if the triplet (α,β,γ) is a cyclic permutation of the indices (e,μ,τ) and zero otherwise.

The traceless Hamiltonian (13) depends on the distance $L = t$ through the set of optical potentials, $\mathbf{q} = (q_e, q_\mu, q_\tau)$, related to the neutrino indices of refraction, $n_\alpha(t)$, for a given medium:

$$q_\alpha(t) = p_\nu[n_\alpha(t) - \langle n(t)\rangle], \quad \langle n(t)\rangle = \frac{1}{3}\sum_\alpha n_\alpha(t). \tag{15}$$

It is seen from Equation (15) that evolution of the neutrino flavors in arbitrary medium depends on no more than two independent potentials $q_\alpha(t)$ due to the identity

$$\sum_\alpha q_\alpha(t) = 0.$$

In general, the indices of refraction $n_\alpha(t)$ and thus optical potentials $q_\alpha(t)$ are complex functions (see below). Owing to radiative electroweak contributions, the real parts of the potentials for different neutrino flavors α differ in magnitude, in both normal cold media [83,84] and hot CP-symmetric plasma (such as the early Universe) [81]. The imaginary parts of the potentials are given by

$$\operatorname{Im} q_\alpha(t) = \frac{1}{2}\left[\frac{1}{\Lambda_\alpha(t)} - \frac{1}{\Lambda(t)}\right], \quad \frac{1}{\Lambda(t)} = \frac{1}{3}\sum_\alpha \frac{1}{\Lambda_\alpha(t)}, \tag{16}$$

and are in general nonzero functions of neutrino energy and distance. This makes the Hamiltonian (13) non-Hermitian.

The neutrino flavor changing oscillation probabilities are just the squared absolute values of the elements of the evolution matrix $\mathbf{S}(t)$,

$$P[\nu_\alpha(0) \to \nu_{\alpha'}(t)] \equiv P_{\alpha\alpha'}(t) = |S_{\alpha'\alpha}(t)|^2. \tag{17}$$

Taking into account Equations (7), (10), (11) and (17) yields

$$P_{\alpha\alpha'}(t) = A(t)\left|\tilde{S}_{\alpha'\alpha}(t)\right|^2, \tag{18}$$

where

$$A(t) = \exp\left[-\int_0^t \frac{dt'}{\Lambda(t')}\right].$$ (19)

This factor accounts for the attenuation (due to inelastic scattering) of all flavors in the mean. It is apparent that in the absence of mixing and refraction (that is an appropriate approximation at superhigh energies),

$$\widetilde{S}_{\alpha'\alpha} = \delta_{\alpha'\alpha} \exp\left[-\int_0^t \operatorname{Im} q_\alpha(t')dt'\right]$$

and, according to Equations (18) and (19), the survival and transition probabilities reduce to the "classical limit":

$$P_{\alpha\alpha'}(t) = \delta_{\alpha\alpha'} \exp\left[-\int_0^t \frac{dt'}{\Lambda_\alpha(t')}\right].$$

Owing to the complex potentials q_α, the Hamiltonian in Equation (13) is non-Hermitian and the evolution matrix $\widetilde{S}(t)$ is non-unitary. It is apparent that the matrix $\mathbf{H}(t)$ becomes Hermitian when one neglects differences in the mean free paths of neutrinos of different flavors. In this case, Equation (12) reduces to one describing the standard Mikheev–Smirnov–Wolfenstein (MSW) mechanism [29–31]. Clearly, this approximation may not be good for very thick environments and/or very high neutrino energies.

At essentially all energies, the CC total cross sections for e or μ production in the neutrino and antineutrino interaction with nucleons are well above the one for the τ-lepton production,

$$\sigma_{\nu_{e,\mu}N}^{CC} > \sigma_{\nu_\tau N}^{CC}, \quad \sigma_{\bar{\nu}_{e,\mu}N}^{CC} > \sigma_{\bar{\nu}_\tau N}^{CC}.$$

This is because of large value of the τ-lepton mass, m_τ, which leads to several consequences (see, e.g., references [85,86] and references therein):

(i) high neutrino energy threshold for τ production;
(ii) sharp shrinkage of the phase spaces for the CC interactions of ν_τ and $\bar{\nu}_\tau$ with protons, neutrons, and nuclei;
(iii) kinematic correction factors ($\propto m_\tau^2$) to the nucleon structure functions (the corresponding structures are negligible for the electron production and small for the muon production);
(iv) the differences $\sigma_{\nu_{e,\mu}N}^{CC} - \sigma_{\nu_\tau N}^{CC}$ and $\sigma_{\bar{\nu}_{e,\mu}N}^{CC} - \sigma_{\bar{\nu}_\tau N}^{CC}$ are relatively slow varying functions of (anti)neutrino energy, having gently sloping maxima in the range of 10–100 PeV and vanishing at super-high energies.

Since the Standard Model NC interactions are universal for all neutrino flavors, it is clear from Equation (16), that the NC contributions to the total cross sections are canceled out from $\operatorname{Im} q_\alpha$ and thus $\operatorname{Im}(q_{e,\mu} - q_\tau) > 0$ at all energies. However, nonstandard NC interactions may be in general different for different flavors and thus contribute to both real and imaginary parts of the potentials q_α. Moreover, flavor-changing interactions (see, e.g., references [87,88] and references therein) would contribute to the non-diagonal elements of the Hamiltonian making these t-dependent.

Similar situation, although in different and rather narrow energy range, holds for $\bar{\nu}_e$ interaction with electrons. This is a particular case for the C-asymmetric media (planets, stars, astrophysical jets, etc.) because of the W-boson resonance formed in the neighborhood of $E_\nu^{res} = m_W^2/2m_e \approx 6.33$ PeV through the reactions

$$\bar{\nu}_e e^- \to W^- \to \text{hadrons} \quad \text{and} \quad \bar{\nu}_e e^- \to W^- \to \bar{\nu}_\ell \ell^- \quad (\ell = e, \mu, \tau).$$

Just at the resonance peak, $\sigma_{\bar{\nu}_e e}^{tot} \approx 250\, \sigma_{\bar{\nu}_e N}^{tot}$ (see, e.g., references [89–91] and references therein).

We conclude this section by explicitly emphasizing that the master equation to be solved is given by Equation (12) and the relevant definitions are given by Equations (8) and (13)–(15).

3. Mixing Matrices In Matter

Solution of the master equation (12) in adiabatic approximation has been found in reference [24]. In the present study we do not use the explicit form of that solution. Moreover, below we will consider an abstract Hamiltonian, which is a 3 × 3 complex matrix $\mathbf{H}$ describing a generic 3-level quantum system with dissipation (through absorption, friction, decay, etc.); such a Hamiltonian may, in particular, be used to describe the nonstandard neutrino interactions and decay. Below, keeping in mind our particular problem (3ν oscillation in absorbing matter) we will use specific notation. In the most general case the Hamiltonian $\mathbf{H}$ depends on time through a set of real parameters $(x_1(t), \ldots, x_s(t)) \equiv \mathbf{x}(t)$. We define these parameters in such a way that $x_k(t) = 0$ in vacuum; in our particular case, $\mathbf{x} = \mathbf{q}$ and this condition holds automatically.

Let us now define the two "mixing matrices" $\mathbf{V}^{(m)}(\mathbf{x})$ and $\overline{\mathbf{V}}^{(m)}(\mathbf{x})$ by the equations

$$\mathbf{H}(\mathbf{x})\mathbf{V}^{(m)}(\mathbf{x}) = \mathbf{V}^{(m)}(\mathbf{x})\mathbf{E}(\mathbf{x}), \quad \mathbf{H}^\dagger(\mathbf{x})\overline{\mathbf{V}}^{(m)}(\mathbf{x}) = \overline{\mathbf{V}}^{(m)}(\mathbf{x})\mathbf{E}^\dagger(\mathbf{x}), \tag{20}$$

with

$$\mathbf{E}(\mathbf{x}) = \mathrm{diag}\left(\mathcal{E}_{N_1}(\mathbf{x}), \mathcal{E}_{N_2}(\mathbf{x}), \mathcal{E}_{N_3}(\mathbf{x})\right). \tag{21}$$

The solution to Equations (20) can be found in two steps. First, one have to find the eigenvalues and eigenvectors of the matrices $\mathbf{H}$ and $\mathbf{H}^\dagger$,

$$\mathbf{H}(\mathbf{x})|N;\mathbf{x}\rangle = \mathcal{E}_N(\mathbf{x})|N;\mathbf{x}\rangle, \quad \mathbf{H}^\dagger(\mathbf{x})\overline{|N;\mathbf{x}\rangle} = \mathcal{E}_N^*(\mathbf{x})\overline{|N;\mathbf{x}\rangle}, \tag{22}$$

where

$$|N;\mathbf{x}\rangle = \begin{pmatrix} U_{Ne}(\mathbf{x}) \\ U_{N\mu}(\mathbf{x}) \\ U_{N\tau}(\mathbf{x}) \end{pmatrix}, \quad \overline{|N;\mathbf{x}\rangle} = \begin{pmatrix} \overline{U}_{Ne}(\mathbf{x}) \\ \overline{U}_{N\mu}(\mathbf{x}) \\ \overline{U}_{N\tau}(\mathbf{x}) \end{pmatrix}, \tag{23}$$

with $N = -1, 0, +1$ or simply $-, 0, +$. For simplicity we will neglect possible degeneracy of the energy levels. Then the eigenvectors form a complete biorthonormal set:

$$\overline{\langle N';\mathbf{x}|}N;\mathbf{x}\rangle = \delta_{NN'}, \quad \sum_N |N;\mathbf{x}\rangle\overline{\langle N;\mathbf{x}|} = \mathbf{I}, \tag{24}$$

or, in the component-wise notation,

$$\sum_\alpha \overline{U}_{N'\alpha}^*(\mathbf{x})U_{N\alpha}(\mathbf{x}) = \delta_{NN'}, \quad \sum_N \overline{U}_{N\alpha}^*(\mathbf{x})U_{N\beta}(\mathbf{x}) = \delta_{\alpha\beta}. \tag{25}$$

Second, from simple algebra it follows that the matrices

$$\begin{aligned}
\mathbf{U}(\mathbf{x}) &\equiv \|U_{\alpha j}(\mathbf{x})\| = (|N_1;\mathbf{x}\rangle, |N_2;\mathbf{x}\rangle, |N_3;\mathbf{x}\rangle), \\
\overline{\mathbf{U}}(\mathbf{x}) &\equiv \|\overline{U}_{\alpha j}(\mathbf{x})\| = \left(\overline{|N_1;\mathbf{x}\rangle}, \overline{|N_2;\mathbf{x}\rangle}, \overline{|N_3;\mathbf{x}\rangle}\right),
\end{aligned} \tag{26}$$

satisfy Equations (20) and thus diagonalize the Hamiltonian matrix $\mathbf{H}$.

The solutions (26) are not however unique. In most general case, the following products

$$\mathbf{V}^{(m)}(\mathbf{x}) = \mathbf{U}(\mathbf{x})\mathbf{D}^\dagger(\mathbf{x}), \quad \overline{\mathbf{V}}^{(m)}(\mathbf{x}) = \overline{\mathbf{U}}(\mathbf{x})\overline{\mathbf{D}}(\mathbf{x}),$$

with arbitrary diagonal and nonsingular matrices $\mathbf{D}(\mathbf{x})$ and $\overline{\mathbf{D}}(\mathbf{x})$, also satisfy Equation (20). This freedom implies that not all elements of the mixing matrices $\mathbf{V}^{(m)}(\mathbf{x})$ and $\overline{\mathbf{V}}^{(m)}(\mathbf{x})$ are physically observable. Recall that the eigenvectors have been built so that

$$\mathbf{U}(0) = \overline{\mathbf{U}}(0) = \mathbf{V}$$

and the following obvious conditions are assumed: $U_{i\alpha}(0) = U_{N_i\alpha}(0) = \overline{U}_{i\alpha}(0) = \overline{U}_{N_i\alpha}(0) = V_{i\alpha}$. Equations (20) and (21) are universal, i.e., they are hold true for any medium and for any value of the neutrino momentum. In particular, they are hold for vacuum. Therefore

$$\mathbf{V}^{(m)}(0) = \overline{\mathbf{V}}^{(m)}(0) = \mathbf{V}, \tag{27}$$

where $\mathbf{V}$ is the vacuum mixing matrix ("correspondence principle"). Hence, according to Equation (27), the matrices $\mathbf{D}(x)$ and $\overline{\mathbf{D}}(x)$ must satisfy the condition

$$\mathbf{D}(0) = \overline{\mathbf{D}}(0) = \mathbf{I}.$$

As a less trivial limiting case, let us consider a medium, where the imaginary part of the optic potentials can be neglected (this standard approximation is true in essence for any media if its thickness is much smaller than the neutrino mean free path). In this case the eigenvalues $\mathcal{E}_N(x)$ are real and the following inequalities are valid [32]:

$$\mathcal{E}_-(x) \leq \mathcal{E}_0(x) \leq \mathcal{E}_+(x).$$

Considering these limiting cases one finds that the numeration of the diagonal elements in (21) (i.e., the one-to-one congruence $N_i \Leftrightarrow i$) is given by the neutrino mass hierarchy. For example, $N_1 = -1$, $N_2 = 0$, $N_3 = +1$ for the "natural hierarchy", $m_1^2 > m_2^2 > m_3^2$ but $N_1 = +1$, $N_2 = 0$, $N_3 = -1$ for the following case: $m_3^2 < m_2^2 < m_1^2$; other cases can be derived similarly. Thus, to simplify formulas, we will use the notation $\mathcal{E}_{N_i}(x) = E_i(x)$, when it is suitable.

According to Equations (25) and (26)

$$\sum_\alpha U_{\alpha i}^* \overline{U}_{\alpha j} = \sum_\alpha \overline{U}_{\alpha i}^* U_{\alpha j} = \delta_{ij}, \tag{28}$$

or, equivalently,

$$\mathbf{U}^\dagger(x)\overline{\mathbf{U}}(x) = \overline{\mathbf{U}}^\dagger(x)\mathbf{U}(x) = \mathbf{I}. \tag{29}$$

It is reasonable to impose the same constraint on the mixing matrices:

$$\left[\mathbf{V}^{(m)}(x)\right]^\dagger \overline{\mathbf{V}}^{(m)}(x) = \left[\overline{\mathbf{V}}^{(m)}(x)\right]^\dagger \mathbf{V}^{(m)}(x) = \mathbf{I}.$$

Then

$$\mathbf{D}^\dagger(x)\overline{\mathbf{D}}(x) = \overline{\mathbf{D}}^\dagger(x)\mathbf{D}^\dagger(x) = \mathbf{I}$$

and therefore

$$\mathbf{D}(x) = \mathrm{diag}\left(e^{-a_1+ib_1}, e^{-a_2+ib_2}, e^{-a_3+ib_3}\right),$$
$$\overline{\mathbf{D}}(x) = \mathrm{diag}\left(e^{+a_1+ib_1}, e^{+a_2+ib_2}, e^{+a_3+ib_3}\right),$$

where $a_k = a_k(x)$ and $b_k = b_k(x)$ are arbitrary real functions which vanish at $x = 0$.

As is generally known (see for example [92]), the vacuum mixing matrix for Majorana neutrinos may be written in the form $\mathbf{V}\mathbf{D}^M$, where

$$\mathbf{D}^M = \mathrm{diag}\left(e^{i\delta_1^M}, e^{i\delta_2^M}, e^{i\delta_3^M}\right)$$

and δ_k^M are the (real) CP-violating parameters (strictly speaking, in the three-neutrino case only two "Majorana parameters" δ_k^M are independent [92,93]). By analogy, one may call the functions $\delta_k(x) = b_k(x) + ia_k(x)$ and $\overline{\delta}_k(x) = b_k(x) - ia_k(x)$ the Majorana phases in matter. Just as in the vacuum case, these phases play no part in neutrino oscillations at relativistic energies [92,93]. Here they merely show the ambiguity in the definition of the mixing matrices in matter. The additional CP-violating Majorana phases are always associated with effects whose magnitude is suppressed by the factor $\left(m_i^M/E_\nu\right)^2$, where E_ν is the neutrino energy in the relevant process and m_i^M is the mass of the Majorana neutrino taking part in the process [93,94].

4. Rephasing Invariant In Matter

Let us introduce two sets of functions

$$J_{\alpha i}^{\pm}(\mathbf{x}) = \frac{1}{2}\eta_\alpha^{\beta\gamma}\eta_i^{jk}V_{\beta j}^{(m)}(\mathbf{x})V_{\gamma k}^{(m)}(\mathbf{x})\left[\overline{V}_{\beta k}^{(m)}(\mathbf{x})\overline{V}_{\gamma j}^{(m)}(\mathbf{x})\right]^*$$
$$\pm \frac{1}{2}\eta_\alpha^{\beta\gamma}\eta_i^{jk}\overline{V}_{\beta j}^{(m)}(\mathbf{x})\overline{V}_{\gamma k}^{(m)}(\mathbf{x})\left[V_{\beta k}^{(m)}(\mathbf{x})V_{\gamma j}^{(m)}(\mathbf{x})\right]^*,$$

which provide the straightforward generalization of the rephasing invariants considered in references [32,33,95,96] (see also reference [34,97] for the Dirac neutrino case or in reference [94] for the Majorana neutrino case (Cheng [94] considered so called second-class rephasing invariant which contains the Majorana phases).

First of all, the functions $J_{\alpha i}^{\pm}(\mathbf{x})$ are independent of the Majorana phases, $\delta_k(\mathbf{x})$, $\overline{\delta}_k(\mathbf{x})$, i.e., these functions are independent of the $\mathbf{D}(\mathbf{x})$ and $\overline{\mathbf{D}}(\mathbf{x})$ matrices. This fact elucidates the term "rephasing invariant". Therefore the functions $J_{\alpha i}^{\pm}(\mathbf{x})$ can be rewritten as

$$J_{\alpha i}^{\pm}(\mathbf{x}) = \frac{1}{2}\eta_\alpha^{\beta\gamma}\eta_i^{jk}U_{\beta j}(\mathbf{x})U_{\gamma k}(\mathbf{x})\overline{U}_{\beta k}^*(\mathbf{x})\overline{U}_{\gamma j}^*(\mathbf{x})$$
$$\pm \frac{1}{2}\eta_\alpha^{\beta\gamma}\eta_i^{jk}\overline{U}_{\beta j}(\mathbf{x})\overline{U}_{\gamma k}(\mathbf{x})U_{\beta k}^*(\mathbf{x})U_{\gamma j}^*(\mathbf{x}). \tag{30}$$

Let us rewrite Equation (29) in the form

$$\mathbf{U}^\dagger(\mathbf{x}) = \overline{\mathbf{U}}^{-1}(\mathbf{x}), \quad \overline{\mathbf{U}}^\dagger(\mathbf{x}) = \mathbf{U}^{-1}(\mathbf{x}),$$

or, in terms of the matrix elements,

$$U_{\alpha i}^*(\mathbf{x}) = |\overline{\mathbf{U}}|^{-1}\eta_\alpha^{\beta\gamma}\eta_i^{jk}\left(\overline{U}_{\beta j}\overline{U}_{\gamma k} - \overline{U}_{\beta k}\overline{U}_{\gamma j}\right),$$
$$\overline{U}_{\alpha i}^*(\mathbf{x}) = |\mathbf{U}|^{-1}\eta_\alpha^{\beta\gamma}\eta_i^{jk}\left(U_{\beta j}U_{\gamma k} - U_{\beta k}U_{\gamma j}\right). \tag{31}$$

Using these identities, one finds from Equation (30) that the real functions

$$\mathrm{Re}J_{\alpha i}^-(\mathbf{x}) = +\frac{1}{2}\mathrm{Re}\left(U_{e1}U_{\mu 2}U_{\tau 3}|\overline{\mathbf{U}}^\dagger| - \overline{U}_{e1}\overline{U}_{\mu 2}\overline{U}_{\tau 3}|\mathbf{U}^\dagger|\right) \equiv R(\mathbf{x}),$$
$$\mathrm{Im}J_{\alpha i}^+(\mathbf{x}) = -\frac{1}{2}\mathrm{Im}\left(U_{e1}U_{\mu 2}U_{\tau 3}|\overline{\mathbf{U}}^\dagger| + \overline{U}_{e1}\overline{U}_{\mu 2}\overline{U}_{\tau 3}|\mathbf{U}^\dagger|\right) \equiv I(\mathbf{x}), \tag{32}$$

as well as their complex combination

$$J(\mathbf{x}) = I(\mathbf{x}) + iR(\mathbf{x}) \tag{33}$$

are independent of indices α and i. Clearly, in the Hermitian case $J_{\alpha i}^- = 0$ and therefore $J = I = -\mathrm{Im}\left(U_{e1}U_{\mu 2}U_{\tau 3}|\mathbf{U}^\dagger|\right)$.

We consider now the following constructions:

$$\mathcal{J}(\mathbf{x}) = \frac{1}{2i}\left[\prod_\alpha \eta_\alpha^{\beta\gamma}H_{\beta\gamma}(\mathbf{x}) - \prod_\alpha \eta_\alpha^{\beta\gamma}H_{\gamma\beta}(\mathbf{x})\right], \tag{34}$$

$$\mathcal{P}(\mathbf{x}) = \prod_\alpha \eta_\alpha^{\beta\gamma} |H_{\beta\gamma}(\mathbf{x})|, \quad \overline{\mathcal{P}}(\mathbf{x}) = \prod_\alpha \eta_\alpha^{\beta\gamma} |H_{\gamma\beta}(\mathbf{x})|, \tag{35}$$

$$\varphi(\mathbf{x}) = \sum_\alpha \eta_\alpha^{\beta\gamma} \arg H_{\beta\gamma}(\mathbf{x}), \quad \overline{\varphi}(\mathbf{x}) = \sum_\alpha \eta_\alpha^{\beta\gamma} \arg H_{\gamma\beta}^*(\mathbf{x}). \tag{36}$$

It is easy to show that
$$\mathcal{J}(\mathbf{x}) = \Im(\mathbf{x}) + i\Re(\mathbf{x}),$$
where
$$\Im(\mathbf{x}) = \frac{1}{2}\left[\mathcal{P}(\mathbf{x})\sin\varphi(\mathbf{x}) + \overline{\mathcal{P}}(\mathbf{x})\sin\overline{\varphi}(\mathbf{x})\right],$$
$$\Re(\mathbf{x}) = \frac{1}{2}\left[\overline{\mathcal{P}}(\mathbf{x})\cos\overline{\varphi}(\mathbf{x}) - \mathcal{P}(\mathbf{x})\cos\varphi(\mathbf{x})\right].$$

In the absence of flavor-changing neutral currents the off-diagonal matrix elements of the Hamiltonian are time independent and thus $\mathcal{J}$ is a complex constant ("dynamic invariant"). In the most general case the following theorem holds true:

$$\mathcal{J}(\mathbf{x}) = \varsigma \mathcal{J}(\mathbf{x}) \prod_L \eta_L^{MN} \left[\mathcal{E}_M(\mathbf{x}) - \mathcal{E}_N(\mathbf{x})\right], \tag{37}$$

where ς is the parity of the cyclic permutation $\begin{pmatrix} -1 & 0 & +1 \\ N_1 & N_2 & N_3 \end{pmatrix}$. The proof of this theorem is given in Appendix A. The obtained identity is very general and does not depend on explicit form of the eigenvalues and eigenvectors, but the full the derivation of these quantities is discussed in detail in reference [24].

To gain a further insight into the identity (37), it is instructive to consider an example of neutrino propagation in matter governed by the Hamiltonian (7). Then, it is seen that $\mathcal{P}(\mathbf{q}) = \overline{\mathcal{P}}(\mathbf{q})$, $\varphi(\mathbf{q}) = \overline{\varphi}(\mathbf{q})$ and these quantities are time independent:

$$\mathcal{P}(\mathbf{q}) = \overline{\mathcal{P}}(\mathbf{q}) = \left|\sum_i V_{\mu i} V_{\tau i}^* \Delta_i\right| \cdot \left|\sum_i V_{ei} V_{\mu i}^* \Delta_i\right| \cdot \left|\sum_i V_{\tau i} V_{ei}^* \Delta_i\right|,$$

$$\varphi(\mathbf{q}) = \overline{\varphi}(\mathbf{q}) = \arg \sum_i \left(V_{\mu i} V_{\tau i}^* + V_{ei} V_{\mu i}^* + V_{\tau i} V_{ei}^*\right) \Delta_i.$$

Therefore, $\Re = 0$ and $\mathcal{J} = \Im = \mathcal{P}\sin\varphi$ in this case. It can be verified that the LHS of Equation (37) is exactly the product the Jarlskog invariant (see Appendix B)

$$J_0 = J(0) = \frac{1}{8}\sin\delta\cos\theta_{13}\sin 2\theta_{12}\sin 2\theta_{23}\sin 2\theta_{13} \tag{38}$$

and the factor

$$\prod_i \eta_i^{jk} \frac{m_j^2 - m_k^2}{2p_\nu} = \prod_i \eta_i^{jk} \frac{\Delta m_{jk}^2}{2p_\nu}.$$

Let us define the effective (complex) masses $\tilde{m}_i = \tilde{m}_i(\mathbf{q})$ of the neutrino mass eigenstates in matter by

$$E_i = \mathcal{E}_{N_i} \stackrel{\text{def}}{=} \frac{\tilde{m}_i^2 - \langle\tilde{m}^2\rangle}{2p_\nu}, \quad \langle\tilde{m}^2\rangle = \frac{1}{3}\sum_i \tilde{m}_i^2,$$

where we used the obvious identity $\sum_i E_i = 0$ and analogy with the vacuum case (see Equation (14)). Then Equation (37) can be written as

$$J(0)\Delta m_{23}^2 \Delta m_{31}^2 \Delta m_{12}^2 = J(\mathbf{q})\Delta\tilde{m}_{23}^2(\mathbf{q})\Delta\tilde{m}_{31}^2(\mathbf{q})\Delta\tilde{m}_{12}^2(\mathbf{q}). \tag{39}$$

The obtained identity is evidently a generalization of the relation (1) to the case of neutrino-absorbing environments. Remarkably that the effective masses are complex functions but the RHS of Equation (39) is proved to be real. The form of Equation (39) confirms that Equations (32) and (33) provide a non-Hermitian extension of the usual rephasing invariant.

5. Summary

In this paper we considered three-neutrino oscillations in thick (including neutrino opaque) media by using the non-Hermitian quantum dynamics framework, which describes the interplay between neutrino mixing, refraction and absorption. We proved an identity which relates (through a product of splitting of the complex energy levels) a rephasing invariant in vacuum and absorbing matter. These findings might be of certain interest in studies of soft-spectrum, high-energy neutrino propagation through Earth or astrophysical objects (jets, blast waves, etc.) whose thickness along the neutrino beam is comparable to or larger than the neutrino mean free path.

Author Contributions: Conceptualization, D.V.N. and V.A.N.; methodology, D.V.N. and V.A.N.; validation, D.V.N., V.A.N., and D.S.S.; writing—original draft preparation, D.V.N., V.A.N., and D.S.S.; writing—review and editing, D.V.N., V.A.N., D.S.S. All authors have read and agreed to the published version of the manuscript.

Funding: This research received no external funding.

Conflicts of Interest: The authors declare no conflict of interest.

Abbreviations

The following abbreviations are used in this manuscript:

PMNS	Pontecorvo-Maki-Nakagawa-Sakata (mixing matrix)
MSW	Mikheev-Smirnov-Wolfenstein (mechanism, equation)
KM	Kobayashi-Maskawa (representation of mixing matrix)
CK	Chau-Keung (representation of mixing matrix)
CC	Charged Current
NC	Neutral Current
AMU	Atomic Mass Unit
CP	Charge Parity
LHS	Left-Hand Side
RHS	Right-Hand Side
QED	Quod Erat Demonstrandum (Lat.)

Appendix A. Proof of The Theorem

Using the definitions for the mixing matrices one can easily show that

$$\mathcal{J} = \frac{1}{2i} \prod_\alpha \eta_\alpha^{\beta\gamma} \sum_i U_{\beta i} \overline{U}^*_{\gamma i} E_i - \frac{1}{2i} \prod_\alpha \eta_\alpha^{\beta\gamma} \sum_i \overline{U}^*_{\beta i} U_{\gamma i} E_i,$$

where we omitted argument **x** for short. Denote

$$G_{ijk} = U_{ei} U_{\mu j} U_{\tau k}, \quad \overline{G}_{ijk} = \overline{U}_{ei} \overline{U}_{\mu j} \overline{U}_{\tau k}.$$

Then $\mathcal{J}$ can be written as

$$\frac{1}{2i} \sum_{ijk} E_i E_j E_k \left(G_{ijk} \overline{G}^*_{kij} - \overline{G}^*_{ijk} G_{kij} \right).$$

It can be shown from here that

$$\mathcal{J} = \frac{1}{2i} \sum_i \eta_i^{jk} E_i^2 \left(C_i^j E_j + C_i^k E_k \right), \tag{A1}$$

where

$$C_i^j = G_{iij}\overline{G}_{jii}^* + G_{iji}\overline{G}_{iij}^* + G_{jii}\overline{G}_{iji}^* - \overline{G}_{iij}^*G_{jii} - \overline{G}_{iji}^*G_{iij} - \overline{G}_{jii}^*G_{iji}$$

and the coefficients C_i^k are defined in a similar way. To derive Equation (A1) it has been taken into account that the terms in the sum over i, j, k with $i = j = k$, as well as with the i, j, and k which unequal to each other, are vanish. This statement is apparent for the term

$$\frac{1}{2i}\sum_i E_i^3 \left(G_{iii}\overline{G}_{iii}^* - \overline{G}_{iii}^*G_{iii}\right).$$

As regards the term

$$\frac{1}{2i}{\sum_{ijk}}' E_i E_j E_k \left(G_{ijk}\overline{G}_{kij}^* - \overline{G}_{ijk}^*G_{kij}\right)$$

(where prime indicates that all indices are different), it can be rewritten in the following form:

$$-\frac{i}{2}|\mathbf{H}|\sum_i \eta_i^{jk}\left(G_{ijk}\overline{G}_{kij}^* + G_{ikj}\overline{G}_{jik}^* - \overline{G}_{ijk}^*G_{kij} - \overline{G}_{ikj}^*G_{jik}\right), \qquad (A2)$$

where it is taken into account that

$$E_1 E_2 E_3 = \mathcal{E}_-\mathcal{E}_0\mathcal{E}_+ = |\mathbf{H}|$$

By applying sequentially the identities (31), one can transform the term (A2) to the following form:

$$-\frac{i}{2}|\mathbf{H}|\sum_i \eta_i^{jk}\left(G_{ijk}|\overline{\mathbf{U}}|^* - \overline{G}_{ijk}^*|\mathbf{U}|\right)\left(U_{\tau j}\overline{U}_{\tau j}^* - U_{\tau k}\overline{U}_{\tau k}^*\right),$$

However, according to Equation (32),

$$\frac{1}{2}\eta_i^{jk}\left(G_{ijk}|\overline{\mathbf{U}}|^* - \overline{G}_{ijk}^*|\mathbf{U}|\right) = R - iI = -iJ.$$

Hence the term (A2) vanishes.
Next, using the identities (28), (31), and definition (32) yields

$$\eta_i^{jk}C_i^j = -2iJ, \quad \eta_i^{jk}C_i^k = 2iJ.$$

Direct substituting into Equation (A1) then gives

$$\mathcal{J} = -J\sum_i \eta_i^{jk}E_i^2(E_j - E_k) = J\prod_i \eta_i^{jk}(E_j - E_k) = \varsigma J\prod_L \eta_L^{MN}(\mathcal{E}_M - \mathcal{E}_N).$$

QED.

Appendix B. Rephasing Invariant In Vacuum

The imaginary part of the rephasing invariant in vacuum (Jarlskog invariant) may be written in terms of the mixing angles and CP-violating Dirac phase dependent of the parametrization of the PMNS mixing matrix. For example, in the Kobayashi–Maskawa (KM) representation [98],

$$\mathbf{V}^{(KM)} = \begin{pmatrix} c_1 & s_1 c_3 & s_1 s_3 \\ -s_1 c_2 & c_1 c_2 c_3 - s_2 s_3 e^{i\delta} & c_1 c_2 s_3 + s_2 c_3 e^{i\delta} \\ -s_1 s_2 & c_1 s_2 c_3 + c_2 s_3 e^{i\delta} & c_1 s_2 s_3 - c_2 c_3 e^{i\delta} \end{pmatrix}$$

(where $s_i = \sin\theta_i$ and $c_i = \cos\theta_i$ for $i = 1,2,3$; $0 < \theta_i < \pi/2$, $-\pi < \delta \leq \pi$, $\det \mathbf{V}^{(KM)} = -e^{i\delta}$),

$$J_0^{(KM)} = \sin\delta \sin\theta_1 \prod_i \sin 2\theta_i$$

In the now more conventional Chau–Keung (CK) representation [99],

$$\mathbf{V}^{(CK)} = \begin{pmatrix} c_{12}c_{31} & s_{12}c_{31} & s_{31}e^{-i\delta} \\ -s_{12}c_{23} - c_{12}s_{23}s_{31}e^{i\delta} & c_{12}c_{23} - s_{12}s_{23}s_{31}e^{i\delta} & s_{23}c_{31} \\ s_{12}s_{23} - c_{12}c_{23}s_{31}e^{i\delta} & -c_{12}s_{23} - s_{12}c_{23}s_{31}e^{i\delta} & c_{23}c_{31} \end{pmatrix}$$

(where $s_{jk} = \sin\theta_{jk}$ and $c_{jk} = \cos\theta_{jk}$ for $j,k = 1,2,3$; $0 < \theta_{jk} < \pi/2$ ($\theta_{jk} \equiv \theta_{jk}$), $0 \leq \delta < 2\pi$, $\det \mathbf{V}^{(CK)} = 1$),

$$J_0^{(CK)} = \sin\delta \cos\theta_{31} \prod_i \eta_i^{jk} \sin 2\theta_{jk}$$

Here the symbol η_i^{jk} has the same sense as $\eta_\alpha^{\beta\gamma}$. Details about the interconnection between the KM and CK representations can be found in references [99–101].

References

1. Pérez de los Heros, C. *Probing Particle Physics with Neutrino Telescopes*; World Scientific: Singapore, 2020. [CrossRef]
2. Cleveland, B.T.; Daily, T.; Davis, R., Jr.; Distel, J.R.; Lande, K.; Lee, C.K.; Wildenhain, P.S.; Ullman, J. Measurement of the solar electron neutrino flux with the Homestake chlorine detector. *Astrophys. J.* **1998**, *496*, 505–526. [CrossRef]
3. Kaether, F.; Hampel, W.; Heusser, G.; Kiko, J.; Kirsten, T. Reanalysis of the GALLEX solar neutrino flux and source experiments. *Phys. Lett. B* **2010**, *685*, 47–54. [CrossRef]
4. Abdurashitov, J.N.; Gavrin, V.N.; Gorbachev, V.V.; Gurkina, P.P.; Ibragimova, T.V.; Kalikhov, A.V.; Khairnasov, N.G.; Knodel, T.V.; Mirmov, I.N.; Shikhin, A.A.; et al. Measurement of the solar neutrino capture rate with gallium metal. III: Results for the 2002–2007 data-taking period. *Phys. Rev. C* **2009**, *80*, 015807. [CrossRef]
5. Fukuda, Y.; Hayakawa, T.; Ichihara, E.; Inoue, K.; Ishihara, K.; Ishino, H.; Itow, Y.; Kajita, T.; Kameda, J.; Kasuga, S.; et al. Evidence for oscillation of atmospheric neutrinos. *Phys. Rev. Lett.* **1998**, *81*, 1562–1567. [CrossRef]
6. Adamson, P.; Anghel, I.; Aurisano, A.; Barr, G.; Bishai, M.; Blake, A.; Bock, G.J.; Bogert, D.; Cao, S.V.; Castromonte, C.M.; et al. Combined analysis of ν_μ disappearance and $\nu_\mu \to \nu_e$ appearance in MINOS using accelerator and atmospheric neutrinos. *Phys. Rev. Lett.* **2014**, *112*, 191801. [CrossRef] [PubMed]
7. Ahn, M.H.; Aoki, S.; Bhang, H.; Boyd, S.; Casper, D.; Choi, J.H.; Fukuda, S.; Fukuda, Y.; Gajewski, W.; Hara, T.; et al. Indications of neutrino oscillation in a 250 km long baseline experiment. *Phys. Rev. Lett.* **2003**, *90*, 041801. [CrossRef]
8. Ashie, Y.; Hosaka, J.; Ishihara, K.; Itow, Y.; Kameda, J.; Koshio, Y.; Minamino, A.; Mitsuda, C.; Miura, M.; Moriyama, S.; et al. Evidence for an oscillatory signature in atmospheric neutrino oscillation. *Phys. Rev. Lett.* **2004**, *93*, 101801. [CrossRef]
9. Abe, S.; Ebihara, T.; Enomoto, S.; Furuno, K.; Gando, Y.; Ichimura, K.; Ikeda, H.; Inoue, K.; Kibe, Y.; Kishimoto, Y.; et al. Precision measurement of neutrino oscillation parameters with KamLAND. *Phys. Rev. Lett.* **2008**, *100*, 221803. [CrossRef]
10. An, F.P.; Bai, J.Z.; Balantekin, A.B.; Band, H.R.; Beavis, D.; Beriguete, W.; Bishai, M.; Blyth, S.; Boddy, K.; Brown, R.L.; et al. Observation of electron-antineutrino disappearance at Daya Bay. *Phys. Rev. Lett.* **2012**, *108*, 171803. [CrossRef]
11. Beuthe, M. Oscillations of neutrinos and mesons in quantum field theory. *Phys. Rept.* **2003**, *375*, 105–218. [CrossRef]
12. Giunti, C.; Kim, C.W. *Fundamentals of Neutrino Physics and Astrophysics*; Oxford University Press: Oxford, UK, 2007. [CrossRef]

13. Kayser, B.; Kopp, J. Testing the wave packet approach to neutrino oscillations in future experiments. *arXiv* **2010**, arXiv:hep-ph/1005.4081.
14. Grimus, W.; Stockinger, P. Real oscillations of virtual neutrinos. *Phys. Rev. D* **1996**, *54*, 3414–3419. [CrossRef] [PubMed]
15. Cardall, C.Y.; Chung, D.J.H. The MSW effect in quantum field theory. *Phys. Rev. D* **1999**, *60*, 073012. [CrossRef]
16. Stockinger, P. Introduction to a field-theoretical treatment of neutrino oscillations. *Pramana* **2000**, *54*, 203–214. [CrossRef]
17. Beuthe, M. Towards a unique formula for neutrino oscillations in vacuum. *Phys. Rev. D* **2002**, *66*, 013003. [CrossRef]
18. Giunti, C.; Kim, C.W.; Lee, J.A.; Lee, U.W. On the treatment of neutrino oscillations without resort to weak eigenstates. *Phys. Rev. D* **1993**, *48*, 4310–4317. [CrossRef]
19. Akhmedov, E.K.; Kopp, J. Neutrino oscillations: Quantum mechanics vs. quantum field theory. *J. High Energy Phys.* **2010**, *1004*, 8. [CrossRef]
20. Naumov, D.V.; Naumov, V.A. A Diagrammatic treatment of neutrino oscillations. *J. Phys. G* **2010**, *37*, 105014. [CrossRef]
21. Berezinsky, V.S.; Gazizov, A.Z.; Zatsepin, G.T.; Rozental, I.L. On penetration of high-energy neutrinos through earth and a possibility of their detection by means of EAS. *Sov. J. Nucl. Phys.* **1986**, *43*, 406.
22. Naumov, V.A.; Perrone, L. Neutrino propagation through dense matter. *Astropart. Phys.* **1999**, *10*, 239–252. [CrossRef]
23. Vincent, A.C.; Argüelles, C.A.; Kheirandish, A. High-energy neutrino attenuation in the Earth and its associated uncertainties. *J. Cosmol. Astropart. Phys.* **2017**, *1711*, 12. [CrossRef]
24. Korenblit, S.E.; Kuznetsov, V.E.; Naumov, V.A. Geometric phases for three-level non-Hermitian system. In Proceedings of the International Workshop on "Quantum Systems: New Trends and Methods", Minsk, Belarus, 23–29 May 1994; Barut, A.O., Feranchuk, I.D., Shnir, Y.M., Tomil'chik, L.M., Eds.; World Scientific: Singapore, 1995; pp. 209–217.
25. Naumov, V.A. High-energy neutrino oscillations in absorbing matter. *Phys. Lett. B* **2002**, *529*, 199–211. [CrossRef]
26. Huang, G.-Y. Sterile neutrinos as a possible explanation for the upward air shower events at ANITA. *Phys. Rev. D* **2018**, *98*, 043019. [CrossRef]
27. Luo, S. Neutrino oscillation in dense matter. *Phys. Rev. D* **2020**, *101*, 033005. [CrossRef]
28. Jarlskog, C. Commutator of the quark mass matrices in the standard electroweak model and a measure of maximal CP nonconservation. *Phys. Rev. Lett.* **1985**, *55*, 1039–1042. [CrossRef]
29. Wolfenstein, L. Neutrino oscillations in matter. *Phys. Rev. D* **1978**, *17*, 2369–2374. [CrossRef]
30. Mikheyev, S.P.; Smirnov, A.Y. Resonance amplification of oscillations in matter and spectroscopy of solar neutrinos. *Sov. J. Nucl. Phys.* **1985**, *42*, 913–917.
31. Mikheev, S.P.; Smirnov, A.Y. Resonant amplification of neutrino oscillations in matter and solar neutrino spectroscopy. *Nuovo Cim. C* **1986**, *9*, 17–26. [CrossRef]
32. Naumov, V.A. Three neutrino oscillations in matter and topological phases. *Sov. Phys. JETP* **1992**, *74*, 1–8.
33. Harrison, P.F.; Scott, W.G. CP and T violation in neutrino oscillations and invariance of Jarlskog's determinant to matter effects. *Phys. Lett. B* **2000**, *476*, 349–355. [CrossRef]
34. Krastev, P.I.; Petcov, S.T. Resonance amplification and T-violation effects in three neutrino oscillations in the Earth. *Phys. Lett. B* **1988**, *205*, 84–92. [CrossRef]
35. Yokomakura, H.; Kimura, K.; Takamura, A. Matter enhancement of T violation in neutrino oscillation. *Phys. Lett. B* **2000**, *496*, 175–184. [CrossRef]
36. Parke, S.J.; Weiler, T.J. Optimizing T violating effects for neutrino oscillations in matter. *Phys. Lett. B* **2001**, *501*, 106–114. [CrossRef]
37. Xing, Z.-Z. Sum rules of neutrino masses and CP violation in the four neutrino mixing scheme. *Phys. Rev. D* **2001**, *64*, 033005. [CrossRef]
38. Yasuda, O. Vacuum mimicking phenomena in neutrino oscillations. *Phys. Lett. B* **2001**, *516*, 111–115. [CrossRef]
39. Guo, W.-L.; Xing, Z.-Z. Rephasing invariants of CP and T violation in the four neutrino mixing models. *Phys. Rev. D* **2002**, *65*, 073020. [CrossRef]

40. Gluza, J.; Zrałek, M. Parameters' domain in three flavor neutrino oscillations. *Phys. Lett. B* **2001**, *517*, 158–166. [CrossRef]
41. Harrison, P.F.; Scott, W.G. Neutrino matter effect invariants and the observables of neutrino oscillations. *Phys. Lett. B* **2002**, *535*, 229–235. [CrossRef]
42. Kimura, K.; Takamura, A.; Yokomakura, H. Exact formula of probability and CP violation for neutrino oscillations in matter. *Phys. Lett. B* **2002**, *537*, 86–94. [CrossRef]
43. Minakata, H.; Nunokawa, H.; Parke, S.J. CP and T trajectory diagrams for a unified graphical representation of neutrino oscillations. *Phys. Lett. B* **2002**, *537*, 249–255. [CrossRef]
44. Kimura, K.; Takamura, A.; Yokomakura, H. Exact formulas and simple CP dependence of neutrino oscillation probabilities in matter with constant density. *Phys. Rev. D* **2002**, *66*, 073005. [CrossRef]
45. Yokomakura, H.; Kimura, K.; Takamura, A. Overall feature of CP dependence for neutrino oscillation probability in arbitrary matter profile. *Phys. Lett. B* **2002**, *544*, 286–294. [CrossRef]
46. Leung, C.N.; Wong, Y.Y.Y. T violation in flavor oscillations as a test for relativity principles at a neutrino factory. *Phys. Rev. D* **2003**, *67*, 056005. [CrossRef]
47. Wong, Y.Y.Y. T violation tests for relativity principles. *J. Phys. G* **2003**, *29*, 1857–1860. [CrossRef]
48. Jacobson, M.; Ohlsson, T. Extrinsic CPT violation in neutrino oscillations in matter. *Phys. Rev. D* **2004**, *69*, 013003. [CrossRef]
49. Harrison, P.F.; Scott, W.G.; Weiler, T.J. Exact matter covariant formulation of neutrino oscillation probabilities. *Phys. Lett. B* **2003**, *565*, 159–168. [CrossRef]
50. Xing, Z.-Z. Flavor mixing and CP violation of massive neutrinos. *Int. J. Mod. Phys. A* **2004**, *19*, 1–80. [CrossRef]
51. Kimura, K.; Takamura, A.; Yokomakura, H. Analytic formulation of neutrino oscillation probability in constant matter. *J. Phys. G* **2003**, *29*, 1839–1842. [CrossRef]
52. Zhang, H.; Xing, Z.-Z. Leptonic unitarity triangles in matter. *Eur. Phys. J. C* **2005**, *41*, 143–152. [CrossRef]
53. Jarlskog, C. Invariants of lepton mass matrices and CP and T violation in neutrino oscillations. *Phys. Lett. B* **2005**, *609*, 323–329. [CrossRef]
54. Xing, Z.-Z.; Zhang, H. Reconstruction of the neutrino mixing matrix and leptonic unitarity triangles from long-baseline neutrino oscillations. *Phys. Lett. B* **2005**, *618*, 131–140. [CrossRef]
55. Takamura, A.; Kimura, K. Large non-perturbative effects of small $\Delta m_{21}^2/\Delta m_{31}^2$ and $\sin\theta_{13}$ on neutrino oscillation and CP violation in matter. *J. High Energy Phys.* **2006**, *1*, 53. [CrossRef]
56. Nunokawa, H.; Parke, S.J.; Valle, J.W.F. CP violation and neutrino oscillations. *Prog. Part. Nucl. Phys.* **2008**, *60*, 338–402. [CrossRef]
57. Kneller, J.P.; McLaughlin, G.C. Three flavor neutrino oscillations in matter: Flavor diagonal potentials, the adiabatic basis and the CP phase. *Phys. Rev. D* **2009**, *80*, 053002. [CrossRef]
58. Chiu, S.H.; Kuo, T.K.; Liu, L.X. Neutrino mixing in matter. *Phys. Lett. B* **2010**, *687*, 184–187. [CrossRef]
59. Oki, H.; Yasuda, O. Sensitivity of the T2KK experiment to the non-standard interaction in propagation. *Phys. Rev. D* **2010**, *82*, 073009. [CrossRef]
60. Asano, K.; Minakata, H. Large-θ_{13} perturbation theory of neutrino oscillation for long-baseline experiments. *J. High Energy Phys.* **2011**, *6*, 22. [CrossRef]
61. Zhou, Y.-L. The Kobayashi-Maskawa parametrization of lepton flavor mixing and its application to neutrino oscillations in matter. *Phys. Rev. D* **2011**, *84*, 113012. [CrossRef]
62. Xing, Z.-Z. Leptonic commutators and clean T violation in neutrino oscillations. *Phys. Rev. D* **2013**, *88*, 017301. [CrossRef]
63. Minakata, H.; Parke, S.J. Simple and compact expressions for neutrino oscillation probabilities in matter. *J. High Energy Phys.* **2016**, *1*, 180. [CrossRef]
64. Xing, Z.-Z.; Zhu, J.-Y. Analytical approximations for matter effects on CP violation in the accelerator-based neutrino oscillations with $E \lesssim 1$ GeV. *J. High Energy Phys.* **2016**, *7*, 11. [CrossRef]
65. Denton, P.B.; Minakata, H.; Parke, S.J. Compact perturbative expressions for neutrino oscillations in matter. *J. High Energy Phys.* **2016**, *6*, 51. [CrossRef]
66. Li, Y.-F.; Zhang, J.; Zhou, S.; Zhu, J.-Y. Looking into analytical approximations for three-flavor neutrino oscillation probabilities in matter. *J. High Energy Phys.* **2016**, *12*, 109. [CrossRef]
67. Zhou, S. Symmetric formulation of neutrino oscillations in matter and its intrinsic connection to renormalization-group equations. *J. Phys. G* **2017**, *44*, 044006. [CrossRef]

68. Yang, Y.; Kneller, J.P.; Perkins, K.M. Multi-flavor effects in stimulated transitions of neutrinos. *arXiv* **2017**, arXiv:hep-ph/1706.01339.
69. Xing, Z.-Z.; Zhou, S.; Zhou, Y.-L. Renormalization-group equations of neutrino masses and flavor mixing parameters in matter. *J. High Energy Phys.* **2018**, *5*, 15. [CrossRef]
70. Xing, Z.-Z.; Zhou, S. Naumov- and Toshev-like relations in the renormalization-group evolution of quarks and Dirac neutrinos. *Chin. Phys. C* **2018**, *42*, 103105. [CrossRef]
71. Petcov, S.T.; Zhou, Y.L. On neutrino mixing in matter and CP and T violation effects in neutrino oscillations. *Phys. Lett. B* **2018**, *785*, 95–104. [CrossRef]
72. Wang, X.; Zhou, S. Analytical solutions to renormalization-group equations of effective neutrino masses and mixing parameters in matter. *J. High Energy Phys.* **2019**, *5*, 35. [CrossRef]
73. Denton, P.B.; Parke, S.J. Simple and precise factorization of the Jarlskog invariant for neutrino oscillations in matter. *Phys. Rev. D* **2019**, *100*, 053004. [CrossRef]
74. Xing, Z.-Z.; Zhu, J.-Y. Sum rules and asymptotic behaviors of neutrino mixing in dense matter. *Nucl. Phys. B* **2019**, *949*, 114803. [CrossRef]
75. Denton, P.B.; Parke, S.J.; Zhang, X. Neutrino oscillations in matter via eigenvalues. *Phys. Rev. D* **2020**, *101*, 093001. [CrossRef]
76. Wang, X.; Zhou, S. On the properties of the effective Jarlskog invariant for three-flavor neutrino oscillations in matter. *Nucl. Phys. B* **2020**, *950*, 114867. [CrossRef]
77. Xing, Z.-Z. Flavor structures of charged fermions and massive neutrinos. *Phys. Rept.* **2020**, *854*, 1–147. [CrossRef]
78. Zhou, S. Continuous and discrete symmetries of renormalization group equations for neutrino oscillations in matter. *arXiv* **2020**, arXiv:hep-ph/2004.10570.
79. Zhu, J.-Y. Radiative corrections to the lepton flavor mixing in dense matter. *J. High Energy Phys.* **2020**, *5*, 97. [CrossRef]
80. Minakata, H. Neutrino amplitude decomposition: Toward observing the atmospheric—Solar wave interference. *arXiv* **2020**, arXiv:hep-ph/2006.16594.
81. D'Olivo, J.C.; Nieves, J.F.; Torres, M. Finite temperature corrections to the effective potential of neutrinos in a medium. *Phys. Rev. D* **1992**, *46*, 1172–1179. [CrossRef]
82. Goldberger, M.L.; Watson, K.M. *Collision Theory*; John Wiley & Sons, Inc.: New York, NY, USA, 1967.
83. Botella, F.J.; Lim, C.S.; Marciano, W.J. Radiative corrections to neutrino indices of refraction. *Phys. Rev. D* **1987**, *35*, 896–901. [CrossRef]
84. Horvat, R.; Pisk, K. Radiative corrections for forward coherent neutrino scattering. *Nuovo Cim. A* **1989**, *102*, 1247–1253. [CrossRef]
85. Paschos, E.A.; Yu, J.Y. Neutrino interactions in oscillation experiments. *Phys. Rev. D* **2002**, *65*, 033002. [CrossRef]
86. Kuzmin, K.S.; Lyubushkin, V.V.; Naumov, V.A. Fine-tuning parameters to describe the total charged-current neutrino-nucleon cross section. *Phys. Atom. Nucl.* **2006**, *69*, 1857–1871. [CrossRef]
87. Blennow, M.; Coloma, P.; Fernandez-Martinez, E.; Hernandez-Garcia, J.; Lopez-Pavon, J. Non-unitarity, sterile neutrinos, and non-standard neutrino interactions. *J. High Energy Phys.* **2017**, *4*, 153. [CrossRef]
88. Capozzi, F.; Chatterjee, S.S.; Palazzo, A. Neutrino mass ordering obscured by nonstandard interactions. *Phys. Rev. Lett.* **2020**, *124*, 111801. [CrossRef]
89. Gandhi, R.; Quigg, C.; Reno, M.H.; Sarcevic, I. Ultrahigh-energy neutrino interactions. *Astropart. Phys.* **1996**, *5*, 81–110. [CrossRef]
90. Gandhi, R.; Quigg, C.; Reno, M.H.; Sarcevic, I. Neutrino interactions at ultrahigh-energies. *Phys. Rev. D* **1998**, *58*, 093009. [CrossRef]
91. Huang, G.; Liu, Q. Hunting the Glashow resonance with PeV neutrino telescopes. *J. Cosmol. Astropart. Phys.* **2020**, *2003*, 5. [CrossRef]
92. Langacker, P.; Petcov, S.T.; Steigman, G.; Toshev, S. On the Mikheev–Smirnov–Wolfenstein (MSW) mechanism of amplification of neutrino oscillations in matter. *Nucl. Phys. B* **1987**, *282*, 589–609. [CrossRef]
93. Bilenky, S.M.; Petcov, S.T. Massive neutrinos and neutrino oscillations. *Rev. Mod. Phys.* **1987**, *59*, 671–754. [CrossRef]
94. Cheng, H.-Y. Cosmological baryon production in spontaneous CP violating models without strong CP problem. *Phys. Rev. D* **1986**, *34*, 3824–3830. [CrossRef]

95. Naumov, V.A. Three neutrino oscillations in matter, CP violation and topological phases. *Int. J. Mod. Phys. D* **1992**, *1*, 379–399. [CrossRef]
96. Naumov, V.A. Berry's phases for three neutrino oscillations in matter. *Phys. Lett. B* **1994**, *323*, 351–359. [CrossRef]
97. Toshev, S. Maximal T violation in matter. *Phys. Lett. B* **1989**, *226*, 335–340. [CrossRef]
98. Kobayashi, M.; Maskawa, T. CP Violation in the renormalizable theory of weak interaction. *Prog. Theor. Phys.* **1973**, *49*, 652–657. [CrossRef]
99. Chau, L.L.; Keung, W.Y. Comments on the parametrization of the Kobayashi–Maskawa matrix. *Phys. Rev. Lett.* **1984**, *53*, 1802–1805. [CrossRef]
100. Fritzsch, H. How to describe weak-interaction mixing and maximal *CP* violation? *Phys. Rev. D* **1985**, *32*, 3058–3061. [CrossRef]
101. Kuznetsov, V.E.; Naumov, V.A. Relationship between the Kobayashi–Maskawa and Chau–Keung presentations of the quark mixing matrix. *Nuovo Cim. A* **1995**, *108*, 1451–1456. [CrossRef]

© 2020 by the authors. Licensee MDPI, Basel, Switzerland. This article is an open access article distributed under the terms and conditions of the Creative Commons Attribution (CC BY) license (http://creativecommons.org/licenses/by/4.0/).

Article

Energy-Momentum Relocalization, Surface Terms, and Massless Poles in Axial Current Matrix Elements

Oleg Teryaev [1,2]

[1] Joint Institute for Nuclear Research, 141980 Dubna, Russia; teryaev@jinr.ru
[2] Institute of Engineering and Physics, Dubna International University, 141980 Dubna, Russia

Received: 15 June 2020; Accepted: 18 August 2020; Published: 24 August 2020

Abstract: The energy-momentum relocalization in classical and quantum theory is addressed with specific impact on non-perturbative QCD and hadronic structure. The relocalization is manifested in the existence of canonical and symmetric (Belinfante and Hilbert) energy momentum tensors (EMT). The latter describes the interactions of hadrons with classical gravity and inertia. Canonical EMT, in turn, is naturally emerging due to the translation invariance symmetry and appears when spin structure of hadrons is considered. Its relation to symmetric Hilbert and Belinfante EMTs requires the possibility to neglect the contribution of boundary terms for the classical fields. For the case of quantum fields this property corresponds to the absence of zero-momentum poles of matrix element of the axial current dual to the spin density. This property is satisfied for quarks manifesting the symmetry counterpart of $U_A(1)$ problem and may be violated for gluons due to QCD ghost pole.

Keywords: gravity; relocalization; topology; boundary; poles

1. Introduction

The space-time symmetry related to the energy-momentum and angular momentum conservation is manifested in field theory as the appearance of energy-momentum and spin currents (see [1] and Ref. therein). Their definition is not unique. The structure of Lagrangian immediately defines, after the application of the Noether theorem, the canonical densities. Passing to the quantum operators and their matrix elements one may analyse how the fundamental fields are manifested in the spin structure of elementary particles.

From the other side, the interaction of particles with gravity involves the symmetruc Hilbert tensor resulting from the variation with respect to metric and the symmetry property naturally emerges after the absorption of spin density to the orbital one using the Belinfante procedure [1].

The interplay of both forms of Energy-Momentum Tensor (EMT) is especially important for hadrons, as in the absence of mathematically rigorous confinement theory their spin structure is an important problem of non-perturbative Quantum Chromodynamics (QCD). The same non-perturbative effects are responsible for the most of the visible mass of the Universe, and, therefore, for its gravitational interaction.

These interactions of hadrons with gravity (and inertia, due to equivalence principle) are encoded in their gravitational formfactors [2–6] (see also [7] and Ref. therein). They define the macrospic properties of all objects, and, as it appeared more recently, the responce of hadrons to fastest ever rotation and acceleration emerging in heavy-ion collisions (see [8] and Ref. therein). Indeed, the angular velocity of quark-gluon matter in the non-central heavy-ion collisions corresponds to the change of the velocity of the order of speed of light c at the distance of order of Compton wavelength l_C, $\omega \sim c/l_C$, which is some 25 orders of magnitude larger than angular velocity of Earth rotation. By coincidence, the acceleration a of this matter which is of the order of $a \sim c^2/l_C$ is larger than Earth's gravity g by almost the same factor.

One may wonder, why the highly non-inertial frame formed by quark-gluon matter in heavy-ion collisions can have any impact on the observables measured by the detector located in the laboratory frame. The non-inertial matter will play a role if its interaction with hadrons may be considered as a quantum measurement which is certainly true if particle spin (essentially quantum object!) or Hawking–Unruh radiation [8] are considered. Let us also note here that the main outcome of equivalence principle (EP) for spin motion in the gravitational field, the equality of classical and quantum rotators (orbital and spin angular momenta) precession frequencies (see [7] and Ref. therein), becomes trivial for the rotating frame (like Earth) if spin is considered just as some vector remaining constant [9] in the inertial frame and rotating in the frame of the Earth like Foucault pendulum. The non-trivial meaning would emerge if the quantum measurement of the spin in the rotating frame is considered and its similarity to pendulum and orbital angular momentum (AM) is the manifestation of EP.

The interaction with gravity, as it was already mentioned, is described by the symmetric Hilbert EMT representing (like Belinfante and any symmetric EMT) the angular momentum (AM) as the orbital one. This form of AM allows one to derive the EP as low-energy theorem (see [7] and Ref. therein) by making use of momentum and angular momentum conservation. Like in QED, the global symmetry puts the restrictions for the interaction (defined by local one) for small momenta. In distinction from QED, where only terms of zero order in momenta are fixed, while the linear ones (momenta) are dynamical, in gravity, because of more complicated gauge group, the momenta are also fixed and anomalous gravitomagnetic moment is absent, which is another formulation [2] of EP.

At the same time, in the analysis of hadronic spin structure the canonical expressions naturally appear. The spin of fundamental fields plays the special role in the physical interpretation of QCD hadronic structure. The interplay of various forms of EMT and AM are discussed in detail in the problems of hadronic structure [10] and heavy-ion collisions [11].

The procedure of relocalization [1] changing the local quantities but preserving the conserved (angular) momenta, requires the possibility to discard the surface terms, which is usually assumed. At the same time, their consideration when generalized to the case of quantum operators was earlier found [12–14] to lead to some non-trivial constraints for matrix elements of axial currents. Here we develop these ideas and put them into modern context. As a result, we find the constraints for the zero-mass poles in matrix elements of singlet axial current (dual to quark spin desnity) leading to an amazing interplay between general symmetry properties of relocalization and very specific topologocal QCD dynamics.

2. Boundary Terms in Coordinate and Momentum Space

Let us start with the following expression for the quark–gluon angular momentum density

$$M^{\mu,\nu\rho} = \frac{1}{2}\epsilon^{\mu\nu\rho\sigma}J_{5,\sigma} + x^\nu T^{\mu\rho} - x^\rho T^{\mu\nu}. \tag{1}$$

The first term in the right hand side (r.h.s.) is just the canonical quark spin tensor dual to singlet axial current. Note that the energy–momentum tensor here accumulates also the quark orbital momentum as well as the total gluon angular momentum. We may proceed further along this way and express the quark spin in the orbital form with the simultaneous change of the energy–momentum tensor to the one suggested by Belinfante long ago

$$M_B^{\mu,\nu\rho} = x^\nu T_B^{\mu\rho} - x^\rho T_B^{\mu\nu}. \tag{2}$$

As the conservation of the angular momentum

$$\partial_\mu M_B^{\mu,\nu\rho} = 0 \tag{3}$$

immediately leads to the symmetry of $T^{\mu\rho}$ (so that symmetric Hilbert tensor may also be considered in such a role), the latter implies that

$$\epsilon_{\mu\nu\rho\alpha} M_B^{\mu,\nu\rho} = 0. \tag{4}$$

One might conclude that the totally antisymmetric quark spin tensor is somehow cancelled and does not contribute to the total angular momentum [15]. This is also the manifestation of the general belief that axial current and angular momentum represent the different aspects of spin structure. Still, it appears possible to extract some quantitative information about their interplay.

The Belinfante (and Hilbert) EMT lead to the same AM as canonical one so that:

$$\int d^3x M_B^{0,\nu\rho} = \int d^3x M^{0,\nu\rho}. \tag{5}$$

We assume (which was done only tacitly in [12]) that also the stronger condition is valid

$$\int d^3x M_B^{\mu,\nu\rho} = \int d^3x M^{\mu,\nu\rho}, \tag{6}$$

so that

$$\epsilon_{\mu\nu\rho\alpha} \int d^3x M^{\mu,\nu\rho} = 0. \tag{7}$$

Substituting here the definition (1) one get

$$\int d^3x (3 J_5^\alpha(x) + 2\epsilon^{\mu\nu\rho\alpha} x_\nu T_{A.\mu\alpha}(x)) = 0, \tag{8}$$

where $T_A^{\mu\alpha} = (T^{\mu\alpha} - T^{\alpha\mu})/2$ is the antisymmetric part of energy momentum tensor responsible for the separate non-conservation of orbital and spin AM. The conservation of total AM results in the relation

$$\frac{1}{4}\epsilon^{\mu\nu\rho\sigma} \partial_\mu J_\sigma^5 = T_A^{\rho\nu}. \tag{9}$$

Note that we consider the weak gravitational fields and the derivatives, as well as quantum states in what follows, correspond to flat space. The consideration of strong fields may be achieved by applying the Dirac equation in curved space (see [7] and Ref. therein).

Set of Equations (8) and (9) allows one to exclude either spin (J_5) or orbital (T_A) AM. The latter is easier, as one can use the local Equation (9). Furthermore, axial current operator is related to many observables.

Excluding EMT antisymmetric part by making use of the conservation of AM density (9), one get in the case of classical fields:

$$(g_{\rho\nu}g_{\alpha\mu} - g_{\rho\mu}g_{\alpha\nu}) \int d^3x \partial^\rho (J_5^\alpha x^\nu) = 0. \tag{10}$$

This is in fact the way to represent the surface terms whose neglect is necessary to apply the Belinfante procedure.

Passing to the most interesting case of quantum operators one should switch the Equations (8) and (after incorporating the AM conservation at operator level) (10) between particle (nucleon) states with the momenta P and $P+q$.

$$(g_{\rho\nu}g_{\alpha\mu} - g_{\rho\mu}g_{\alpha\nu}) \int d^3x \langle P|\partial^\rho (J_5^\alpha x^\nu)|P+q\rangle = 0. \tag{11}$$

Expressing the local operator $J_5^\alpha(x) = exp(i\hat{P}x) J_5^\alpha(0) exp(-i\hat{P}x)$ by action of shift operator $exp(i\hat{P}x)$ allows one to perform the integration resulting in appearance of $\delta(\vec{q})$. Now $\frac{\partial}{\partial x^\mu}$ is substituted by $-iq_\mu$ and x^μ by $i\frac{\partial}{\partial q_\mu}$ acting on that $\delta^3(\vec{q})$. The latter is, by definition, equal, up to a sign, to the derivative acting on the matrix element. As a result, one obtains the following constraint:

$$q^\mu \frac{\partial}{\partial q^\alpha}|_0 \langle P|J_5^\alpha|P+q\rangle = q^\alpha \frac{\partial}{\partial q^\alpha}|_0 \langle P|J_5^\mu|P+q\rangle. \tag{12}$$

It is a quantum counterpart of (10) and it is natural that surface terms in coordinate space correspond to zero momenta. To make it more clear, let us multiply both sides by q_μ:

$$q^2 \frac{\partial}{\partial q^\alpha}\langle P|J_5^\alpha|P+q\rangle = (q^\beta \frac{\partial}{\partial q^\beta} - 1)q_\gamma \langle P|J_5^\gamma|P+q\rangle. \tag{13}$$

This equality is obviously valid up to the second and higher powers of q. Note that the differential operator in the r.h.s. subtracts the terms linear in q from the divergence matrix element proportional to sq for the pure kinematical reasons.

What can be dangerous is the pole for $q^2 \to 0$ which naturally appears for anomalous axial current already in perturbation theory for massless fermions [16]. For massless quarks, due to t'Hooft consistency principle, in the case of non-singlet currents these poles correspond to the exchange of the massless Goldstone mesons.

The exception is provided by singlet channel where η' remains massive manifesting the famous $U_A(1)$ problem and the correspondent pole is absent (see, e.g., [17] and Ref. therein):

$$\langle P,S|J_{5,\mu}(0)|P+q,S\rangle = 2MS_\mu G_1 + q_\mu(Sq)G_2, \tag{14}$$

$$q^2 G_2|_0 = 0. \tag{15}$$

The G_2 pole term, if present, would provide a contribution linear in q to the l.h.s.

Therefore, solution of $U_A(1)$ problem provides simulateneously the necessary dynamical mechanism for relocalization of massless quarks spin. This, in turn, leads to relation of conservation laws and canonical EMT with Belinfante and Hilbert EMT, supporting the emergence of equivalence principle as low-energy theorem [7].

3. Problems with Relocalization for Gluons

The situation is changed in the case of gluons. The relevant matrix element of topological current

$$\langle P,S|K_\mu^5(0)|P+q,S\rangle = 2MS_\mu \tilde{G}_1(q^2) + q_\mu(Sq)\tilde{G}_2(q^2), \tag{16}$$

$$q^2 \tilde{G}_2(q^2)|_0 \neq 0, \tag{17}$$

contains the contribution $\tilde{G}_2$ of the relevant Kogut–Susskind ghost (or instanton [18]) pole [19] which is fully responsible [20] for the value of the forward matrix element of anomaly-free quark gluon current $J_5^\mu - K^\mu$.

The consideration of topological current as dual to spin is naturally supported by the studies of bosonic anomalies in gravitational field [21] which may be relevant also for consideration of rotating quark-gluon matter in heavy-ion collisions [22]. Therefore one has a contradiction between kinematics of Relocalization Invariance (RI) requiring the absence of surface terms (corresponding to zero-mass poles of matrix elements) and instanton-type dynamics requiring their presence. The possible outcomes are the following

(i) If RI is indeed violated the coupling of nucleons to gravity (described by the formfactors of Belinfante EMT [13,23]) may be unconstrained by the form of conservation laws in terms of canonical EMT. In extreme case, assuming that just the canoncical form is related to translational invariance, this might result in the violation of Equivalence Principle for nucleons at several percent level which may be tested experimentally and is probably already excluded by the data.

(ii) One may assume "Hadronic censorship" leading to the absence of the ghost pole: in this case the matrrix element

$$< P, S | J_5^\mu - K^\mu | P, S > = 0$$

in the chiral limit. Bearing in mind the smallness of gluon spin one should mostly attribute the quark spin to the (predominantly strange) quark mass. This may explain the relative smallness of quark spin ("Spin Crisis") and may be checked, say, by lattice calculations of pseudoscalar quark densities.

(iii) The simplest solution would be the impossibility to separate spin and orbital momenta of gluons in the meaningful way.

4. Discussion

We found the relation between general space-time symmetry responsible for interactions of hadrons with gravity and the specific QCD dynamics. As a result, the quark spin relocalization is supported by solution of $U_A(1)$ problem and the same non-perturbative dynamics may spoil the extraction of totally antisymmetric gluon spin density.

The future studies, besides the exploration of mentioned in the previous section alternatives may include following developments:

(i) investigation of boundary terms in hydrodynamic approximation;
(ii) exploration of the role of boundary terms (spoiling the transition of spin to orbital AM) for twisted states, which might be obtained also at high energies (see [24] and Ref. therein) and provide the complementary description of Transverse Momentum Dependent parton correlators.

Funding: This research was funded by RFBR grant 18-02-01107.

Acknowledgments: I am intebted to Yu.N. Obukhov, G.Yu. Prokhorov, A.Ya. Silenko and V.I. Zakharov for useful discussions.

Conflicts of Interest: The author declares no conflict of interest.

References

1. Hehl, F.W.; Macias, A.; Mielke, E.W.; Obukhov, Y.N. On the structure of the energy-momentum and the spin currents in Dirac's electron theory. In *On Einstein's Path*; Springer: New York, NY, USA, 1999; pp. 257–274.
2. Kobzarev, I.Y.; Okun, L.B. Gravitational Interaction of Fermions. *Zh. Eksp. Teor. Fiz.* **1962**, *43*, 1904–1909.
3. Kobzarev, I.Y.; Zakharov, V.I. Consequences of the transversality of the graviton emission amplitude. *Ann. Phys.* **1970**, *60*, 448–463. [CrossRef]
4. Pagels, H. Energy-Momentum Structure Form Factors of Particles. *Phys. Rev.* **1966**, *144*, 1250–1260. [CrossRef]
5. Polyakov, M.V.; Son, H.D. Nucleon gravitational form factors from instantons: Forces between quark and gluon subsystems. *J. High Energy Phys.* **2018**, *9*, 156. [CrossRef]
6. Lorcé, C. The light-front gauge-invariant energy-momentum tensor. *J. High Energy Phys.* **2015**, *8*, 45. [CrossRef]
7. Teryaev, O. Gravitational form factors and nucleon spin structure. *Front. Phys.* **2016**, *11*, 111207. [CrossRef]
8. Prokhorov, G.Y.; Teryaev, O.V.; Zakharov, V.I. Effects of rotation and acceleration in the axial current: Density operator vs Wigner function. *J. High Energy Phys.* **2019**, *2*, 146. [CrossRef]
9. Kobzarev, I.Y.; Zakharov, V.I. Spin precession in a gravitational field. *Ann. Phys.* **1966**, *37*, 1–6. [CrossRef]
10. Leader, E.; Lorcé, C. The angular momentum controversy: What's it all about and does it matter? *Phys. Rept.* **2014**, *541*, 163–248. [CrossRef]
11. Becattini, F. Polarization in relativistic fluids: A quantum field theoretical derivation. *arXiv* **2020**, arXiv:2004.04050.
12. Teryaev, O. The EMC spin crisis: Comparing proton and photon. *arXiv* **1993**, arXiv:hep-ph/9303228.
13. Teryaev, O. Equivalence principle and partition of angular momenta in the nucleon. *AIP Conf. Proc.* **2007**, *915*, 260–263. [CrossRef]
14. Teryaev, O. Can gluon spin contribute to that of nucleon? *Phys. Part. Nucl.* **2014**, *45*, 57–58. [CrossRef]

15. Jaffe, R.; Manohar, A. The G(1) Problem: Fact and Fantasy on the Spin of the Proton. *Nucl. Phys. B* **1990**, *337*, 509–546. [CrossRef]
16. Dolgov, A.; Zakharov, V. On Conservation of the axial current in massless electrodynamics. *Nucl. Phys. B* **1971**, *27*, 525–540. [CrossRef]
17. Khlebtsov, S.; Klopot, Y.; Oganesian, A.; Teryaev, O. Dispersive approach to non-Abelian axial anomaly. *Phys. Rev. D* **2019**, *99*, 016008. [CrossRef]
18. Hooft, G. How Instantons Solve the U(1) Problem. *Phys. Rept.* **1986**, *142*, 357–387. [CrossRef]
19. Diakonov, D.; Eides, M.I. Massless Ghost Pole in Chromodynamics and the Solution of the U(1) Problem. *Sov. Phys. JETP* **1981**, *54*, 232–240.
20. Efremov, A.; Soffer, J.; Teryaev, O. Spin Structure of Nucleon and the Axial Anomaly. *Nucl. Phys. B* **1990**, *346*, 97–114. [CrossRef]
21. Dolgov, A.; Khriplovich, I.; Vainshtein, A.; Zakharov, V.I. Photonic Chiral Current and Its Anomaly in a Gravitational Field. *Nucl. Phys. B* **1989**, *315*, 138–152. [CrossRef]
22. Prokhorov, G.; Teryaev, O.; Zakharov, V. CVE for photons: Black-hole vs. flat-space derivation. *arXiv* **2020**, arXiv:2003.11119.
23. Teryaev, O. Spin structure of nucleon and equivalence principle. *arXiv* **1999**, arXiv:hep-ph/9904376.
24. Silenko, A.J.; Teryaev, O.V. Siberian snake-like behavior for an orbital polarization of a beam of twisted (vortex) electrons. *Phys. Part. Nucl. Lett.* **2020**, *16*, 77–78. [CrossRef]

© 2020 by the authors. Licensee MDPI, Basel, Switzerland. This article is an open access article distributed under the terms and conditions of the Creative Commons Attribution (CC BY) license (http://creativecommons.org/licenses/by/4.0/).

MDPI
St. Alban-Anlage 66
4052 Basel
Switzerland
Tel. +41 61 683 77 34
Fax +41 61 302 89 18
www.mdpi.com

Symmetry Editorial Office
E-mail: symmetry@mdpi.com
www.mdpi.com/journal/symmetry

www.ingramcontent.com/pod-product-compliance
Lightning Source LLC
LaVergne TN
LVHW070630100526
838202LV00012B/768